Lecture Notes in Mathematics 1787

Editors:
J.-M. Morel, Cachan
F. Takens, Groningen
B. Teissier, Paris

Springer
Berlin
Heidelberg
New York
Barcelona
Hong Kong
London
Milan
Paris
Tokyo

Stefaan Caenepeel
Gigel Militaru
Shenglin Zhu

Frobenius and Separable Functors for Generalized Module Categories and Nonlinear Equations

 Springer

Authors

Stefaan Caenepeel
Faculty of Applied Sciences
Vrije Universiteit Brussel, VUB
Pleinlaan 2
1050 Brussels
Belgium
e-mail: scaenepe@vub.ac.be
http://homepages.vub.ac.be/~scaenepe/welcome.html

Gigel Militaru
Faculty of Mathematics
University of Bucharest
Strada Academiei 14
70109 Bucharest 1
Romania
e-mail: gmilit@al.math.unibuc.ro

Shenglin ZHU
Institute of Mathematics
Fudan University
Shanghai 200433
China
e-mail: zhuslk@online.sh.cn

Cataloging-in-Publication Data applied for

Die Deutsche Bibliothek - CIP-Einheitsaufnahme
Caenepeel, Stefaan:
Frobenius and separable functors for generalized module categories and
nonlinear equations / Stefaan Caenepeel ; Gigel Militaru ; Shenglin Zhu. -
Berlin ; Heidelberg ; New York ; Barcelona ; Hong Kong ; London ; Milan ;
Paris ; Tokyo : Springer, 2002
  (Lecture notes in mathematics ; Vol. 1787)
  ISBN 3-540-43782-7

Mathematics Subject Classification (2000):
PRIMARY 16W30, SECONDARY 16D90, 16W50, 16B50

ISSN 0075-8434
ISBN 3-540-43782-7 Springer-Verlag Berlin Heidelberg New York

Springer-Verlag Berlin Heidelberg New York a member of BertelsmannSpringer
Science + Business Media GmbH

http://www.springer.de

© Springer-Verlag Berlin Heidelberg 2002
Printed in Germany

Typesetting: Camera-ready TeX output by the author

SPIN: 10878510   41/3142/ du - 543210 - Printed on acid-free paper

Dedicated to Gilda, Lieve and Xiu

Preface

One of the key tools in classical representation theory is the fact that a representation of a group can also be viewed as an action of the group algebra on a vector space. This has been (one of) the motivations to introduce algebras, and modules over algebras. During the past century, it has become clear that several different notions of module can be introduced, with a variety of applications in different mathematical disciplines. For example, actions by group algebras can also be used to develop Galois descent theory, with its applications in number theory. Graded modules originated from projective algebraic geometry. In fact a group grading can be considered as a coaction by the group algebra, i.e. the dual of an action. One may then consider various types of modules over bialgebras and Hopf algebras: Hopf modules (in integral theory), relative Hopf modules (in Hopf-Galois theory), dimodules (when studying the Brauer group). Perhaps the most important ones are the Yetter-Drinfeld modules, that have been studied in connection with the theory of quantum groups, the quantum Yang-Baxter equation, braided monoidal categories, and knot theory.

Frobenius fuctors generalize the classical concept of Frobenius algebra that appeared first 100 years ago in the work of Frobenius on representation theory. The study of Frobenius algebras has seen a revival during the passed five years, serving as an important tool in problems arising from different fields: Jones theory of subfactors of von Neumann algebras ([98], [100]), topological quantum field theory ([3], [8]), geometry of manifolds and quantum cohomology ([79], [129] and the references indicated there), the quantum Yang-Baxter equation ([15], [42]), and Yetter-Drinfeld modules ([49], [88]).

Separable functors are a generalization of the theory of separable field extensions, and of separable algebras. Separability plays a crucial role in several topics in algebra, number theory and algebraic geometry, for example in classical Galois theory, ramification theory, Azumaya algebras and the Brauer group theory, Hochschild cohomology and étale cohomology. A more recent application can be found in the Jones theory of subfactors of von Neumann algebras, already mentioned above with respect to Frobenius algebras.

In this monograph, we present - from a purely algebraic point of view - a unification program for actions and coactions and their properties, where we are mainly interested in generalizations of Frobenius and separability properties. The unification theory takes place at four different levels.

First, we have a unification on the level of categories of modules: *Doi-Koppinen modules* were introduced first, and all modules mentioned above can be viewed as special cases. *Entwined modules* arose from noncommutative geometry; they are at the same time more general and easier to deal with, and provide new fields of applications. Secondly, there is a unification at the level of functors between module categories: one can introduce morphisms of entwining structures, and then associate such a morphism a pair of adjoint functors. Many "classical" pairs of adjoint functors (the induction functor, forgetful functors, restriction of (co)scalars, functors forgetting a grading, and their adjoints) are in fact special cases of this construction. A third unification takes place at the level of the properties of these pairs of adjoint functors. Here the inspiration comes from two at first sight completely different algebraic notions, having their roots in representation theory: separable algebras and Frobenius. We give a categorical approach, leading to the introduction of *separable functors* and *Frobenius functors*. Not only this explains the at first sight mysterious fact that both separable and Frobenius algebras can be characterized using Casimir elements, it also enables us to prove Frobenius and separability type properties in a unified framework, with several new versions of Maschke's Theorem as a consequence.

The fourth unification is based on the theory of Yetter-Drinfeld modules, their relation with the quantum Yang-Baxter equation, and the FRT Theorem. The pentagon equation has appeared in the theory of duality for von Neumann algebras, in connection with C^*-algebras. Here we explain how they are related to Hopf modules. In a similar way, another nonlinear equation which we called the Long equation is related to the category of Long dimodules, that finds its origin in generalizations of the Brauer-Wall group. Finally, the FS equation can be used to characterize Frobenius algebras, as well as separable algebras, providing yet another explanation of the relationship between the two notions. For all these equations, we have a version of the FRT Theorem.

In Chapter 1, some preliminary results are given. We have included a Section about coalgebras and bialgebras, and one about adjoint functors. Section 1.2 deals with a classical treatment of Frobenius and separable algebras over fields, and we explain how they are connected to classical representation theory.

Chapter 2 provides a discussion of entwining structures and their representations, entwined modules, and we discuss how they generalize other types of modules and how they are related to the smash (co)product and the factorization problem of an algebra through two subalgebras. We also give the general pair of adjoint functors mentioned earlier. First properties of the category of entwined modules are discussed, for example we discuss when the category of entwined modules is a monoidal category. We use entwining structures mainly as a tool to unify all kinds of modules, but we want to point out that they were originally introduced with a completely different motivation, coming from noncommutative geometry: one can generalize the notion of principal bundles to a very general setting in which the role of coordinate functions on the base is played by a general noncommutative algebra A, and the fibre of the principal bundle by a coalgebra C, where A and C are related by a map $\psi : A \otimes C \to C \otimes A$, called the entwining map, that has to satisfy certain compatibility conditions (see [32] and [33]). Entwined modules, as representations of an entwining structure, were introduced by Brzeziński [23], and he proved that Doi-Koppinen Hopf modules and, a fortiori, graded modules, Hopf modules and Yetter-Drinfeld modules are special cases. Entwined modules can also be applied to introduce coalgebra Galois theory, we come back to this in Section 4.8, where we also explain the link to descent theory.

The starting points of Chapter 3 are *Maschke's Theorem* from Representation Theory (a group algebra is semisimple if and only if the order of the group does not divide the characteristic of the base field), and the classical result that a finite group algebra is *Frobenius*. Larson and Sweedler have given Hopf algebraic generalizations of these two results, using integrals.

Both the Maschke and Frobenius Theorem can be restated in categorical terms. Let us first look at Maschke's Theorem. If we replace the base field k by a commutative ring, then we obtain the following result: if the order of the group G is invertible in k, then every exact sequence of kG-modules that splits as a sequence of k-modules is split as a sequence of kG-modules. If k is field, this implies immediately that kG is semisimple; in fact it turns out that all variations of Maschke's Theorem that exist in the literature admit such a formulation. In fact we have more: the kG-splitting maps are constructed deforming the k-splitting maps in a *functorial* way. A proper definition of functors that have this functorial Maschke property was given by Năstăsescu, Van den Bergh, and Van Oystaeyen [145]. They called these functors *separable functors* because for a given ring extension $R \to S$, the restriction of scalars functor is separable if and only if S/R is separable in the classical sense. A Theorem of Rafael [158] gives necessary and sufficient conditions for a functor with an adjoint to be separable: the unit (or counit) of the adjunction has to be split (or cosplit). We will see that the separable functor philosophy can be applied successfully to any adjoint pair of functors between categories of entwined modules. We will focus mainly on the functors forgetting the action and the coaction, as this is more transparent and leads to several interesting results.

A similar functorial approach can be used for the Frobenius property. It is well-known that a k-algebra S is Frobenius if and only if the restriction of scalars functors is at the same time a left and right adjoint of the induction functor. This has lead to the introduction of *Frobenius functors*: this is a functor which has a left adjoint that is also a right adjoint. An adjoint pair of Frobenius functors is called a Frobenius pair.

Let $\eta : 1 \to GF$ be the unit of an adjunction; as we have seen, to conclude that F is separable, we need a natural transformation $\nu : GF \to 1$. Our strategy will be to describe *all* natural transformations $GF \to 1$; we will see that they form a k-algebra, and that the natural transformations that split the unit are idempotents (*separability idempotents*) in this algebra.

A look at the definition of adjoint pairs of functors tells us that we have to investigate natural transformations $GF \to 1$ and $1 \to FG$; the difference is that the normalizing properties for the separability property and the Frobenius property are not the same. But still we can handle both problems in a unified framework, and this is what we will do in Chapter 3. In Chapter 4, we will apply the results from Chapter 3 in some important subcases. We have devoted Sections to relative Hopf modules and Hopf-Galois theory, graded modules, Yetter-Drinfeld modules and the Drinfeld double, and Long dimodules. For example, we prove that, for a finitely generated projective Hopf algebra H, the Drinfeld double $D(H)$ is a Frobenius extension of H if and only if H is unimodular.

Part I tells us that Hopf modules, Yetter-Drinfeld modules and Long dimodules over a Hopf algebra H can be regarded as special cases as Doi-Koppinen Hopf modules and entwined modules, and that a unified theory can be developed. In Part II, we look at these three types of modules from a different point of view: we will see how they are connected to three different nonlinear equations. The celebrated FRT Theorem shows us the close relationship between Yetter-Drinfeld modules and the *quantum Yang-Baxter* equation (QYBE) (see e.g. [115], [108], [128]). We will discuss how the two other types of modules, Hopf modules and Long dimodules, are related to other nonlinear equations. It comes as a surprise that the nonlinear equation related to the category of Hopf modules $_H\mathcal{M}^H$ is the *pentagon* (or *fusion*) equation, which is even older, and somehow more basic then the quantum Yang-Baxter equation. Using Hopf modules, we will present two different approaches to solving this equation: a first approach is to prove an FRT type Theorem for the pentagon equation; a second, completely different, approach was developed by Baaj and Skandalis for unitary operators on Hilbert spaces ([10]) and, more recently, by Davydov ([65]) for arbitrary vector spaces. We will conclude Chapter 6 with a few open problems that may have important consequences: from a philosophical point of view the theory presented herein views a finite dimensional Hopf algebra H simply as an invertible matrix $R \in \mathcal{M}_{n^2}(k) \cong \mathcal{M}_n(k) \otimes \mathcal{M}_n(k)$ that is a solution for the pentagon equation $R^{12}R^{13}R^{23} = R^{23}R^{12}$. Furthermore, in this case $\dim(H)|n$. This point of view could be crucial in reducing the problem of classifying finite dimensional Hopf algebras (currently in full development and using different and complex techniques) to the elementary theory of matrices from linear algebra. At this point a new Jordan theory (we called it *restricted Jordan theory*) has to be developed.

In Chapter 8, we will focus on the Frobenius-separability equation, all solutions of which are also solutions of the braid equation. An FRT type theorem will enable us to clarify the structure of two fundamental classes of algebras, namely separable algebras and Frobenius algebras. The fact that separable algebras and Frobenius algebras are related to the same nonlinear equation is related to the fact that separability and Frobenius properties studied in Chapters 3 and 4 are based on the same techniques.

As we already indicated, the quantum Yang-Baxter equation has been intensively studied in the literature. For completeness sake, and to illustrate the similarity with our other nonlinear equations, we decided that to devote a special Chapter to it. This will also allow us to present some recent results, see Section 5.5.

The three authors started their common research on Doi-Koppinen Hopf modules in 1995, with a three month visit by the second and third author to Brussels. The research was continued afterwards within the framework of the bilateral projects "Hopf algebras and (co)Galois theory" and "Hopf algebras in Algebra, Topology, Geometry and Physics" of the Flemish and Romanian governments, and "New computational, geometric and algebraic methods applied to quantum groups and differential operators" of the Flemish and Chinese governments.

We benefitted greatly from direct and indirect contributions from - in alphabetical order - Margaret Beattie, Tomasz Brzeziński, Sorin Dăscălescu, Jose (Pepe) Gómez Torrecillas, Bogdan Ichim, Bogdan Ion, Lars Kadison, Claudia Menini, Constantin Năstăsescu, Şerban Raianu, Peter Schauenburg, Mona Stanciulescu, Dragos Ştefan, Lucien Van hamme, Fred Van Oystaeyen, Yinhuo Zhang, and Yonghua Xu. Chapters 2 and 3 are based on an seminar given by the first author in Brussels during the spring of 1999. The first author wishes to thank Sebastian Burciu, Corina Calinescu and Erwin De Groot for their useful comments. Finally we wish to thank Paul Taylor for his kind permission to use his "diagrams" software.

A few words about notation: in Part I, we work over a commutative ring k; unadorned Hom, $\otimes$, $\mathcal{M}$ etc. are assumed to be taken over k. In Part II, we are always assuming that we work over a field k. For k-modules M and N, I_M will be the identity map on M, and $\tau : N \otimes M \to M \otimes N$ will be the switch map mapping $m \otimes n$ to $n \otimes m$. Also it is possible to read part II without reading part I first: one needs the generalities of Chapter 1, and the definitions in the first Sections of Chapter 2.

Brussels, Bucharest, Shanghai,
February 2002

Stefaan Caenepeel
Gigel Militaru
Shenglin Zhu

Table of Contents

Part II Nonlinear equations

Part I

Entwined modules and Doi-Koppinen Hopf modules

1 Generalities

1.1 Coalgebras, bialgebras, and Hopf algebras

In this Section, we give a brief introduction to Hopf algebras. A more detailed discussion can be found in the literature, see for example [1], [63], [140] or [172].

Throughout, k will be a commutative ring. In some specific cases, we will assume that k is a field. $_kM = \mathcal{M}$ will denote the category of (left) k-modules (we omit the index k if no confusion is possible). $\otimes$ and Hom will be shorter notation for $\otimes_k$ and Hom_k. Let M and N be k-modules. $I_M : M \to M$ will be the identity map, and $\tau_{M,N} : M \otimes N \to N \otimes M$ the switch map. Indices will be omitted if no confusion is possible.

$M^* = \mathrm{Hom}(M, k)$ is the dual of the k-module M. For $m \in M$ and $m^* \in M^*$, we will often use the duality notation

$$\langle m^*, m \rangle = m^*(m)$$

Let M be a finitely generated and projective k-module. Then there exists a (finite) dual basis

$$\{m_i, m_i^* \mid i = 1, \cdots, n\}$$

for M. This means that

$$m = \sum_{i=1}^{n} \langle m_i^*, m \rangle m_i \quad \text{and} \quad m^* = \sum_{i=1}^{n} \langle m^*, m_i \rangle m_i^*$$

for all $m \in M$ and $m^* \in M^*$.

Algebras and coalgebras Recall that a k-algebra (with unit) is a k-module together with a multiplication map $m = m_A : A \otimes A \to A$ and a unit element $1_A \in A$ satisfying the conditions

$$m \circ (m \otimes I) = m \circ (I \otimes m)$$

$$m(a \otimes 1_A) = m(1_A \otimes a) = a$$

for all $a \in A$. The map $\eta = \eta_A : k \to A$ mapping $x \in k$ to $x1_A$ is called the unit map of A and satisfies the condition

$$m \circ (\eta \otimes I) = m \circ (I \otimes \eta) = I$$

The opposite A^{op} of an algebra A, is equal to A as a k-module, with multiplication $m_{A^{\mathrm{op}}} = m_A \circ \tau$. A is commutative if $A = A^{\mathrm{op}}$, or $m \circ \tau = m$.
k-alg will be the category of k-algebras, and multiplicative maps.
Coalgebras are defined in a similar way: a k-coalgebra C is a k-module together with k-linear maps

$$\Delta = \Delta_C : C \to C \otimes C \quad \text{and} \quad \varepsilon = \varepsilon_C : C \to k$$

satisfying

$$(\Delta \otimes I) \circ \Delta = (I \otimes \Delta) \circ \Delta \tag{1.1}$$

$$(\varepsilon \otimes I) \circ \Delta = (I \otimes \varepsilon) \circ \Delta = I \tag{1.2}$$

Δ is called the *comultiplication* or the *diagonal map*, and ε is called the *counit* or *augmentation map*. (1.2) tells us that the comultiplication is *coassociative*. We will use the Sweedler-Heyneman notation for the comultiplication: for $c \in C$, we write

$$\Delta(c) = \sum_{(c)} c_{(1)} \otimes c_{(2)} = c_{(1)} \otimes c_{(2)}$$

The summation symbol $\sum$ will usually be omitted. The coassociativity can then be reformulated as follows:

$$c_{(1)(1)} \otimes c_{(1)(2)} \otimes c_{(2)} = c_{(1)} \otimes c_{(2)(1)} \otimes c_{(2)(2)}$$

and therefore we write

$$\Delta^2(c) = (\Delta \otimes I)(\Delta(c)) = (I \otimes \Delta)(\Delta(c)) = c_{(1)} \otimes c_{(2)} \otimes c_{(3)}$$

and, in a similar way,

$$\Delta^3(c) = c_{(1)} \otimes c_{(2)} \otimes c_{(3)} \otimes c_{(4)}$$

The counit property (1.2) can be restated as

$$\varepsilon(c_{(1)})c_{(2)} = \varepsilon(c_{(2)})c_{(1)} = c$$

The co-opposite C^{cop} of a coalgebra C is equal to C as a k-module, with comultiplication $\Delta_{C^{\mathrm{cop}}} = \tau \circ \Delta_C$. C is called *cocommutative* if $C = C^{\mathrm{cop}}$, or $\tau \circ \Delta = \tau$, or

$$c_{(1)} \otimes c_{(2)} = c_{(2)} \otimes c_{(1)}$$

for all $c \in C$.
A k-linear map $f : C \to D$ between two coalgebras C and D is called a morphism of k-coalgebras if

$$\Delta_D \circ f = (f \otimes f)\Delta_C \quad \text{and} \quad \varepsilon_D \circ f = \varepsilon_C$$

or

$$f(c)_{(1)} \otimes f(c)_{(2)} = f(c_{(1)}) \otimes f(c_{(2)})$$

and

$$\varepsilon_D(f(c)) = \varepsilon_C(c)$$

for all $c \in C$. We also say that f is *comultiplicative*. The category of k-coalgebras and comultiplicative map is denoted by k-coalg. The tensor product of two coalgebras C and D is again a coalgebra. The comultiplication and counit are given by the formulas

$$\Delta_{C \otimes D} = (I_C \otimes \tau_{C,D} \otimes I_D) \circ (\Delta_C \otimes \Delta_D) \quad \text{and} \quad \varepsilon_{C \otimes D} = \varepsilon_C \otimes \varepsilon_D$$

Example 1. Let X be an arbitrary set, and $C = kX$ the free k-module with basis X. On C we define a comultiplication and counit as follows:

$$\Delta_C(x) = x \otimes x \quad \text{and} \quad \varepsilon_C(x) = 1$$

for all $x \in X$. kX is called the grouplike coalgebra.

The convolution product Let C be a coalgebra, and A an algebra. Then we can define a multiplication on $\text{Hom}(C, A)$ in the following way: for $f, g : C \to A$, we let $f * g = m_A \circ (f \otimes g) \circ \Delta_C$, that is,

$$(f * g)(c) = f(c_{(1)})g(c_{(2)})$$

This multiplication is called the *convolution*. $\eta_A \circ \varepsilon_C$ is a unit for the convolution.

In particular, if $A = k$, we find that C^* is a k-algebra, with unit ε, and comultiplication given by

$$\langle c^* * d^*, c \rangle = \langle c^*, c_{(1)} \rangle \langle d^*, c_{(2)} \rangle$$

In fact, the multiplication on C^* is the dual of the comultiplication on C. If A is an algebra, which is finitely generated and projective as a k-module, then A^* is a coalgebra. The comultiplication is given by

$$A^* \xrightarrow{m_A^*} (A \otimes A)^* \cong A^* \otimes A^*$$

This means that $\Delta(a^*) = a^*_{(1)} \otimes a^*_{(2)}$ if and only if

$$\langle a^*, ab \rangle = \langle a^*_{(1)}, a \rangle \langle a^*_{(2)}, b \rangle$$

for all $a \in A$ and $b \in B$. The comultiplication can be described in terms of a dual basis $\{a_i, a^*_i \mid i = 1, \cdots, n\}$ of A:

$$\Delta(a^*) = \sum_{i,j=1}^{n} \langle a^*, a_i a_j \rangle a^*_i \otimes a^*_j \tag{1.3}$$

for all $a^* \in A^*$. From (1.3), it also follows that

$$\sum_{i,j=1}^{n} a_i a_j \otimes a_i^* \otimes a_j^* = \sum_{i=1}^{n} a_i \otimes \Delta(a_i^*) \tag{1.4}$$

For later use, we rewrite this formula in terms of coalgebras: put $C = A^*$, and let $\{c_i, c_i^* \mid i = 1, \cdots, n\}$ be a finite dual basis for C. Then

$$\sum_i \Delta(c_i) \otimes c_i^* = \sum_{i,j} c_i \otimes c_j \otimes c_i^* * c_j^* \tag{1.5}$$

Bialgebras and Hopf algebras

Proposition 1. *For a k-module H that is at once a k-algebra and a k-coalgebra, the following assertions are equivalent:*

1. *m_H and η_H are comultiplicative;*

2. *Δ_H and ε_H are multiplicative;*

3. *for all $h, g \in H$, we have*

$$\Delta(gh) = g_{(1)} h_{(1)} \otimes g_{(2)} h_{(2)} \tag{1.6}$$
$$\varepsilon(gh) = \varepsilon(g) \varepsilon(h) \tag{1.7}$$
$$\Delta(1) = 1 \otimes 1 \tag{1.8}$$
$$\varepsilon(1) = 1 \tag{1.9}$$

In this situation, we call H a bialgebra. A map between bialgebras that is multiplicative and comultiplicative is called a morphism of bialgebras.

Proof. This follows from the following observations:
m_H is comultiplicative $\iff$ (1.6) and (1.8) hold;
η_H is comultiplicative $\iff$ (1.7) and (1.9) hold
Δ_H is multiplicative $\iff$ (1.6) and (1.7) hold;
ε_H is multiplicative $\iff$ (1.8) and (1.9) hold

Definition 1. *A bialgebra H is called a Hopf algebra if the identity I_H has an inverse S in the convolution algebra $\mathrm{Hom}(H, H)$. Thus we need a map $S : H \to H$ satisfying*

$$S(h_{(1)}) h_{(2)} = h_{(1)} S(h_{(2)}) = \eta(\varepsilon(h)) \tag{1.10}$$

The map S is called the antipode of H.

Let $f : H \to K$ be a morphism of bialgebras between two Hopf algebras H and K. It is well-known that f also preserves the antipode, that is,

$$S_K \circ f = f \circ S_H$$

and f is called a morphism of Hopf algebras.

Example 2. Let G be a semigroup. Then kG is a coalgebra (see Example 1), and a k-algebra. It is easy to see that kG is a bialgebra. If G is a group, then kG is a Hopf algebra. The antipode is given by $S(g) = g^{-1}$, for all $g \in G$.

If H is bialgebra, then H^{op}, H^{cop} and H^{opcop} are also bialgebras. If H has an antipode S, then S is also an antipode for H^{opcop}. An antipode $\overline{S}$ for H^{op} is also an antipode for H^{cop}, and is called a twisted antipode. $\overline{S}$ has to satisfy the property

$$\overline{S}(h_{(2)})h_{(1)} = h_{(2)}\overline{S}(h_{(1)}) = \eta(\varepsilon(h)) \tag{1.11}$$

for all $h \in H$.

Proposition 2. *Let H be a Hopf algebra. Then S is a bialgebra morphism from H to H^{opcop}. If S is bijective, then S^{-1} is a twisted antipode. If H is commutative or cocommutative, then $S \circ S = I_H$, and consequently $S = \overline{S}$.*

Proof. Consider the maps $\nu, \rho : H \otimes H \to H$ given by

$$\nu(h \otimes k) = S(k)S(h) \quad \text{and} \quad \rho(h \otimes k) = S(hk)$$

It is easy to prove that both ν and ρ are convolution inverses of the multiplication map m, and $\nu = \rho$, and $S(hk) = S(k)S(h)$ for all $h, k \in H$. Furthermore

$$1 = \eta(\varepsilon(1)) = (I * S)(1) = I(1)S(1) = S(1)$$

and we find that $S : H \to H^{\mathrm{op}}$ is multiplicative.

In a similar way, we prove that $S : H \to H^{\mathrm{cop}}$ is comultiplicative: the maps $\psi, \varphi : H \to H \otimes H$ given by

$$\psi(h) = \Delta(S(h)) \quad \text{and} \quad \varphi(h) = S(h_{(2)}) \otimes S(h_{(1)})$$

are both convolution inverses of Δ_H, and therefore $\psi = \varphi$ and

$$\Delta(S(h)) = S(h_{(2)}) \otimes S(h_{(1)})$$

for all $h \in H$. Finally

$$\varepsilon(h) = \varepsilon((\eta \circ \varepsilon)(h)) = \varepsilon(S(h_{(1)})h_{(2)}) = \varepsilon(S(h_{(1)}))\varepsilon(h_{(2)}) = \varepsilon(S(h))$$

Assume that S is bijective. Then $S^{-1}(hk) = S^{-1}(k)S^{-1}(h)$, and $S^{-1}(1) = 1$. Applying S^{-1} to (1.10), we find (1.11), and S^{-1} is a twisted antipode. Finally, if H is commutative or cocommutative, then S is also a twisted antipode, and we have for all $h \in H$ that

$$\begin{aligned}
(S * (S \circ S))(h) &= S(h_{(1)}S(S(h_{(2)}))) \\
&= S(S(h_{(2)})h_{(1)}) \\
&= S((\eta \circ \varepsilon)(h)) = (\eta \circ \varepsilon)(h)
\end{aligned}$$

proving that $S \circ S$ is a convolution inverse for S, and $S \circ S = I$.

Modules Let A be a k-algebra. A left A-module M is a k-module, together with a map

$$\psi = \psi_M^l : \ A \otimes M \to M, \ \psi(a \otimes m) = am$$

such that

$$a(bm) = (ab)m \ \text{ and } \ 1m = m$$

for all $a, b \in A$ and $m \in M$. We say that ψ is a left A-action on M, or that A acts on M from the left. Let M and N be two left A-modules. A k-linear map $f : \ M \to N$ is called left A-linear if $f(am) = af(m)$, for all $a \in A$ and $m \in M$. The category of left A-modules and A-linear maps is denoted by $_A\mathcal{M}$. In a similar way, we can introduce right A-modules, and the category of right A-modules $\mathcal{M}_A$. Let B be another k-algebra. A k-module M that is at once a left A-module and a right B-module such that

$$a(mb) = (am)b$$

for all $a \in A$, $b \in B$ and $m \in M$ is called an (A, B)-bimodule. $_A\mathcal{M}_B$ will be the category of (A, B)-bimodules. Observe that we have isomorphisms of categories

$$_A\mathcal{M}_B \cong {}_{A \otimes B^{\mathrm{op}}}\mathcal{M} \cong \mathcal{M}_{A^{\mathrm{op}} \otimes B}$$

Take $M \in \mathcal{M}_A$ and $N \in {}_A\mathcal{M}$. The tensor product $M \otimes_A N$ is by definition the coequalizer of the maps $I_M \otimes \psi_N^l$ and $\psi_M^r \otimes I_N$, that is, we have an exact sequence

$$M \otimes A \otimes N \rightrightarrows M \otimes N \longrightarrow M \otimes_A M \longrightarrow 0$$

If H is a bialgebra, then the tensor product of two (left) H-modules M and N is again an H-module. The action on $M \otimes N$ is given by

$$h(m \otimes n) = h_{(1)}m \otimes h_{(2)}n$$

We also write

$$M^H = \{m \in M \mid hm = \varepsilon(h)m, \text{ for all } h \in H\}$$

Module algebras and module coalgebras Assume that H is a bialgebra. Let A be a left H-module, and a k-algebra. We call A a left H-*module algebra* if the unit and multiplication are left H-linear, or

$$h(ab) = (h_{(1)}a)(h_{(2)}b) \ \text{ and } \ h1_A = \varepsilon(h)1_A \tag{1.12}$$

for all $h \in H$, and $a, b \in A$. In a similar way, we introduce right H-module algebras. If A is a left H-module algebra, then A^{op} is a right H^{opcop}-module algebra.

A k-coalgebra that is also a left H-module is called a left H-*module coalgebra* if the counit and the comultiplication are left H-linear. This is equivalent to

$$\Delta_C(hc) = h_{(1)}c_{(1)} \otimes h_{(2)}c_{(2)} \ \text{ and } \ \varepsilon_C(hc) = \varepsilon_H(h)\varepsilon_C(c) \tag{1.13}$$

for all $h \in H$ and $c \in C$. We can also introduce right module coalgebras, and if C is a left H-module coalgebra, then C^{cop} is a right H^{opcop}-module coalgebra.

If C is a right H-module coalgebra, then C^* is a left H-module algebra. The left H-action on C^* is given by the formula

$$\langle h \cdot c^*, c \rangle = \langle c^*, ch \rangle \tag{1.14}$$

In a similar way, if C is a left H-module coalgebra, then C^* is a right H-module algebra, with

$$\langle c^* \cdot h, c \rangle = \langle c^*, hc \rangle \tag{1.15}$$

Example 3. Let G be a group, and X a right G-set. This means that we have a map

$$X \times G \to X : (x, g) \mapsto xg$$

such that $(xg)h = x(gh)$, for all $g, h \in G$. Then the coalgebra kX is a right kG-module coalgebra.

Comodules Let C be a coalgebra. A right C-comodule M is a k-module together with a map

$$\rho = \rho_M^r : M \to M \otimes C$$

such that

$$(\rho \otimes I_C) \circ \rho = (I_M \otimes \Delta_C) \circ \rho \quad \text{and} \quad (I_C \otimes \varepsilon_C) \circ \rho = I_M \tag{1.16}$$

We will say that C acts from the right on M. We will use the Sweedler-Heyneman notation

$$\rho(m) = m_{[0]} \otimes m_{[1]}$$

and

$$(\rho \otimes I_C)(\rho(m)) = (I_M \otimes \Delta_C)(\rho(m)) = m_{[0]} \otimes m_{[1]} \otimes m_{[2]}$$

The second identity in (1.16) can be rewritten as

$$\varepsilon(m_{[1]}) m_{[0]} = m$$

for all $m \in M$. A map $f : M \to N$ between two right comodules is called a morphism of C-comodules, or a right C-colinear map if

$$\rho_N^r \circ f = (f \otimes I_C) \circ \rho_M^r$$

or

$$f(m)_{[0]} \otimes f(m)_{[1]} = f(m_{[0]}) \otimes m_{[1]}$$

for all $m \in M$. $\mathcal{M}^C$ will be the category of right C-comodules and right C-colinear maps.

Example 4. Let $C = kX$, with X an arbitrary set. Let M be a k-module graded by X, that is

$$M = \bigoplus_{x \in X} M_x$$

where every M_x is a k-module. Then M is a kX-comodule, the coaction is given by

$$\rho^r(m) = m_x \otimes x$$

if $m = m_x$ with $m_x \in M_x$. Conversely, every kX-comodule M is graded by X, one defines the grading by

$$M_x = \{m \in M \mid \rho(m) = m \otimes x\}$$

Thus we have an equivalence between $\mathcal{M}^{kX}$ and the category of X-graded modules.

We have a functor

$$F: \ \mathcal{M}^C \to {}_{C^*}\mathcal{M}$$

defined as follows: for a right C-comodule M, we let $F(M) = M$, with left C^*-action given by

$$c^* \cdot m = \langle c^*, m_{[1]} \rangle m_{[0]}$$

for all $c^* \in C^*$ and $m \in M$; if $f : M \to N$ is right C-colinear, then it is easy to prove that f is also left C^*-linear, and we let $F(f) = f$.

Proposition 3. *The functor $F: \ \mathcal{M}^C \to {}_{C^*}\mathcal{M}$ is faithful. If C is projective as a k-module, then F is fully faithful. If C is finitely generated and projective, then F is an isomorphism of categories.*

Proof. Take two right C-comodules M and N. Obviously $\operatorname{Hom}^C(M, N) \to {}_{C^*}\operatorname{Hom}(F(M), F(N))$ is injective, so F is faithful.
Assume that C is k-projective, and let $\{c_i, c_i^* \mid i \in I\}$ be a dual basis. Let M and N be C-comodules, and assume that $f : M \to N$ is left C^*-linear. We claim that f is also right C-colinear. Indeed, for all $m \in M$, we have

$$
\begin{aligned}
f(m_{[0]}) \otimes m_{[1]} &= \sum_{i \in I} f(m_{[0]}) \otimes \langle c_i^*, m_{[1]} \rangle c_i \\
&= \sum_{i \in I} f(c_i^* \cdot m) \otimes c_i \\
&= \sum_{i \in I} c_i^* \cdot f(m) \otimes c_i \\
&= \sum_{i \in I} \langle c_i^*, f(m)_{[1]} \rangle f(m)_{[0]} \otimes c_i \\
&= f(m)_{[0]} \otimes f(m)_{[1]}
\end{aligned}
$$

Assume moreover that C is finitely generated, and let $\{c_i, c_i^* \mid i = 1, \cdots, n\}$ be a dual basis for C. We define a functor $G : {}_{C^*}\mathcal{M} \to \mathcal{M}^C$ as follows: $G(M) = M$ as a k-module, with right C-coaction

$$\rho(m) = \sum_{i=1}^n c_i^* \cdot m \otimes c_i$$

We will show that ρ defines a coaction, and leave all other verifications to the reader. We obviously have

$$(I_M \otimes \varepsilon)(\rho(m)) = \sum_{i=1}^n \varepsilon(c_i) c_i^* \cdot m = \varepsilon \cdot m = m$$

Next we want to prove that

$$(\rho \otimes I_C) \circ \rho = (I_M \otimes \Delta_C) \circ \rho \tag{1.17}$$

For all $c^*, d^* \in C^*$, we have

$$
\begin{aligned}
&((I_M \otimes c^* \otimes d^*) \circ (\rho \otimes I_C) \circ \rho)(m) \\
&= (I_M \otimes c^* \otimes d^*)(\sum_{i,j}(c_j^* * c_i^*) \cdot m \otimes c_j \otimes c_i) \\
&= \sum_{i,j} \langle c^*, c_j \rangle \langle d^*, c_i \rangle (c_j^* * c_i^*) \cdot m \\
&= (c^* * d^*) \cdot m \\
&= \langle c^* * d^*, c_i \rangle c_i^* \cdot m \\
&= (I_M \otimes c^* \otimes d^*)(c_i^* \cdot m \otimes \delta(c_i)) \\
&= (I_M \otimes c^* \otimes d^*)(((I_M \otimes \Delta_C) \circ \rho^r)(m))
\end{aligned}
$$

and (1.17) follows after we apply Lemma 1

Lemma 1. *Let M, N be k-modules, and assume that N is finitely generated and projective. Take $\sum_j m_j \otimes p_j$ and $\sum_k m_k' \otimes p_k'$ in $M \otimes N$. If*

$$\sum_j \langle n^*, p_j \rangle m_j = \sum_k \langle n^*, p_k' \rangle m_k'$$

for all $n^* \in N^*$, then

$$\sum_j m_j \otimes p_j = \sum_k m_k' \otimes p_k'$$

Proof. Let $\{n_i, n_i^* \mid i = 1, \cdots, n\}$ be a dual basis for N. Then

$$\sum_j m_j \otimes p_j = \sum_{i,j} m_j \otimes \langle n_i^*, p_j \rangle n_i = \sum_{i,k} m_k' \otimes \langle n_i^*, p_k' \rangle n_i = \sum_k m_k' \otimes p_k'$$

Let H be a bialgebra. If M and N are right H-comodules, then $M \otimes N$ is again a right H-comodule. The H-coaction is given by

$$\rho^r_{M \otimes N}(m \otimes n) = m_{[0]} \otimes n_{[0]} \otimes m_{[1]}n_{[1]}$$

We call

$$M^{coH} = \{m \in M \mid \rho(m) = m \otimes 1\}$$

the submodule of *coinvariants* of M.

We can also introduce left C-comodules. For a left C-comodule M, the Sweedler-Heyneman notation takes the following form:

$$\rho^l_M(m) = m_{[-1]} \otimes m_{[0]} \in C \otimes M$$

The category of left C-comodules and left C-colinear maps is denoted by $^C\mathcal{M}$. We have an isomorphism of categories

$$^C\mathcal{M} \cong \mathcal{M}^{C^{cop}}$$

If M is at once a left C-comodule and a right D-comodule in such a way that

$$(\rho^l \otimes I_D) \circ \rho^r = (I_C \otimes \rho^r) \circ \rho^l$$

then we say that M is a (C, D)-bicomodule. We then write, following the Sweedler-Heyneman philosophy:

$$(m_{[0]})_{[-1]} \otimes (m_{[0]})_{[0]} \otimes m_{[1]} = m_{[-1]} \otimes (m_{[0]})_{[0]} \otimes (m_{[0]})_{[1]}$$
$$= m_{[-1]} \otimes m_{[0]} \otimes m_{[1]} = \rho^{lr}(m)$$

Observe that C itself is a (C, C)-bicomodule. $^C\mathcal{M}^D$ is the category of (C, D)-bicomodules and left C-colinear right C-colinear maps. We have isomorphisms

$$^C\mathcal{M}^D \cong {}^{C \otimes D^{cop}}\mathcal{M} \cong \mathcal{M}^{C^{cop} \otimes D}$$

Proposition 4. *Let C be a coalgebra, and M a finitely generated projective k-module. Right C-coaction on M are in bijective correspondence with left C-coactions on M^*.*

Proof. Let $\{m_i, m_i^* \mid i = 1, \cdots, n\}$ be a dual basis for M, and let $\rho^r : M \to M \otimes C$ be a right C-coaction. We define $\rho^l = \alpha(\rho^r) : M^* \to C \otimes M^*$ by

$$\rho^l(m^*) = \sum_{i=1}^{n} m_{i[1]} \otimes \langle m^*, m_{i[0]} \rangle m_i^* \tag{1.18}$$

This is a coaction on M^* since

$$(I_C \otimes \rho^l)(\rho^l(m^*)) = \sum_{i,j=1}^{n} m_{i[1]} \otimes m_{j[1]} \otimes \langle m^*, m_{i[0]} \rangle \langle m_i^*, m_{j[0]} \rangle m_j^*$$

$$= \sum_{j=1}^{n} m_{j[1]} \otimes m_{j[2]} \otimes \langle m^*, m_{j[0]} \rangle m_j^*$$

$$= (\Delta_C \otimes I_{M^*})(\rho^l(m^*))$$

$$\sum_{i=1}^{n} \varepsilon(m_{[-1]}^*) m_{[0]}^* = \sum_{i=1}^{n} \langle \varepsilon, m_{i[1]} \rangle \langle m^*, m_{i[0]} \rangle m_i^*$$

$$= \sum_{i=1}^{n} \langle m^*, m_i \rangle m_i^* = m^*$$

Conversely, given $\rho^l : M^* \to C \otimes M^*$, we define $\rho^r = \tilde{\alpha}(\rho^l) : M \to M \otimes C$ by

$$\rho^r(m) = \sum_{i=1}^{n} \langle m_{i[0]}^*, m \rangle m_i \otimes m_{i[-1]}^*$$

An easy computation shows that α and $\tilde{\alpha}$ are each others inverses.

The category of comodules over a coalgebra over a field k is a Grothendieck category. Over a commutative ring, we have the following generalization of this result, due to Wisbauer [187].

Proposition 5. *Let C be a coalgebra over a commutative ring k. The following assertions are equivalent:*

1. *C is flat as a k-module;*
2. *$\mathcal{M}^C$ is a Grothendieck category and the forgetful functor $\mathcal{M}^C \to \mathcal{M}$ is exact;*
3. *$\mathcal{M}^C$ is an abelian category and the forgetful functor $\mathcal{M}^C \to \mathcal{M}$ is exact.*

Proof. 1. $\Rightarrow$ 2. It is clear that $\mathcal{M}^C$ is additive. Let $f : M \to N$ be a map in $\mathcal{M}^C$. To prove that $\mathrm{Ker}\,(f)$ is a C-comodule, we need to show, for any $m \in \mathrm{Ker}\,(f)$:

$$\rho(m) \in \mathrm{Ker}\,(f) \otimes C = \mathrm{Ker}\,(f \otimes I_C)$$

(using the fact that C is k-flat). This is obvious, since

$$(f \otimes I_C)\rho(m) = f(m_{[0]}) \otimes m_{[1]} = \rho(f(m)) = 0$$

On $\mathrm{Coker}\,(f)$, we put a C-comodule structure as follows:

$$\rho(\overline{n}) = \overline{n_{[0]}} \otimes n_{[1]}$$

for all $n \in N$. This is well-defined: if $n = f(m)$, then

$$\overline{n_{[0]}} \otimes n_{[1]} = \overline{f(m)_{[0]}} \otimes f(m)_{[1]} = \overline{f(m_{[0]})} \otimes m_{[1]} = 0$$

It is clear that every monic in $\mathcal{M}^C$ is the kernel of its cokernel, and that every epic is the cokernel of its cokernel, so $\mathcal{M}^C$ is an abelian category. Let us next see that $\mathcal{M}^C$ is an AB3-category. If $\{M_\lambda \mid \lambda \in \Lambda\}$ is a family in $\mathcal{M}^C$, then $M = \oplus_\lambda M_\lambda$ is again a comodule: we have maps

$$M_\lambda \xrightarrow{\rho_\lambda} M_\lambda \otimes C \xrightarrow{i_\lambda \otimes I_C} M \otimes C$$

and therefore a unique map $\rho : M \to M \otimes C$ making M into a comodule, and i_λ into a right C-colinear map. The fact that $\mathcal{M}^C$ is an AB5-category follows easily since $\mathcal{M}$ is AB5, and the functor forgetting the C-coaction is exact.

Let us finally show that $\mathcal{M}^C$ has a family of generators. First observe that every right C-comodule of the form $M \otimes C$, with C-coaction induced by C, is generated by C. Indeed, for any k-module M, we can find an epimorphism $k^{(\lambda)} \to M$ in $\mathcal{M}$, and therefore an epimorphism

$$k^{(\lambda)} \otimes C = C^{(\lambda)} \to M \otimes C$$

in $\mathcal{M}^C$. Now we claim that the C-subcomodules of C form a family of generators of $\mathcal{M}^C$. It suffices to show that for every right C-comodule M and $m \in M$, there exists a C-subcomodule D of C and a C-colinear map $f : D \to M$ such that $m \in \mathrm{Im}\,(f)$.

$\rho : M \to M \otimes C$ is a monomorphism in $\mathcal{M}^C$, so M is isomorphic to $\rho(M) = \{n_{[0]} \otimes n_{[1]} \mid n \in M\}$. C generates $M \otimes C$, so there exists a C-colinear map $f : C \to M \otimes C$ and $c \in C$ such that $f(m) = \rho(m)$. Now let

$$D = \{d \in C \mid f(d) \in \rho(M)\}$$

Indeed, for $d \in D$, we can find $n \in N$ such that $f(d) = n_{[0]} \otimes n_{[1]}$, and we see that

$$(\rho \otimes I_C)(f(d)) = n_{[0]} \otimes n_{[1]} \otimes n_{[2]} \in \rho(M) \otimes C$$

Now look at the diagram with exact rows that defines D:

$$
\begin{array}{ccccc}
1 & \longrightarrow & D & \longrightarrow & C \\
 & & \downarrow{\scriptstyle f} & & \downarrow{\scriptstyle f} \\
1 & \longrightarrow & \rho(M) & \longrightarrow & M \otimes C
\end{array}
$$

C is flat, so we have a commutative diagram with exact rows

$$
\begin{array}{ccccc}
1 & \longrightarrow & D \otimes C & \longrightarrow & C \otimes C \\
 & & \downarrow{\scriptstyle f \otimes I_C} & & \downarrow{\scriptstyle f \otimes I_C} \\
1 & \longrightarrow & \rho(M) \otimes C & \longrightarrow & M \otimes C \otimes C
\end{array}
$$

and
$$D \otimes C = \{x \in C \otimes C \mid (f \otimes I_C)(x) \in \rho(M) \otimes C$$

It follows that $\rho(d) \in D \otimes C$, and D is a right C-comodule. We now have $f : D \to \rho(M) \cong M$ in $\mathcal{M}^C$, and $f(c) = m_{[0]} \otimes m_{[1]} \cong m$.

2. $\Rightarrow$ 3. is trivial.

3. $\Rightarrow$ 1. The forgetful functor $F : \mathcal{M}^C \to \mathcal{M}$ is a left adjoint of $\bullet \otimes C : \mathcal{M} \to \mathcal{M}^C$. The unit and counit of the adjunction are given by

$$\rho : M \to M \otimes C \; ; \; \rho(m) = m_{[0]} \otimes m_{[1]}$$

$$\varepsilon_N : N \otimes C \to N \; : \; \varepsilon_N(n \otimes c) = \varepsilon(c)n$$

for all $M \in \mathcal{M}^C$ and $N \in \mathcal{M}$. It is well-known that a functor between abelian categories that is a right adjoint of a covariant functor is left exact (see e.g. [11, I.7.1]), and it follows that $\bullet \otimes C : \mathcal{M} \to \mathcal{M}^C$ is exact. Now the forgetful functor $\mathcal{M}^C \to \mathcal{M}$ is also left exact, by assumption, so the composition $\bullet \otimes C : \mathcal{M} \to \mathcal{M}$ is left exact, and C is flat, as needed.

Remark 1. The assumption that the forgetful functor is exact, in the second and third condition of the Proposition, means the following: for a C-colinear map $f : \mathcal{M}^C \to \mathcal{M}^C$, the (co)kernel of f in $\mathcal{M}^C$ has to be equal as a k-module to the kernel of f viewed as a map between k-modules. J. Gómez Torrecillas kindly pointed out to us that this condition is missing in Wisbauer's paper [187]. For an example of a coalgebra C such that $\mathcal{M}^C$ is abelian, while C is not flat, and the functor forgetting the coaction is not exact, we refer to [80].

The cotensor product Take $M \in \mathcal{M}^C$ and $N \in {}^C\mathcal{M}$. The cotensor product $M\square_C N = M \otimes^C N$ is defined as the equalizer

$$0 \longrightarrow M\square_C N \longrightarrow M \otimes N \rightrightarrows M \otimes C \otimes N$$

Example 5. Let $C = kX$, and M and N X-graded modules. Then

$$M\square_C N = \bigoplus_{x \in X} M_x \otimes N_x$$

For a fixed right C-comodule M, we have a functor

$$M\square_C \bullet : {}^C\mathcal{M} \to \mathcal{M}$$

If M is flat as a k-module, then $M \otimes \bullet$ is an exact functor, and it follows easily that $M\square_C \bullet$ is left exact, but not necessarily right exact.

Definition 2. *A right C-comodule M is called right C-coflat if it is flat as a k-module, and if $M\square_C \bullet$ is an exact functor. A similar definition applies to left C-comodules.*

Now take $M \in \mathcal{M}^C$, $N \in {}^C\mathcal{M}$, and $P \in \mathcal{M}$. We then have a natural map

$$f : (M\square_C N) \otimes P \to M\square_C(N \otimes P)$$

given by $f((\sum_i m_i \otimes n_i) \otimes p) = \sum_i m_i \otimes (n_i \otimes p)$.

Lemma 2. *With notation as above, the natural map*

$$f : (M\square_C N) \otimes P \to M\square_C(N \otimes P)$$

is an isomorphism in each of the following cases:

1. *P is k-flat (e.g. if k is a field);*
2. *M is right C-coflat.*

Proof. 1. $M\square_C N$ is defined by the exact sequence

$$0 \longrightarrow M\square_C N \longrightarrow M \otimes N \rightrightarrows M \otimes C \otimes N$$

Using the fact that P is k-flat, we obtain a commutative diagram with exact rows

$$
\begin{array}{ccccccc}
0 & \longrightarrow & (M\square_C N) \otimes P & \longrightarrow & M \otimes N \otimes P & \rightrightarrows & M \otimes C \otimes N \otimes P \\
 & & \downarrow{\scriptstyle f} & & \downarrow{\scriptstyle \cong} & & \downarrow{\scriptstyle \cong} \\
0 & \longrightarrow & M\square_C(N \otimes P) & \longrightarrow & M \otimes N \otimes P & \rightrightarrows & M \otimes C \otimes N \otimes P
\end{array}
$$

and the result follows from the Five Lemma (see e.g. [123, Sec. VIII.4]).

2. Recall the definition of the tensor product: $N \otimes P = N \times P/I$, where I is the ideal generated by elements of the form

$$(n, p + q) - (n, p) - (n, q) \;\; ; \;\; (n + m, p) - (n, p) - (m, p) \;\; ; \;\; (nx, p) - (n, xp)$$

and we have an exact sequence of left C-comodules

$$0 \longrightarrow I \longrightarrow N \times P \longrightarrow N \otimes P \longrightarrow 0$$

and, using the right C-coflatness of M, we find a commutative diagram with exact rows

$$
\begin{array}{ccccccccc}
0 & \longrightarrow & M\square_C I & \longrightarrow & M\square_C(N \times P) & \longrightarrow & M\square_C(N \otimes P) & \longrightarrow & 0 \\
 & & \downarrow{\scriptstyle =} & & \downarrow{\scriptstyle \cong} & & \uparrow{\scriptstyle f} & & \\
0 & \longrightarrow & J & \longrightarrow & (M\square_C N) \times P & \longrightarrow & (M\square_C N) \otimes P & \longrightarrow & 0
\end{array}
$$

and the result follows again from Five Lemma.

Assume that A is a k-algebra, C a k-coalgebra, $P \in {}_A\mathcal{M}$, $M \in \mathcal{M}^C$ and $N \in {}^C\mathcal{M}_A$. By this we mean that N is a left C-comodule and a right A-module such that the right A-action is left C-colinear, i.e.

$$\rho^l(na) = n_{[-1]} \otimes n_{[0]}a$$

for all $n \in N$ and $a \in A$.

Lemma 3. *With notation as above, the natural map*

$$f : (M\square_C N) \otimes_A P \to M\square_C(N \otimes_A P)$$

is an isomorphism in each of the following situations:

1. *P is left A-flat;*
2. *M is right C-coflat.*

Proof. 1) The proof is identical to the proof of the first part of Lemma 2

2) The right A-action on $M\square_C N$ is given by

$$\left(\sum_i m_i \otimes n_i\right)a = \sum_i m_i \otimes n_i a \in M\square_C N$$

for every $\sum_i m_i \otimes n_i \in M\square_C N$. Now $(M\square_C N) \otimes_A P$ is the equalizer of

$$(M\square_C N) \otimes A \otimes P \rightrightarrows (M\square_C N) \otimes P$$

which is by Lemma 2 isomorphic to the equalizer of

$$M\square_C(N \otimes A \otimes P) \rightrightarrows M\square_C(N \otimes P)$$

and this equalizer is isomorphic to $M\square_C(N \otimes_A P)$ because M is right C-coflat. ∎

In some situations, the cotensor product can be computed explicitely.

Proposition 6. *Let M and N be right C-comodules, and assume that M is finitely generated and projective as a k-module. Then we have a natural isomorphism*

$$\mathrm{Hom}^C(M, N) \cong N\square_C M^*$$

Proof. We use notation as in Proposition 4. We know from (1.18) that M^* is a left C-comodule. From (1.18), we deduce that

$$\langle m^*_{[0]}, m\rangle m^*_{[-1]} = \langle m^*, m_{[0]}\rangle m_{[1]} \tag{1.19}$$

M is finitely generated projective, so we have an isomorphism

$$\alpha : \mathrm{Hom}(M, N) \to N \otimes M^*$$

given by

$$\alpha(f) = \sum_{i=1}^n f(m_i) \otimes m^*_i \quad \text{and} \quad \alpha^{-1}(n \otimes m^*)(m) = \langle m^*, m\rangle n$$

We will show that α restricts to the required isomorphism. Assume first that f is right C-colinear. Using (1.18) we find that

$$\sum_i f(m_i) \otimes m^*_{i[-1]} \otimes m^*_{i[0]} = \sum_{i,j} f(m_i) \otimes m_{j[1]} \otimes \langle m^*_i, m_{j[0]} \rangle m^*_j$$

$$= \sum_j f(m_{j[0]}) \otimes m_{j[1]} \otimes m^*_j$$

$$= \sum_j f(m_j)_{[0]} \otimes f(m_j)_{[1]} \otimes m^*_j$$

and it follows that $\alpha(f) \in N \square_C M^*$. Now take $\sum_k n_k \otimes m^*_k \in N \square_C M^*$, and let $f = \alpha^{-1}(\sum_k n_k \otimes n^*_k)$. f is then right C-colinear, since for all $m \in M$, we have

$$f(m_{[0]}) \otimes m_{[1]} = \sum_k \langle n^*_k, m_{[0]} \rangle n_k \otimes m_{[1]}$$

$$(1.19) \qquad = \sum_k \langle n^*_{k[0]}, m \rangle n_k \otimes n^*_{k[-1]}$$

$$= \sum_k \langle n^*_k, m \rangle n_{k[0]} \otimes n_{k[1]}$$

$$= \rho(f(m))$$

Coflatness versus injectivity Let C be a coalgebra over a field. We will show that a C-comodule is an injective object in the category of C-comodules if and only if it is C-coflat. Our proof is based on the approach presented in [63]. First we need some Lemmas.

Lemma 4. *Let C be a coalgebra over a field k, and M a right C-comodule. For every $m \in M$, there exists a finite dimensional subcomodule M' of M containing m. Consequently there exists an index set J and a set $\{M_j \mid j \in J\}$ consisting of finite dimensional right C-comodules, and an epimorphism $\phi: \oplus_{j \in J} M_j \to M$ in $\mathcal{M}^C$.*

Proof. Let $\{c_i \mid i \in I\}$ be a basis for C as a k-vector space, and write

$$\rho(m) = \sum_{i \in I} m_i \otimes c_i$$

where only a finite number of the m_i are nonzero - for a change, we do not use the Sweedler notation. Let M' be the k-subspace of M generated by the m_k. M' is finite dimensional, and

$$m = \sum_{i \in I} \varepsilon(c_i) m_i \in M'$$

We can write

$$\Delta(c_i) = \sum_{j,l \in I} a_i^{jl} c_l \otimes c_m$$

where only a finite number of the $a_i^{jl} \in k$ are different from 0. We now compute that

$$\sum_{i \in I} \rho(m_i) \otimes c_i = \sum_{i \in I} m_i \otimes \Delta(c_i)$$

$$= \sum_{i,j,l \in I} a_i^{jl} m_i \otimes c_l \otimes c_m$$

$$= \sum_{i,j,l \in I} a_l^{ji} m_l \otimes c_l \otimes c_i$$

Since the c_i form a basis of C, we have

$$\rho(m_i) = \sum_{j,l \in I} a_l^{ji} m_l \otimes c_l \in M' \otimes C$$

for all $i \in I$, and this proves that M' is a subcomodule of M.

Consider two right C-comodules M and Q. We say that Q is M-injective if for every subcomodule $M' \subset M$, the canonical map

$$\mathrm{Hom}^C(M, Q) \to \mathrm{Hom}^C(M', Q)$$

is surjective. Clearly Q is an injective comodule (i.e. an injective object of $\mathcal{M}^C$) if and only if Q is M-injective for every $M \in \mathcal{M}^C$.

Lemma 5. *If $\{M_i \mid i \in I\}$ is a collection of C-comodules, and $Q \in \mathcal{M}^C$ is M_i-injective for all $i \in I$, then Q is also $\oplus_{i \in I} M_i$-injective.*

Proof. Write $M = \oplus_{i \in I} M_i$. Let M' be a subcomodule of M, and $f : M' \to Q$ C-colinear. Consider

$$\mathcal{P} = \{(L, g) \mid M' \subset L \subset M \text{ in } \mathcal{M}^C, \ g : L \to Q \text{ in } \mathcal{M}^C, \ g_{|M'} = f\}$$

$\mathcal{P}$ is nonempty since $(M', f) \in \mathcal{P}$, and $\mathcal{P}$ is ordered: $(L, g) \leq (L', g')$ if $L \subset L'$ and $g'_{|L} = g$. It is easy to show that this ordering is inductive, so $\mathcal{P}$ has a maximal element, by Zorn's Lemma. We call this element (L_0, g_0), and we claim that $M_i \subset L_0$, for all $i \in I$.

Assume M_i is not contained in L_0, and consider

$$h = g_{0|M_i \cap L_0} : M_i \cap L_0 \to Q$$

Since Q is M_i-injective, we have a C-colinear map

$$\overline{h} : M_i \to Q \text{ such that } \overline{h}_{|M_i \cap L_0} = h$$

Now define $g : M_i + L_0 \to Q$ as follows

$$g(x + y) = \overline{h}(x) + g_0(y)$$

for $x \in M_i$ and $y \in L_0$. g is well-defined, since $\bar{h}$ and g_0 coincide on $M_i + L_0$. Now $g_{|L_0} = g_0$ and $M_i + L_0$ strictly contains L_0, so

$$(L_0, g_0) < (M_i + L_0, g) \text{ in } \mathcal{P}$$

which is a contradiction.
We conclude that $M_i \subset L_0$, so $M = \oplus_{i \in I} M_i \subset L_0$, and $g_0 : M = L_0 \to Q$ extends f.

Theorem 1. *Let C be a coalgebra over a field k. For a right C-comodule Q, the following assertions are equivalent.*

1. *Q is injective as a C-comodule;*
2. *Q is M-injective, for every finite dimensional C-comodule M;*
3. *Q is right C-coflat.*

Proof. $\underline{1. \Rightarrow 3.}$ Assume that Q is injective. The coaction ρ_Q is monomorphic, so we have a C-colinear map $\nu_Q : Q \otimes C \to Q$ splitting ρ_Q. Let $f : X \to Y$ be a surjective morphism of left C-comodules, and take $\sum_i q_i \otimes y_i \in Q\square_C Y$. As f is surjective, we find $x_i \in X$ such that $f(x_i) = y_i$, and our problem is that we don't know whether $\sum_i q_i \otimes x_i \in Q\square_C X$. We have

$$\sum_i q_{i[0]} \otimes q_{i[1]} \otimes f(x_i) = \sum_i q_i \otimes x_{i[-1]} \otimes f(x_{i[0]})$$

so

$$\sum_i q_i \otimes y_i = \sum_i \nu_M(q_{i[0]} \otimes q_{i[1]}) \otimes f(x_i)$$
$$= (I_M \otimes f)(\nu_M(q_i \otimes x_{i[-1]}) \otimes x_{i[0]})$$

Using the fact that ν_Q is C-colinear, we find

$$(\rho_Q \otimes I_X)(\nu_Q(q_i \otimes x_{i[-1]}) \otimes x_{i[0]}) = \nu_Q(q_i \otimes x_{i[-2]}) \otimes x_{i[-1]} \otimes x_{i[0]}$$
$$= (I_Q \otimes \rho_X)(\nu_Q(q_i \otimes x_{i[-1]}) \otimes x_{i[0]})$$

so $\nu_Q(q_i \otimes x_{i[-1]}) \otimes x_{i[0]} \in M\square_C X$, and this shows that $I_Q\square_C f : Q\square_C X \to Q\square_C Y$ is surjective.
$\underline{3. \Rightarrow 2.}$ Let $M \in \mathcal{M}^C$ be finite dimensional, and take a subcomodule $M' \subset M$. Then M^* and M'^* are left C-comodules, and Proposition 6 implies that

$$Q\square_C M^* \cong \text{Hom}^C(M, Q) \text{ and } Q\square_C M'^* \cong \text{Hom}^C(M', Q)$$

Now $M^* \to M'^*$ is surjective, so $Q\square_C M^* \to Q\square_C M'^*$ is also surjective since Q is C-coflat, and we find that $\text{Hom}^C(M, Q) \to \text{Hom}^C(M', Q)$ is surjective, as needed.
$\underline{2. \Rightarrow 1.}$ Take an arbitrary $N \in \mathcal{M}^C$. From Lemma 4, we know that there

exists a collection $\{M_i \mid i \in I\}$ of finite dimensional C-comodules and a C-colinear surjection $\phi: \oplus_{i \in I} M_i \to N$. Let $P = \operatorname{Ker} \phi$.

Now take a subcomodule $N' \subset N$, and let $M' = \phi^{-1}(N')$. Then $P \subset M'$, so we have the following commutative diagram with exact rows in $\mathcal{M}^C$:

$$
\begin{array}{ccccccccc}
0 & \longrightarrow & P & \longrightarrow & M & \overset{\phi}{\longrightarrow} & N & \longrightarrow & 0 \\
  &   & \big\Vert &   & \big\uparrow C & & \big\uparrow C & & \\
0 & \longrightarrow & P & \longrightarrow & M' & \overset{\phi}{\longrightarrow} & N' & \longrightarrow & 0
\end{array}
$$

Applying $\operatorname{Hom}^C(\bullet, Q)$ to this diagram, we find

$$
\begin{array}{ccccccc}
0 & \longrightarrow & \operatorname{Hom}^C(N, Q) & \longrightarrow & \operatorname{Hom}^C(M, Q) & \longrightarrow & \operatorname{Hom}^C(P, Q) \\
  &   & \big\downarrow &   & \big\downarrow &   & \big\uparrow = \\
0 & \longrightarrow & \operatorname{Hom}^C(N', Q) & \longrightarrow & \operatorname{Hom}^C(M', Q) & \longrightarrow & \operatorname{Hom}^C(P, Q)
\end{array}
$$

$\operatorname{Hom}^C(M, Q) \to \operatorname{Hom}^C(M', Q)$ is surjective, by Lemma 5. An easy diagram argument shows that $\operatorname{Hom}^C(N, Q) \to \operatorname{Hom}^C(N', Q)$ is surjective, as needed.

Comodule algebras and comodule coalgebras Let H be a bialgebra. A right H-comodule A that is also a k-algebra is called a right H-*comodule algebra* , if the unit and multiplication are right H-colinear, that is

$$\rho^r(ab) = a_{[0]}b_{[0]} \otimes a_{[1]}b_{[1]} \quad \text{and} \quad \rho^r(1_A) = 1_A \otimes 1_H \tag{1.20}$$

for all $a, b \in A$. Left H-comodule algebras are introduced in a similar way, and if A is a right H-comodule algebra, then A^{op} is a left H^{opcop}-comodule algebra.

A k-coalgebra C that is also a right H-comodule is called a right H-*comodule coalgebra* if the comultiplication and the counit are right H-colinear, or

$$c_{[0](1)} \otimes c_{[0](2)} \otimes c_{[1]} = c_{(1)[0]} \otimes c_{(2)[0]} \otimes c_{(1)[1]}c_{(2)[1]} \tag{1.21}$$

and

$$\varepsilon_C(c_{[0]})c_{[1]} = \varepsilon_C(c)1_H \tag{1.22}$$

for all $c \in C$.

Example 6. Let G be a (semi)group, and take $H = kG$. Then a kG-comodule algebra is nothing else then a G-graded k-algebra (see [146] for an extensive study of graded rings). A kG-comodule coalgebra is a G-graded coalgebra (see [144]).

Proposition 7. *Let C be a coalgebra which is finitely generated and pro-jective as a k-module. There is a bijective correspondence between right H-comodule coalgebra structures on C and left H-comodule algebra structures on C^*.*

Proof. Let $\{c_i, c_i^* \mid i = 1, \cdots, n\}$ be a finite dual basis of C, and assume that C is a right H-comodule coalgebra. We know from Proposition 4 that C^* is a left H-comodule, with left H-coaction given by

$$\rho^l(c^*) = \sum_{i=1}^n c_{i[1]} \otimes \langle c^*, c_{i[0]} \rangle c_i^*$$

This makes C^* into a left H-comodule algebra since

$$c_{[-1]}^* d_{[-1]}^* \otimes c_{[0]}^* d_{[0]}^* = \sum_{i,j=1}^n c_{i[1]} c_{j[1]} \otimes \langle c^*, c_{i[0]} \rangle \langle d^*, c_{j[0]} \rangle c_i^* * c_j^*$$

$$(1.5) \quad = \sum_{i=1}^n c_{i(1)[1]} c_{i(2)[1]} \otimes \langle c^*, c_{i(1)[0]} \rangle \langle d^*, c_{i(2)[0]} \rangle c_i^*$$

$$(1.21) \quad = \sum_{i=1}^n c_{i[1]} \otimes \langle c^* * d^*, c_{i[0]} \rangle c_i^*$$

$$= \rho^l(c^* * d^*)$$

$$\rho^l(\varepsilon_C) = \sum_{i=1}^n c_{i[1]} \otimes \langle \varepsilon_C, c_{i[0]} \rangle c_i^*$$

$$(1.22) \quad = \sum_{i=1}^n 1_H \otimes \langle \varepsilon_C, c_i \rangle c_i^* = 1_H \otimes \varepsilon_C$$

The further details of the proof are left to the reader.

1.2 Adjoint functors

We give a brief discussion of properties of pairs of adjoint functors; of course these results are well-known, but we have organized them in such a way that they can be applied easily to Frobenius and separable functors in Chapter 3.

We will occasionally use the *Godement product* of two natural transforma-tions. Let us introduce the Godement product briefly, refering the reader to [21] for more detail. Let $\mathcal{C}$, $\mathcal{D}$ and $\mathcal{E}$ be categories, and consider functors

$$F, G: \ \mathcal{C} \to \mathcal{D} \ \text{ and } \ H, K: \ \mathcal{D} \to \mathcal{E}$$

and natural transformations

$$\alpha : F \to G \quad \text{and} \quad \beta : H \to K$$

The Godement product

$$\beta * \alpha : HF \to KG$$

is defined by

$$(\beta * \alpha)_C = \beta_{G(C)} \circ H(\alpha_C) = K(\alpha_C) \circ \beta_{F(C)} : HF(C) \to KG(C)$$

If $F = G$, and $\alpha = 1_F$, then we find

$$(\beta * 1_F)_C = \beta_{F(C)}$$

If $H = K$, and $\beta = 1_H$, then we find

$$(1_H * \alpha)_C = H(\alpha_C)$$

Now consider, in addition, functors

$$L : \mathcal{C} \to \mathcal{D} \quad \text{and} \quad M : \mathcal{D} \to \mathcal{E}$$

and natural transformations

$$\gamma : G \to L \quad \text{and} \quad \delta : K \to M$$

then we have the following formula:

$$(\delta * \gamma) \circ (\beta * \alpha) = (\delta \circ \beta) * (\gamma \circ \alpha)$$

Pairs of adjoint functors Let $\mathcal{A}$, $\mathcal{B}$, $\mathcal{C}$ and $\mathcal{D}$ be categories, and consider functors

$$F : \mathcal{A} \to \mathcal{C}, \ G : \mathcal{B} \to \mathcal{C}, \ H : \mathcal{A} \to \mathcal{D}, \text{and} \ K : \mathcal{B} \to \mathcal{D}$$

We have functors

$$\mathrm{Hom}_{\mathcal{C}}(F, G), \ \mathrm{Hom}_{\mathcal{D}}(H, K) : \mathcal{A}^{\mathrm{op}} \times \mathcal{B} \to \underline{\mathrm{Sets}}$$

and we can consider natural transformations

$$\theta : \mathrm{Hom}_{\mathcal{C}}(F, G) \to \mathrm{Hom}_{\mathcal{D}}(H, K)$$

The naturality of θ can be expressed as follows: given $a : A' \to A$ in $\mathcal{A}$, $b : B \to B'$ in $\mathcal{B}$, and $f : F(A) \to G(B)$ in $\mathcal{C}$, we have

$$\theta_{A',B'}(G(b) \circ f \circ F(a)) = K(b) \circ \theta_{A,B}(f) \circ H(a) \qquad (1.23)$$

Proposition 8. *For two functors $F : \mathcal{C} \to \mathcal{D}$ and $G : \mathcal{D} \to \mathcal{C}$, we have the following isomorphisms of classes of natural transformations:*

$$\underline{\mathrm{Nat}}(1_{\mathcal{C}}, GF) \cong \underline{\mathrm{Nat}}(\mathrm{Hom}_{\mathcal{D}}(F, \bullet), \mathrm{Hom}_{\mathcal{C}}(\bullet, G)) \qquad (1.24)$$

$$\underline{\mathrm{Nat}}(FG, 1_{\mathcal{D}}) \cong \underline{\mathrm{Nat}}(\mathrm{Hom}_{\mathcal{C}}(\bullet, G), \mathrm{Hom}_{\mathcal{D}}(F, \bullet)) \qquad (1.25)$$

Proof. (Sketch) Consider a natural transformation $\eta : 1_{\mathcal{C}} \to GF$. The corresponding natural transformation $\theta : \text{Hom}_{\mathcal{D}}(F, \bullet) \to \text{Hom}_{\mathcal{C}}(\bullet, G)$ is defined by

$$\theta_{C,D}(f) = G(f) \circ \eta_C \tag{1.26}$$

for all $f : F(C) \to D$ in $\mathcal{D}$. Conversely, given θ, the corresponding η is given by

$$\eta_C = \theta_{C,F(C)}(I_{F(C)})$$

for all $C \in \mathcal{C}$.

Lemma 6. *Let F and G be as in Proposition 8, and consider natural transformations $\theta : \text{Hom}_{\mathcal{D}}(F, \bullet) \to \text{Hom}_{\mathcal{C}}(\bullet, G)$ and $\psi : \text{Hom}_{\mathcal{C}}(\bullet, G) \to \text{Hom}_{\mathcal{D}}(F, \bullet)$. let $\eta : 1_{\mathcal{C}} \to GF$ and $\varepsilon : FG \to 1_{\mathcal{D}}$ be natural transformations from Proposition 8.*

1. $\psi \circ \theta$ is the identity natural transformation if and only if

$$(\varepsilon * F) \circ (F * \eta) = 1_F \tag{1.27}$$

2. $\theta \circ \psi$ is the identity natural transformation if and only if

$$(G * \varepsilon) \circ (\eta * G) = 1_G \tag{1.28}$$

Proof. 1. Take $f : F(C) \to D$ in $\mathcal{D}$. We easily compute that

$$\psi_{C,D}(\theta_{C,D}(f)) = \varepsilon_D \circ FG(f) \circ F(\eta_C)$$

Now take $D = F(C)$ and $f = I_{F(C)}$. Then

$$\psi_{C,F(C)}(\theta_{C,F(C)}(I_{F(C)})) = \varepsilon_{F(C)} \circ F(\eta_C)$$

and, under the assumption that $\psi \circ \theta$ is the identity natural transformation, we find (1.27). Conversely, assume that (1.27) holds. ε is natural, so we have the following commutative diagram for any $f : F(C) \to D$ in $\mathcal{D}$:

$$
\begin{array}{ccc}
FGF(C) & \xrightarrow{FG(f)} & FG(D) \\
\downarrow{\varepsilon_{F(C)}} & & \downarrow{\varepsilon_D} \\
F(C) & \xrightarrow{FG(f)} & D
\end{array}
$$

and we find that

$$f = f \circ \varepsilon_{F(C)} \circ F(\eta_C)$$
$$= \varepsilon_D \circ FG(f) \circ F(\eta_C) = \psi_{C,D}(\theta_{C,D}(f))$$

The proof of 2. is similar.

Recall that (F, G) is an *adjoint pair of functors* if $\operatorname{Hom}_\mathcal{D}(F, \bullet)$ and $\operatorname{Hom}_\mathcal{C}(\bullet, G)$ are naturally isomorphic, or, equivalently, if there exists natural transformations $\eta : 1_\mathcal{C} \to GF$ and $\varepsilon : FG \to 1_\mathcal{D}$ satisfying (1.27-1.28). In this case, F is called a left adjoint of G, and G is called an adjoint of F. η is called the unit of the adjunction, while ε is called the counit.

It is well-known that the left or right adjoint of a functor is unique up to natural isomorphism; we include a proof for completeness sake.

Proposition 9. (Kan) [101]. *If G and G' are both adjoints of a functor $F : \mathcal{C} \to \mathcal{D}$, then G and G' are naturally isomorphic.*

Proof. We have two adjunctions (F, G) and (F, G'). Let (η, ε) and (η', ε') be the unit and counit of both adjunctions, and consider the natural transformations

$$\gamma = (G' * \varepsilon) \circ (\eta' * G) : \ G \to G'$$
$$\gamma' = (G * \varepsilon') \circ (\eta * G') : \ G' \to G$$

η is natural, so for any $D \in \mathcal{D}$, we have a commutative diagram

$$
\begin{array}{ccc}
G'FG(D) & \xrightarrow{\ G'(\varepsilon_D)\ } & G'(D) \\
{\scriptstyle \eta_{G'FG(D)}}\downarrow & & \downarrow{\scriptstyle \eta_{G'(D)}} \\
GFG'FG(D) & \xrightarrow[\ GFG'(\varepsilon_D)\ ]{} & GFG'(D)
\end{array}
$$

or

$$(\eta * G') \circ (G' * \varepsilon) = (GFG' * \varepsilon) \circ (\eta * G'FG)$$

Now η is natural, and we have a commutative diagram

$$
\begin{array}{ccc}
G(D) & \xrightarrow{\ \eta'_{G(D)}\ } & G'FG(D) \\
{\scriptstyle \eta_{G(D)}}\downarrow & & \downarrow{\scriptstyle \eta_{G'FG(D)}} \\
GFG(D) & \xrightarrow[\ GF(\eta'_{G(D)})\ ]{} & GFG'FG(D)
\end{array}
$$

or

$$(\eta * G'FG) \circ (\eta' * G) = (GF * \eta' * G) \circ (\eta * G)$$

The naturality of ε' gives a commutative diagram

$$
\begin{array}{ccc}
FG'FG(D) & \xrightarrow{\ FG'(\varepsilon_D)\ } & FG'(D) \\
{\scriptstyle \varepsilon'_{FG(D)}}\downarrow & & \downarrow{\scriptstyle \varepsilon'_D} \\
FG(D) & \xrightarrow[\ \varepsilon_D\ ]{} & D
\end{array}
$$

or

$$\varepsilon' \circ (FG' * \varepsilon) = \varepsilon \circ (\varepsilon' * FG)$$

and it follows that

$$(G * \varepsilon') \circ (GFG' * \varepsilon) = (G * \varepsilon) \circ (G * \varepsilon' * FG)$$

Combining all these formulas, we find

$$
\begin{aligned}
\gamma' \circ \gamma &= (G * \varepsilon') \circ (\eta * G') \circ (G' * \varepsilon) \circ (\eta' * G) \\
&= (G * \varepsilon') \circ (GFG' * \varepsilon) \circ (\eta * G'FG) \circ (\eta' * G) \\
&= (G * \varepsilon) \circ (G * \varepsilon' * FG) \circ (GF * \eta' * G) \circ (\eta * G) \\
&= (G * \varepsilon) \circ \big(G * ((\varepsilon' * F) \circ (F * \eta')) * G\big) \circ (\eta * G) \\
&= (G * \varepsilon) \circ \big(G * 1_F * G\big) \circ (\eta * G) \\
&= (G * \varepsilon) \circ (\eta * G) = 1_G
\end{aligned}
$$

In a similar way, we obtain that $\gamma \circ \gamma' = 1_{G'}$, and it follows that G and G' are naturally isomorphic.

Recall the following properties of adjoint pairs:

Theorem 2. *Let (F, G) be an adjoint pair of functors.*
F preserves colimits, and, in particular, coproducts, initial objects and cokernels.
G preserves limits, and, in particular, products, final objects and kernels.
If $\mathcal{C}$ and $\mathcal{D}$ are abelian categories, then F is right exact, and G is left exact. If F is exact, then G preserves injective objects. If G is exact, then F preserves projective objects.

Here is another well-known property of adjoint functors that will be useful in the sequel.

Proposition 10. *Let (F, G) be an adjoint pair functors, then we have isomorphisms*

$$\underline{\mathrm{Nat}}(F, F) \cong \underline{\mathrm{Nat}}(G, G) \cong \underline{\mathrm{Nat}}(1_{\mathcal{C}}, GF) \cong \underline{\mathrm{Nat}}(FG, 1_{\mathcal{D}})$$

Proof. We will show that $\underline{\mathrm{Nat}}(G, G) \cong \underline{\mathrm{Nat}}(1_{\mathcal{C}}, GF)$, the proof of the other assertions is left to the reader. For a natural transformation $\theta : 1_{\mathcal{C}} \to GF$, we define $\alpha = X(\theta) : G \to G$ by

$$\alpha_D = G(\varepsilon_D) \circ \theta_{G(D)} \tag{1.29}$$

Conversely, for $\alpha : G \to G$, $\theta = X^{-1}(\alpha) : 1_{\mathcal{C}} \to GF$ is defined by

$$\theta_C = \alpha_{F(C)} \circ \eta_C \tag{1.30}$$

We are done if we can show that X and X^{-1} are each others inverses. First take $\alpha : G \to G$, and $\theta = X^{-1}(\alpha)$. The diagram

$$
\begin{array}{ccccc}
G(D) & \xrightarrow{\eta_{G(D)}} & GFG(D) & \xrightarrow{G(\varepsilon_D)} & G(D) \\
& \searrow \theta_{G(D)} & \downarrow \alpha_{FG(D)} & & \downarrow \alpha_D \\
& & GFG(D) & \xrightarrow{G(\varepsilon_D)} & G(D)
\end{array}
$$

commutes: the triangle is commutative because of (1.30), and the square commutes because α is natural. From (1.27), it follows that the composition of the two maps in the top row is $I_{G(N)}$, and then we see from the diagram that $\alpha = X(\theta)$.

Conversely, take $\theta : 1_C \to GF$, and let $\alpha = X(\theta)$. Then $\theta = X^{-1}(\alpha)$ because the following diagram commutes:

$$
\begin{array}{ccccc}
C & \xrightarrow{\eta_C} & GF(C) & & \\
\downarrow \theta_C & & \downarrow \theta_{GF(C)} & \searrow \alpha_{F(C)} & \\
GF(C) & \xrightarrow{GF(\eta_C)} & GFGF(C) & \xrightarrow{G(\varepsilon_{F(C)})} & GF(C)
\end{array}
$$

A result of the same type is the following:

Proposition 11. *Let (F, G) be an adjoint pair of functors. Then we have isomorphisms*

$$\underline{\mathrm{Nat}}(GF, 1_C) \cong \underline{\mathrm{Nat}}(\mathrm{Hom}_{\mathcal{D}}(F, F), \mathrm{Hom}_{\mathcal{C}}(\bullet, \bullet)) \tag{1.31}$$

$$\underline{\mathrm{Nat}}(1_{\mathcal{D}}, FG) \cong \underline{\mathrm{Nat}}(\mathrm{Hom}_{\mathcal{C}}(G, G), \mathrm{Hom}_{\mathcal{D}}(\bullet, \bullet)) \tag{1.32}$$

Proof. We outline the proof of the first statement. Given a natural transformation $\nu : GF \to 1_C$, we define

$$\theta = \alpha(\nu) : \mathrm{Hom}_{\mathcal{D}}(F, F) \to \mathrm{Hom}_{\mathcal{C}}(\bullet, \bullet)$$

as follows: take $g : F(C) \to F(C')$ in $\mathcal{D}$, and put

$$\theta_{C, C'}(g) = \nu_{C'} \circ G(g) \circ \eta_C$$

Straightforward arguments show that θ is natural. Conversely, given

$$\theta : \mathrm{Hom}_{\mathcal{D}}(F, F) \to \mathrm{Hom}_{\mathcal{C}}(\bullet, \bullet)$$

we define $\alpha^{-1}(\theta) = \nu : GF \to 1_C$ by

$$\nu_C = \theta_{GF(C),C}(\varepsilon_{F(C)}) : \; GF(C) \to C$$

We leave it as an exercise to show that ν is natural, as needed, and that α and α^{-1} are inverses.

The proof of the second statement is similar. Let us just mention that, given $\zeta : 1_{\mathcal{D}} \to FG$, we define

$$\beta(\zeta) = \psi : \; \mathrm{Hom}_{\mathcal{C}}(G, G) \to \mathrm{Hom}_{\mathcal{D}}(\bullet, \bullet)$$

as follows: given $f : \; G(D) \to G(D')$ in $\mathcal{C}$, we put

$$\psi_{D,D'}(f) = \varepsilon_{D'} \circ F(f) \circ \zeta_D$$

1.3 Separable algebras and Frobenius algebras

In this Section, we give the classical Definition, and elementary properties of separable and Frobenius algebras. We will refer to it in Chapter 3, where we will introduce separable and Frobenius functors, and show that they are generalizations of the classical concepts. The Section on separable algebras is based on [109], and the one on Frobenius algebras on [113].

Separable algebras Let k be a commutative ring, A a k-algebra and M an A-bimodule. Recall that M can be viewed as a left A^e-module, where $A^e = A \otimes A^{op}$ is the enveloping algebra of A. A *derivation* of A in M is a k-linear map $D : \; A \to M$ such that

$$D(ab) = D(a)b + aD(b) \tag{1.33}$$

for all $a, b \in A$. $\mathrm{Der}_k(A, M)$ will be the k-module consisting of all derivation of A into M. For any $m \in M$, we have a derivation D_m, given by

$$D_m : \; A \to M, \quad D_m(a) = am - ma$$

called the *inner derivation* asociated to m. It is clear that $D_m = 0$ if and only if $m \in M^A = \{m \in M \mid am = ma, \forall a \in A\}$, so we have an exact sequence

$$0 \to M^A \to M \to \mathrm{Der}_k(A, M) \tag{1.34}$$

We also note that

$$M^A \cong \mathrm{Hom}_{A^e}(A, M), \quad M \cong \mathrm{Hom}_{A^e}(A^e, M) \tag{1.35}$$

The multiplication m_A on A induces an epimorphism $A \otimes A^{op} \to A$ of left A^e-modules, still denoted by m_A, and we have another exact sequence

$$0 \to I(A) = \mathrm{Ker}\,(m_A) \to A \otimes A^{op} \to A \to 0 \tag{1.36}$$

We have a derivation

$$\delta : \ A \to I(A), \quad \delta(a) = a \otimes 1 - 1 \otimes a$$

for all $a \in A$. It is clear that $\delta(a) \in I(A)$ and

$$A\delta(A) = I(A) = \delta(A)A$$

Indeed, take $x = \sum_i a_i \otimes b_i \in I(A)$, then

$$x = \sum_i a_i(1 \otimes b_i - b_i \otimes 1) = -\sum_i a_i\delta(b_i) \in A\delta(A)$$

Lemma 7. *Let M be an A-bimodule over a k-algebra A. Then we have an isomorphism of k-modules*

$$\mathrm{Hom}_{A^e}(I(A), M) \cong \mathrm{Der}_k(A, M) \tag{1.37}$$

Proof. We define

$$\phi : \ \mathrm{Hom}_{A^e}(I(A), M) \to \mathrm{Der}_k(A, M), \quad \phi(f) = f \circ \delta$$

ϕ^{-1} is given by

$$\phi^{-1}(D)(\sum_i a_i \otimes b_i) = -\sum_i a_i D(b_i)$$

We show that $\phi^{-1}(D)$ is left A^e-linear, and leave the other details to the reader.

$$\phi^{-1}(D)((a \otimes b)(\sum_i a_i \otimes b_i)) = \phi^{-1}(D)(\sum_i aa_i \otimes b_ib)$$

$$= -\sum_i aa_i D(b_ib) = -\sum_i aa_i D(b_i)b - \sum_i aa_ib_i D(b)$$

$$= (a \otimes b)\phi^{-1}(D)(\sum_i a_i \otimes b_i)$$

Applying the functor $\mathrm{Hom}_{A^e}(\bullet, M)$ to the exact sequence (1.36) and taking (1.35) and (1.37) into account, we find a long exact sequence

$$0 \to M^A \to M \to \mathrm{Der}_k(A, M) \to \mathrm{Ext}^1_{A^e}(A, M) \to 0 \tag{1.38}$$

extending (1.34). Indeed, $\mathrm{Ext}^1_{A^e}(A^e, M) = 0$, since A^e is projective as a left A^e-module.

$$H^1(A, M) = \mathrm{Ext}^1_{A^e}(A, M)$$

is another notation, and $H^1(A, M)$ is called the first *Hochschild cohomology group* of A with coefficients in M. For more information on Hochschild cohomology, we refer to [55, Ch. IX]. Thus (1.38) tells us that

$$H^1(A, M) \cong \mathrm{Der}_k(A, M)/\mathrm{InnDer}_k(A, M)$$

A is called a *separable k-algebra* if it satisfies the equivalent conditions of the following theorem:

Theorem 3. *For a k-algebra A the following statements are equivalent:*

1. *A is projective as a left A^e-module;*
2. *the exact sequence (1.36) splits as a sequence of left A^e-modules;*
3. *there exists $e = \sum e^1 \otimes e^2 \in A \otimes A$ such that*

$$ae = ea \quad \text{and} \quad \sum e^1 e^2 = 1 \tag{1.39}$$

   for all $a \in A$.
4. *$H^1(A, M) = 0$, for any A-bimodules M.*
5. *the derivation $\delta : A \to I(A)$, $\delta(a) = a \otimes 1 - 1 \otimes a$ is inner;*
6. *every derivation $D : A \to M$ is inner, for any A-bimodule M.*

An element $e \in A^e$ satisfying $ae = ea$ for all $a \in A$ is called a *Casimir element*. If, in addition, $\sum e^1 e^2 = 1$, then e is an idempotent, and it is called a *separability idempotent*.

Proof. 1. $\Leftrightarrow$ 2. is obvious.
2. $\Rightarrow$ 3. If $\psi : A \to A^e$ is a left A^e-module map and section of m_A then, $e = \psi(1)$ satisfies (1.39).
3. $\Rightarrow$ 2. Define $\psi : A \to A^e$, $\psi(a) = ae = ea$. ψ is left A^e-module map and $m_A \psi(a) = \sum ae^1 e^2 = a$.
1. $\Leftrightarrow$ 4. is obvious.
4. $\Leftrightarrow$ 6. follows from the exact sequence (1.38).
6. $\Rightarrow$ 5. is trivial.
5. $\Rightarrow$ 6. Let $D : A \to M$ be a derivation. From the above Lemma we know that there is a $f \in \text{Hom}_{A^e}(I(A), M)$ such that $D = f \circ \delta$. δ is inner, so we can write $\delta = D_x$, with $x \in I(A)$. Now,

$$D(a) = f(\delta(a)) = f(ax - xa) = af(x) - f(x)a = D_{f(x)}(a)$$

i.e. D is inner.

Let us now prove some immediate properties of separable algebras.

Proposition 12. *Any projective separable algebra A over a commutative ring k is finitely generated.*

Proof. We take a dual basis $\{s_i, s_i^* \mid i \in I\}$ for A. This means that, for all $s \in A$, the set

$$I(s) = \{i \in I \mid \langle s_i^*, s \rangle \neq 0\}$$

is finite, and

$$s = \sum_{i \in I} \langle s_i^*, s \rangle s_i$$

For all $i \in I$, we define $\phi_i : A \otimes A^{\text{op}} \to A$ by

$$\phi_i(s \otimes t) = \langle s_i^*, t \rangle s$$

such that
$$\phi_i(s's \otimes t) = \langle s_i^*, t \rangle s's = s'\phi_i(s \otimes t)$$

and ϕ_i is left A-linear. We now claim that $\{z_i = 1 \otimes s_i, \phi_i \mid i \in I\}$ is a dual basis of $A \otimes A^{\mathrm{op}}$ as a left A-module.

Take $z = s \otimes t \in A \otimes A^{\mathrm{op}}$. If $\phi_i(z) = \langle s_i^*, t \rangle s \neq 0$, then $\langle s_i^*, t \rangle \neq 0$, so $i \in I(t)$, and we conclude that

$$I(z) = \{i \in I \mid \phi_i(z) \neq 0\} \subset I(t)$$

is finite. Moreover

$$s \otimes t = \sum_{i \in I} s \otimes \langle s_i^*, t \rangle s_i = \sum_{i \in I} \langle s_i^*, t \rangle s \otimes s_i$$
$$= \sum_{i \in I} \phi_i(s \otimes t) \otimes s_i = \sum_{i \in I} \phi_i(s \otimes t)(1 \otimes s_i)$$

A is separable, so we have a separability idempotent $e = e^1 \otimes e^2 \in A \otimes A^{\mathrm{op}}$. Our next claim is that
$$I(et) \subset I(e)$$

for all $t \in A$. Indeed, we compute

$$\phi_i(et) = \phi_i(e^1 \otimes e^2 t) = \phi_i(te^1 \otimes e^2) = t\phi_i(e)$$

so $i \in I(et)$, or $\phi_i(et) \neq 0$, implies $\phi_i(e) \neq 0$ and $i \in I(e)$.
For all $t \in A$, we finally compute

$$t = 1t = m(e)t = m(et) = m\Big(\sum_{i \in I(e)} \phi_i(et)z_i \Big)$$
$$= m\Big(\sum_{i \in I(e)} \phi_i(e^1 \otimes e^2 t)z_i \Big) = m\Big(\sum_{i \in I(e)} \langle s_i^*, e^2 t \rangle e^1 z_i \Big)$$
$$= \sum_{i \in I(e)} \langle s_i^*, e^2 t \rangle e^1 m(z_i) = \sum_{i \in I(e)} \langle s_i^*, e^2 t \rangle e^1 s_i$$

Write $e = \sum_{j=1}^r e_j \otimes e'_j$. We have shown that

$$\{e_j s_i, \langle s_i^*, e'_j \bullet \rangle \mid i \in I(e), j = 1, \cdots, r\}$$

is a finite dual basis for A.

Proposition 13. *A separable algebra A over a field k is semisimple.*

Proof. Let $e = \sum e^1 \otimes e^2 \in A \otimes A$ be a separability idempotent and N an A-submodule of a right A-module M. As k is a field, the inclusion $i : N \to M$ splits in the category of k-vector spaces. Let $f : M \to N$ be a k-linear map such that $f(n) = n$, for all $n \in N$. Then

$$\tilde{f} : \ M \to N, \quad \tilde{f}(m) := \sum f(me^1)e^2$$

is a right A-module map that splits the inclusion i. Thus N is an A-direct factor of M, and it follows that M is completely reducible. This shows that A is semisimple.

Examples 1. 1. Let k be a field of characteristic p, and $a \in k \setminus k^p$. $l = k[X]/(X^p - a)$ is then a purely inseparable field extension of k, and l is not a separable k-algebra in the above sense. Indeed,

$$\frac{\mathrm{d}}{\mathrm{d}X} : \ l \to l$$

is a derivation that is not inner. More generally, one can prove that a finite field extension l/k is separable in the classical sense if and only if l is separable as a k-algebra, see [66, Proposition III.3.4].
2. Let k be a field. It can be show that a separable k-algebra is of the form

$$A = M_{n_1}(D_1) \times \cdots \times M_{n_r}(D_r) \tag{1.40}$$

where D_i is a division algebra with center a finite separable field extension l_i of k. See [66, Theorem III.3.1] for details.
3. Any matrix ring $M_n(k)$ is separable as a k-algebra: for any $i = 1, \cdots, n$, $e_i = \sum_{j=1}^{n} e_{ji} \otimes e_{ij}$ is a separablity idempotent. More generally, any Azumaya algebra A is separable as a k-algebra.

Frobenius algebras In this Section we will recall the classical definition of a Frobenius algebra, thus showing how it came up in representation theory. We will work over a a field k. For a k-algebra A, the k-dual $A^* = \mathrm{Hom}_k(A, k)$ is an A-bimodule via the actions

$$\langle r^* \cdot r, r' \rangle = \langle r^*, rr' \rangle, \quad \langle r \cdot r^*, r' \rangle = \langle r^*, r'r \rangle \tag{1.41}$$

for all $r, r' \in A$ and $r^* \in A^*$.

Definition 3. *A finite dimensional k-algebra A is called a Frobenius algebra if $A \cong A^*$ as right A-modules.*

Remarks 1. 1. A finite dimensional k-algebra A is Frobenius if and only if there exists a k-linear map $\lambda : \ A \to k$ such that for any $\psi \in A^*$ there exists a unique element $r = r_\psi \in A$ such that

$$\psi(x) = \lambda(rx)$$

for all $x \in A$. In particular, the matrix algebra $M_n(k)$ is Frobenius: take $\lambda = \mathrm{Tr}$, the trace map.
2. The concept of Frobenius algebra is left-right symmetric: that is $A \cong A^*$ in $\mathcal{M}_A$ if and only if $A \cong A^*$ in $_A\mathcal{M}$.
It suffices to observe that there exists a one to one correspondence between the following data:

- the set of all isomorphisms of right A-modules $f : A \to A^*$;
- the set of all bilinear, nondegenerate and associative maps $B : A \times A \to k$;
- the set of all isomorphisms of left A-modules $g : A \to A^*$, given by the formulas

$$f(x)(y) = B(x, y) = g(y)(x) \tag{1.42}$$

for all x, $y \in A$.

Let us now explain how the original problem of Frobenius arises naturally in representation theory, as explained in the book of Lam [113]. We fix a basis $\{e_1, \cdots, e_n\}$ of a finite dimensional algebra A. Then for any $r \in A$ we can find scalars $a_{ij}^{(r)}$ and $b_{ij}^{(r)}$ such that

$$e_i r = \sum_{j=1}^n a_{ij}^{(r)} e_j, \quad r e_i = \sum_{j=1}^n b_{ji}^{(r)} e_j \tag{1.43}$$

for all $i = 1, \cdots, n$. Hence we have constructed k-linear maps

$$\alpha, \beta : A \to M_n(k), \quad \alpha(r) = (a_{ij}^{(r)}), \quad \beta(r) = (b_{ij}^{(r)}) \tag{1.44}$$

for all $r \in A$. It is straightforward to prove that α and β are algebra maps, i.e. they are representations of the k-algebra A.

The problem of Frobenius: When are the above representations α and β equivalent? We recall that two representations α, $\beta : A \to M_n(k)$ are equivalent if there exists an invertible matrix $U \in M_n(k)$ such that

$$\beta(r) = U\alpha(r)U^{-1},$$

for all $r \in A$.

Before giving the answer to the problem we present one more construction: let $(c_{ij}^l)_{i,j,l=1,n}$ be the structure constants of the algebra A, that is

$$e_i e_j = \sum_{k=1}^n c_{ij}^k e_k$$

for all i, $j = 1, \cdots, n$. For $a = (a_1, \cdots, a_n) \in k^n$, let $P_a \in M_n(k)$ be the matrix given by

$$(P_a)_{i,j} = \sum_{k=1}^n a_k c_{ij}^k$$

The matrix P_a is called the *paratrophic matrix*.
In the next Theorem, the equivalence 2. $\Leftrightarrow$ 3. was the original theorem of Frobenius, while the equivalence 1. $\Leftrightarrow$ 2. translate the problem from representation theory into the language of modules.

Theorem 4. *For an n-dimensional algebra A, the following statements are equivalent:*

1. A is Frobenius;
2. the representations α and β : $A \to M_n(k)$ constructed in (1.44) are equivalent;
3. there exists $a \in k^n$ such that the paratrophic matrix P_a is invertible;
4. there exists a bilinear, nondegenerate and associative map B : $A \times A \to k$, i.e. $B(xy, z) = B(x, yz)$, for all x, y, $z \in A$;
5. there exists a hyperplane of A that does not contain a nonzero right ideal of A;
6. there exists a pair (ε, e), called a Frobenius pair , where $\varepsilon \in A^*$ and $e = \sum e^1 \otimes e^2 \in A \otimes A$ such that

$$ae = ea, \quad \text{and} \quad \sum \varepsilon(e^1)e^2 = \sum e^1\varepsilon(e^2) = 1. \qquad (1.45)$$

Before proving the Theorem, let us recall some well-known facts. First of all, let V be a n-dimensional vector space with basis $\mathcal{B} = \{v_1, \cdots, v_n\}$. Let

$$can_V : \ \text{End}_k(V)^{\text{op}} \to M_n(k), \quad can_V(f) = M_{\mathcal{B}}(f)$$

be the canonical isomorphism of algebras; here, for $f \in \text{End}_k(V)$, $M_{\mathcal{B}}(f) = (a_{ij})$, is the matrix asociated to f with respect to the basis $\mathcal{B}$ written as follows

$$f(v_i) = \sum_{j=1}^{n} a_{ij}v_j$$

for all $i = 1, \cdots, n$.

Secondly, a k-vector space M has a structure of right A-module if and only if there exists an algebra map

$$\varphi_M : \ A \to \text{End}_k(M)^{\text{op}}$$

φ_M is called the representation associated to M. The correspondence between the action " $\cdot$ " and the representation is given by $\varphi_M(r)(m) = m \cdot r$. In particular, if $\dim_k(M) = n$, M has a structure of right A-module if and only if there exists an algebra map

$$\tilde{\varphi}_M (= can_M \circ \varphi_M) : A \to M_n(k).$$

Finally, let M and N be two right A-modules and φ_M : $A \to \text{End}_k(M)$, φ_N : $A \to \text{End}_k(N)$ the associated representations. Then $M \cong N$ (as right A-modules) if and only if there exists an isomorphism of k-vector spaces θ : $M \to N$ such that

$$\varphi_M(r) = \theta^{-1} \circ \varphi_N(r) \circ \theta$$

for all $r \in A$. Indeed, a k-linear map $\theta : M \to N$ is a right A-module map if and only if $\theta(m \cdot r) = \theta(m) \cdot r$ for all $m \in M$, $r \in A$. This is equivalent to

$$\theta(\varphi_M(r)(m)) = \varphi_N(r)(\theta(m))$$

or

$$\theta \circ \varphi_M(r) = \varphi_N(r) \circ \theta$$

for all $r \in A$.

Proof. (of Theorem 4). 1. $\Leftrightarrow$ 2. This follows from the remarks made above if we can prove that

$$\alpha = \tilde{\varphi}_A, \quad \beta = \tilde{\varphi}_{A^*}.$$

Let us prove first that $\alpha = \tilde{\varphi}_A$, where $A \in \mathcal{M}_A$, via right multiplication. The representation associated to this structure is

$$\varphi_A : \ A \to \operatorname{End}_k(A), \quad \varphi_A(r)(r') = r'r$$

hence,

$$\varphi_A(r)(e_i) = e_i r = \sum_{j=1}^{n} a_{ij}^{(r)} e_j$$

i.e. $\alpha = \tilde{\varphi}_A$. Let us show next that $\beta = \tilde{\varphi}_{A^*}$. Let $\{e_i^*\}$ be the dual basis of $\{e_i\}$ and

$$\varphi_{A^*} : \ A \to \operatorname{End}_k(A^*), \quad \varphi_{A^*}(r)(r^*) = r^*(r).$$

Now $\beta(r) = \tilde{\varphi}_{A^*}(r)$ if and only if

$$e_i^* \cdot r = \sum_{j=1}^{n} b_{ij}^{(r)} e_j^*$$

or

$$\langle e_i^*, r e_k \rangle = \langle \sum_{j=1}^{n} b_{ij}^{(r)} e_j^*, e_k \rangle$$

for all k. Both sides are equal to $b_{ik}^{(r)}$.

1. $\Leftrightarrow$ 3. Any right A-module map $f : \ A \to A^*$ has the form $f(r) = \lambda \cdot r$, for some $\lambda \in A^*$. Thus, there exists $a_1, \cdots, a_n \in k$ such that

$$f(r) = (a_1 e_1^* + \cdots + a_n e_n^*) \cdot r$$

for any $r \in A$. Using the dual basis formula we have

$$\langle e_k^* \cdot e_i, e_j \rangle = \langle e_k^*, e_i e_j \rangle = c_{ij}^k$$

Hence $e_k^* \cdot e_i = \sum_{j=1}^{n} c_{ij}^k e_j^*$, and it follows that

$$f(e_i) = \sum_{k=1}^{n} a_k e_k^* \cdot e_i = \sum_{j=1}^{n} (\sum_{k=1}^{n} c_{ij}^k a_k) e_j^*$$

for all $i = 1, \cdots, n$. This means that the matrix associated to f in the pair of basis $\{e_i, e_i^*\}$ is just the paratrophic matrix P_a, where $a = (a_1, \cdots, a_n) \in k^n$.

1. $\Leftrightarrow$ 4. follows from (1.42).

4. $\Rightarrow$ 5. $H = \{a \in A \mid B(1, a) = 0\}$ is a k-subspace of A of codimension 1. Assume that J is a right ideal of A and $J \subset H$, and take $x \in J$. using the fact that $xA \subset J \subset H$, and that B is associative, we obtain

$$0 = B(1, xA) = B(x, A)$$

As B is nondegenerate we obtain that $x = 0$.

5. $\Rightarrow$ 1. Let H be a such a hyperplane. As k is a field, we can pick a k-linear map $\lambda: A \to k$ such that $\text{Ker}(\lambda) = H$. Then

$$f = f_\lambda: A \to A^*, \quad \langle f(x), y \rangle = \lambda(xy)$$

for all $x, y \in A$, is an injective right A-linear map. Indeed, for $x, y, z \in A$ we have

$$\langle f(xy), z \rangle = \lambda(xyz) = \langle f(x), yz \rangle = \langle f(x) \cdot y, z \rangle$$

On the other hand, from $f(x) = 0$ it follows that $\lambda(xA) = 0$, hence $xA \subset \text{Ker}(\lambda) = H$. We obtain, that $xA = 0$, i.e. $x = 0$. Thus, f is an injective right A-module map, that is an isomorphism as A and A^* have the same dimension.

1. $\Rightarrow$ 6. Let (e_i, e_i^*) be a dual basis of A and $f: A \to A^*$ an isomorphism of right A-modules. Then $(\varepsilon = f(1), e = \sum_i e_i \otimes f^{-1}(e_i^*))$ is a Frobenius pair. This is an elementary computation left to the reader at this point; in Theorem 28, we give the same proof in a more general situation.

6. $\Rightarrow$ 1. If $(\varepsilon, e = \sum e^1 \otimes e^2)$ is a Frobenius pair, then

$$f: A \to A^*, \quad \langle f(x), y \rangle = \varepsilon(xy)$$

is an isomorphism of right A-modules with inverse

$$f^{-1}: A^* \to A, \quad f^{-1}(a^*) = \sum \langle a^*, e^1 \rangle e^2$$

for all $a^* \in A^*$.

Examples 2. 1. Theorem 4 gives an elementary way to check whether an algebra A is Frobenius. Let $A = k[X, Y]/(X^2, Y^2)$. Then A has a basis $e_1 = 1$, $e_2 = x$, $e_3 = y$ and $e_4 = xy$. Through a trivial computation we find that the paratrophic matrix is

$$P_a = \begin{pmatrix} a_1 & a_2 & a_3 & a_4 \\ a_2 & 0 & a_4 & 0 \\ a_3 & a_4 & 0 & 0 \\ a_4 & 0 & 0 & 0 \end{pmatrix}$$

Thus, if a_4 is non-zero, then P_a is invertible, so A is a Frobenius algebra.

2. A similar computation shows that the k-algebra

$$A = k[X,Y]/(X^2, XY^2, Y^3)$$

is not Frobenius.

3. Using the criterium 5) given by Theorem 4 we can see that any finite dimensional division k-algebra D is a Frobenius algebra. It can be proved that $M_n(D)$ is also a Frobenius k-algebra. Using (1.40) and the fact that a product of Frobenius algebras is Frobenius algebra, we obtain that any separable algebra over a field is Frobenius.

2 Doi-Koppinen Hopf modules and entwined modules

In this Chapter, we introduce entwining structures and entwined modules. We show how various kinds of modules that appear in ring theory are special cases of entwined modules. We also show that there is a close analogy, based on duality arguments, with the factorization problem for algebras, and the smash product of algebras. Entwined modules themselves can be viewed as special cases of comodules over corings. Pairs of adjoint functors between categories of entwined modules are investigated, and it is discussed how one can make the category of entwined modules into a monoidal category.

2.1 Doi-Koppinen structures and entwining structures

Entwining structures Throughout this Section, k is a commutative ring. A (right-right) *entwining structure* on k consists of a triple (A, C, ψ), where A is a k-algebra, C a k-coalgebra, and $\psi : C \otimes A \to A \otimes C$ a k-linear map satisfying the relations

$$(ab)_\psi \otimes c^\psi = a_\psi b_\Psi \otimes c^{\psi\Psi} \tag{2.1}$$

$$(1_A)_\psi \otimes c^\psi = 1_A \otimes c \tag{2.2}$$

$$a_\psi \otimes \Delta_C(c^\psi) = a_{\psi\Psi} \otimes c_{(1)}^\Psi \otimes c_{(2)}^\psi \tag{2.3}$$

$$\varepsilon_C(c^\psi)a_\psi = \varepsilon_C(c)a \tag{2.4}$$

Here we used the sigma notation

$$\psi(c \otimes a) = a_\psi \otimes c^\psi = a_\Psi \otimes c^\Psi$$

A morphism $(\alpha, \gamma) : (A, C, \psi) \to (A', C', \psi')$ consists of an algebra map $\alpha : A \to A'$ and a coalgebra map $\gamma : C \to C'$ such that

$$(\alpha \otimes \gamma) \circ \psi = \psi' \circ (\gamma \otimes \alpha) \tag{2.5}$$

or, equivalently,

$$\alpha(a_\psi) \otimes \gamma(c^\psi) = \alpha(a)_{\psi'} \otimes \gamma(c)^{\psi'} \tag{2.6}$$

$\mathbb{E}_\bullet^\bullet(k)$ will denote the category of entwining structures. The category $\mathbb{E}_\bullet^\bullet(k)$ is monoidal. $\mathbb{E}^*{}_\bullet^\bullet(k)$ is the full subcategory of $\mathbb{E}_\bullet^\bullet(k)$ consisting of entwining structures (A, C, ψ) with ψ invertible.

Left-right, right-left, and left-left versions can also be introduced. For example, ${}_{\bullet}\mathbb{E}^{\bullet}(k)$ is the category with objects (A, C, ψ), where now

$$\psi : \ A \otimes C \to A \otimes C, \ \psi(a \otimes c) = a_\psi \otimes c^\psi$$

is a map satisfying (2.2,2.3,2.4) and

$$(ab)_\psi \otimes c^\psi = a_\psi b_\Psi \otimes c^{\Psi\psi} \tag{2.7}$$

In ${}^{\bullet}\mathbb{E}_{\bullet}(k)$, we need maps $\psi : \ C \otimes A \to C \otimes A$, satisfying (2.1,2.2,2.4) and

$$a_\psi \otimes \Delta_C(c^\psi) = a_{\psi\Psi} \otimes (c_{(1)})^\psi \otimes (c_{(2)})^\Psi \tag{2.8}$$

In ${}_{\bullet}^{\bullet}\mathbb{E}(k)$, we will need maps $\psi : \ A \otimes C \to C \otimes A$ satisfying (2.2,2.4,2.7) and (2.8).

Proposition 14. *The categories* $\mathbb{E}_{\bullet}^{\bullet}(k)$, ${}_{\bullet}\mathbb{E}^{\bullet}(k)$, ${}^{\bullet}\mathbb{E}_{\bullet}(k)$, *and* ${}_{\bullet}^{\bullet}\mathbb{E}(k)$ *are isomorphic.*

Proof. It is easy to see that the isomorphism between $\mathbb{E}_{\bullet}^{\bullet}(k)$ and ${}_{\bullet}\mathbb{E}^{\bullet}(k)$ is given by sending (A, C, ψ) to $(A^{\mathrm{op}}, C, \psi \circ \tau)$. The other isomorphisms are left to the reader.

Obviously the isomorphisms in Proposition 15 restrict to the subcategories consisting of structures with invertible ψ. For these subcategories, there exists alternative isomorphisms.

Proposition 15. *The categories* $\mathbb{E}^{*\bullet}_{\bullet}(k)$ *and* ${}_{\bullet}^{\bullet}\mathbb{E}^{*}(k)$ *are isomorphic via the functor* $\mathcal{S}$ *given by*

$$\mathcal{S}(A, C, \psi) = (A, C, \psi^{-1}) \tag{2.9}$$

Proof. Asssume that $\psi : \ C \otimes A \to A \otimes C$ satisfies (2.1-2.4). We have to show that $\varphi = \psi^{-1}$ satisfies (2.2,2.4, 2.7) and (2.8). (2.2) and (2.4) are obvious. (2.1) is equivalent to commutativity of the diagram

$$
\begin{array}{ccccc}
C \otimes A \otimes A & \xrightarrow{\psi \otimes I_A} & A \otimes C \otimes A & \xrightarrow{I_A \otimes \psi} & A \otimes A \otimes C \\
\downarrow{\scriptstyle I_C \otimes m_A} & & & & \downarrow{\scriptstyle m_A \otimes I_C} \\
C \otimes A & & \xrightarrow{\hspace{3cm}\psi\hspace{3cm}} & & A \otimes C
\end{array}
$$

This is equivalent to commutativity of the following diagram

$$
\begin{array}{ccccc}
A \otimes A \otimes C & \xrightarrow{I_A \otimes \varphi} & A \otimes C \otimes A & \xrightarrow{\varphi \otimes I_A} & C \otimes A \otimes A \\
\downarrow{\scriptstyle m_A \otimes I_C} & & & & \downarrow{\scriptstyle I_C \otimes m_A} \\
A \otimes C & & \xrightarrow{\hspace{3cm}\varphi\hspace{3cm}} & & C \otimes A
\end{array}
$$

which is equivalent to

$$c^{\varphi\phi} \otimes a_\phi b_\varphi = c^\varphi \otimes (ab)_\varphi$$

and this tells us that φ satisfies (2.7). In a similar way (2.3) implies that φ satisfies (2.8).

Doi-Koppinen structures Let H be a bialgebra, A a right H-comodule algebra, and C a right H-module coalgebra. We call (H, A, C) a right-right *Doi-Koppinen structure* or *DK structure* over k. A morphism between two DK structures consists of a triple $\varphi = (\hbar, \alpha, \gamma) : (H, A, C) \to (H', A', C')$, where $\hbar : H \to H'$, $\alpha : A \to A'$, and $\gamma : C \to C'$ are respectively a bialgebra map, an algebra map, and a coalgebra map such that

$$\rho_A(\alpha(a)) = \alpha(a_{[0]}) \otimes \hbar(a_{[1]}) \tag{2.10}$$
$$\gamma(ch) = \gamma(c)\hbar(h) \tag{2.11}$$

for all $a \in A$, $c \in C$, and $h \in H$. The category of right-right Doi-Hopf structures over k is denoted by $\mathbb{DK}_\bullet^\bullet(k)$. $\mathbb{DK}_\bullet^\bullet(k)$ is a monoidal category, if we define

$$(H, A, C) \otimes (H', A', C') = (H \otimes H', A \otimes A', C \otimes C')$$

with the obvious structure maps. The unit element is (k, k, k). We will also consider the full subcategories $\mathbb{H}_\bullet^\bullet(k)$, $\mathbb{HA}_\bullet^\bullet(k)$, and $\mathbb{HC}_\bullet^\bullet(k)$ of $\mathbb{DK}_\bullet^\bullet(k)$, consisting of objects respectively of the form

$$(H, H, H), \ (H, A, H), \ (H, H, C)$$

The subcategory of $\mathbb{DK}_\bullet^\bullet(k)$ consisting of objects (H, A, C) and morphisms $(\hbar, \alpha, \gamma)$ where H has a twisted antipode $\overline{S}$, and where $\hbar$ preserves the twisted antipode, is denoted by $\mathbb{DK}s_\bullet^\bullet(k)$.
In a similar way, we introduce the categories $_\bullet\mathbb{DK}^\bullet(k)$, $^\bullet\mathbb{DK}_\bullet(k)$, and $_\bullet^\bullet\mathbb{DK}(k)$, and their various subcategories. For example, $_\bullet\mathbb{DK}^\bullet(k)$ has objects (H, A, C), where A is a right H-comodule algebra, and C is a left H-module coalgebra. In the definition of $_\bullet\mathbb{DK}^{\bullet\bullet}(k)$ and $^\bullet\mathbb{DK}^\bullet_\bullet(k)$, we require that the bialgebra H in each object is a Hopf algebra (i.e., it has an antipode). In the left-left case, we want a twisted antipode.

Proposition 16. *The categories* $\mathbb{DK}_\bullet^\bullet(k)$, $_\bullet\mathbb{DK}^\bullet(k)$, $^\bullet\mathbb{DK}_\bullet(k)$, *and* $_\bullet^\bullet\mathbb{DK}(k)$ *are isomorphic. Similar statements hold for the respective subcategories introduced above.*

Proof. Let $(H, A, C) \in \mathbb{DK}_\bullet^\bullet(k)$. Then the opposite algebra A^{op} with the original right H-coaction is a right H^{op}-comodule algebra. The coalgebra C with left H^{op}-action defined by

$$h^{\mathrm{op}} \cdot c = ch$$

is a left H^{op}-module coalgebra. The functor $\mathbb{DK}_\bullet^\bullet(k) \to {}_\bullet\mathbb{DK}^\bullet(k)$, mapping (H, A, C) to $(H^{\mathrm{op}}, A^{\mathrm{op}}, C)$ is easily seen to be an isomorphism of categories. Observe also that H^{op} has an antipode in case H has a twisted antipode, so we also find an isomorphism between the categories $\mathbb{DK}s_\bullet^\bullet(k)$ and ${}_\bullet\mathbb{DK}s^\bullet(k)$. The other statements follow in a similar way, let us mention that the objects corresponding to $(H, A, C) \in \mathbb{DK}_\bullet^\bullet(k)$ are $(H^{\mathrm{cop}}, A, C^{\mathrm{cop}}) \in {}^\bullet\mathbb{DK}_\bullet$ and $(H^{\mathrm{opcop}}, A^{\mathrm{op}}, C^{\mathrm{cop}}) \in {}_\bullet^\bullet\mathbb{DK}(k)$.

Proposition 17. *We have faithful functors*

$$F: \ \mathbb{DK}_\bullet^\bullet(k) \to \mathbb{E}_\bullet^\bullet(k)$$

and

$$F: \ \mathbb{DK}^*{}_\bullet^\bullet(k) \to \mathbb{E}^*{}_\bullet^\bullet(k)$$

Proof. We define F by

$$F(H, A, C) = (A, C, \psi) \ ; \ \ F(\hbar, \alpha, \gamma) = (\alpha, \gamma)$$

with $\psi: \ C \otimes A \to A \otimes C$ given by

$$\psi(c \otimes a) = a_{[0]} \otimes ca_{[1]}$$

We leave it to the reader to check that ψ satisfies (2.1-2.4), and that (α, γ) satisfies (2.5). If $(H, A, C) \in \mathbb{DK}s_\bullet^\bullet(k)$, then H has a twisted antipode $\overline{S}$, and the inverse of ψ is given by the formula

$$\psi^{-1}(a \otimes c) = c\overline{S}(a_{[1]}) \otimes a_{[0]}$$

Alternative Doi-Koppinen structures These structures were recently introduced by Schauenburg [163]. A left-right *alternative Doi-Koppinen structure* consists of a triple (H, A, C), where H is a bialgebra, A is left H-module algebra, and C is a right H-comodule coalgebra. We write ${}_\bullet a\mathbb{DK}^\bullet(k)$ for the category left-right alternative Doi-Koppinen structures. The morphisms are defined in the obvious way, and analogous definitions can be given in the right-left, left-left and right-right situations. The alternative version of Proposition 17 is the following:

Proposition 18. *We have a faithful functor*

$$F_a: \ {}_\bullet a\mathbb{DK}^\bullet(k) \to {}_\bullet\mathbb{E}^\bullet(k)$$

Proof. We put

$$F_a(H, A, C) = (A, C, \psi) \ \text{ and } \ F_a(\hbar, \alpha, \gamma) = (\alpha, \gamma)$$

with

$$\psi: \ A \otimes C \to A \otimes C, \ \ \psi(a \otimes c) = c_{[1]}a \otimes c_{[0]}$$

A straightforward computation shows that (A, C, ψ) is a left-right entwining structure.

Doi-Koppinen structures versus entwining structures An obvious question is the following: is an entwining structure (A, C, ψ) defined by a Doi-Koppinen structure, i.e. can we find a bialgebra H, an H-coaction on A, and an H-action on C such that

$$(A, C, \psi) = F(H, A, C)$$

We will see that a sufficient condition is that A is finitely generated and projective as a k-module. If C is finitely generated and projective, then every entwining structure comes from an alternative Doi-Koppinen structure. A recent counterexample due to Schauenburg shows that there exist entwining structures that do not arise from Doi-Koppinen structures.

We start with a construction due to Sweedler [172, p.155], reformulated by Tambara [178]. Let A be finitely generated projective, with dual basis $\{a_i, a_i^* \mid i = 1, \cdots, n\}$, and write

$$H = H(A) = T(A^* \otimes A)/I$$

the tensor algebra of $A^* \otimes A$ divided by the ideal I generated by elements of the form

$$\langle a^*, 1_A \rangle - a^* \otimes 1_A \tag{2.12}$$

$$a^* \otimes ab - (a_{(1)}^* \otimes a) \otimes (a_{(2)}^* \otimes b) \tag{2.13}$$

where $a^* \in A^*$ and $a, b \in A$. We write $[a^* \otimes a]$ for the class represented by $a^* \otimes a$.

Proposition 19. *Let A be a finitely generated projective k-algebra. Then $H = H(A)$ is a bialgebra, with comultiplication and counit given by*

$$\Delta_H[a^* \otimes a] = \sum_{i=1}^{n} [a^* \otimes a_i] \otimes [a_i^* \otimes a] \quad \text{and} \quad \varepsilon_H[a^* \otimes a] = \langle a^*, a \rangle$$

Proof. A straightforward calculation; we will show that Δ_H is well-defined, i.e. $\Delta_H = 0$ on I. First,

$$\Delta_H(a^* \otimes 1_A) = \sum_{i=1}^{n} [a^* \otimes a_i] \otimes [a_i^* \otimes 1_A]$$

$$= \sum_{i=1}^{n} [a^* \otimes a_i] \otimes \langle a_i^*, 1_A \rangle = \langle a^*, 1_A \rangle 1 \otimes 1$$

Next

$$\Delta_H([(a_{(1)}^* \otimes a) \otimes (a_{(2)}^* \otimes b)])$$

$$= \left(\sum_{i=1}^{n} [a_{(1)}^* \otimes a_i] \otimes [a_i^* \otimes a] \right) \left(\sum_{j=1}^{n} [a_{(2)}^* \otimes a_j] \otimes [a_j^* \otimes b] \right)$$

$$= \sum_{i,j=1}^{n} [a^*_{(1)} \otimes a_i][a^*_{(2)} \otimes a_j] \otimes [a^*_i \otimes a][a^*_j \otimes b]$$

$$= \sum_{i,j=1}^{n} [a^* \otimes a_i a_j] \otimes [a^*_i \otimes a][a^*_j \otimes b]$$

$$(1.4) \quad = \sum_{i=1}^{n} [a^* \otimes a_i] \otimes [a^*_{i(1)} \otimes a][a^*_{i(1)} \otimes b]$$

$$= \sum_{i=1}^{n} [a^* \otimes a_i] \otimes [a^*_i \otimes ab]$$

$$= \Delta_H [a^* \otimes ab]$$

Remark 2. As above, let A be finitely generated and projective, and consider the functor

$$F : \ k\text{-Alg} \to k\text{-Alg} \ ; \quad F(B) = A \otimes B$$

Tambara [178] observes that F has a right adjoint G, and, as a k-algebra, $H(A) = G(A)$. For any k-algebra B, we write

$$G(B) = a(A, B) = T(A^* \otimes B)/I$$

with I defined as above (1_A is replaced by 1_B, and $a, b \in B$). The unit and counit of the adjunction are given by

$$\eta_B : \ B \to a(A, A \otimes B) \ ; \quad \eta_B(b) = \sum_{i=1}^{n} [a^*_i \otimes (a_i \otimes b)]$$

$$\varepsilon_B : \ A \otimes a(A, B) \to B \ ; \quad \varepsilon_B(a \otimes [a^* \otimes b]) = \langle a^*, a \rangle b$$

The comultiplication and counit on $a(A, A) = H(A)$ can be defined using the adjunction properties.

Proposition 20. *Let A be a finitely generated projective k-algebra, and $H = H(A)$. Then A is a right H-comodule algebra, and A^* is a left H-comodule coalgebra. The structure maps are*

$$\rho^r(a) = \sum_{i=1}^{n} a_i \otimes [a^*_i \otimes a] \tag{2.14}$$

$$\rho^l(a^*) = \sum_{i=1}^{n} [a^* \otimes a_i] \otimes a^*_i \tag{2.15}$$

Proof. A is a right H-comodule since

$$(\rho^r \otimes I_H)(\rho^r(a)) = \sum_{i,j=1}^{n} a_j \otimes [a^*_j \otimes a_i] \otimes [a^*_i \otimes a] = (I_A \otimes \Delta_H)(\rho^r(a))$$

A is a right H-comodule algebra since

$$\rho^r(ab) = \sum_{i=1}^{n} a_i \otimes [a_i^* \otimes ab]$$

$$(2.13) \qquad = \sum_{i=1}^{n} a_i \otimes [a_{i(1)}^* \otimes a][a_{i(2)}^* \otimes b]$$

$$(1.4) \qquad = \sum_{i,j=1}^{n} a_i a_j \otimes [a_i^* \otimes a][a_j^* \otimes b]$$

$$= \rho^r(a)\rho^r(b)$$

$$\rho^r(1_A) = \sum_{i=1}^{n} a_i \otimes [a_i^* \otimes 1_A]$$

$$(2.12) \qquad = \sum_{i=1}^{n} a_i \otimes \langle a_i^*, 1_A \rangle = 1_A \otimes 1_H$$

From Proposition 7, it follows that A^* is a left H-comodule coalgebra, with

$$\rho^l(a^*) = \sum_{i=1}^{n} a_{i[1]} \otimes \langle a^*, a_{i[0]} \rangle a_i^*$$

$$= \sum_{i,j=1}^{n} [a_j^* \otimes a_i] \otimes \langle a^*, a_j \rangle a_i^*$$

$$= \sum_{i=1}^{n} [a^* \otimes a_i] \otimes a_i^*$$

Theorem 5. *Let A be a finitely generated projective algebra, and C a coalgebra. There is a bijective correspondence between left $H(A)$-module coalgebra structures on C, and left-right entwining structures of the form (A, C, ψ). Consequently every entwining structure (A, C, ψ) with A finitely generated and projective can be derived from a Doi-Koppinen structure.*

Proof. First consider an entwining structure $(A, C, \psi) \in \bullet\mathbb{E}^\bullet(k)$. As before, we write $H = H(A)$. On C, we define the following left H-action:

$$[a^* \otimes a] \cdot c = \langle a^*, a_\psi \rangle c^\psi \qquad (2.16)$$

This action is well-defined since

$$[a^* \otimes 1_A] \cdot c = \langle a^*, (1_A)_\psi \rangle c^\psi = \langle a^*, 1_A \rangle c$$

and

$$[a^* \otimes ab] \cdot c = \langle a^*, (ab)_\psi \rangle c^\psi = \langle a^*, a_\psi b_\Psi \rangle c^{\Psi\psi}$$
$$= \langle a_{(1)}^*, a_\psi \rangle \langle a_{(2)}^*, b_\Psi \rangle c^{\Psi\psi}$$
$$= [a_{(1)}^*, a] \cdot ([a_{(2)}^*, b] \cdot c)$$

The comultiplication and counit of C are left H-linear since

$$\sum_{i=1}^{n}[a^* \otimes a_i] \cdot c_{(1)} \otimes [a_i^* \otimes a] \cdot c_{(2)} = \sum_{i=1}^{n}\langle a^*, a_{i\psi}\rangle c_{(1)}^{\psi} \otimes \langle a_i^*, a_{\Psi}\rangle c_{(2)}^{\Psi}$$

$$= \langle a^*, a_{\Psi\psi}\rangle c_{(1)}^{\psi} \otimes c_{(2)}^{\Psi} = \langle a^*, a_{\psi}\rangle \Delta(c^{\psi}) = \Delta([a^* \otimes a] \cdot c)$$

and

$$\varepsilon_C([a^* \otimes a] \cdot c) = \varepsilon_C(\langle a^*, a_{\psi}\rangle c^{\psi}) = \langle a^*, a\rangle \varepsilon_C(c) = \varepsilon_H([a^* \otimes a])\varepsilon_C(c)$$

Conversely, let C be a left H-module coalgebra. We know from Proposition 20 that A is a right H-comodule algebra, so we have $(H, A, C) \in {}_\bullet\mathbb{DK}^\bullet(k)$ and $(A, C, \psi) = F(H, A, C) \in {}_\bullet\mathbb{E}^\bullet(k)$. Recall that $\psi : A \otimes C \to A \otimes C$ is given by

$$\psi(a \otimes c) = a_{[0]} \otimes a_{[1]}c$$

Let us check that we have a bijective correspondence, as needed. Let (A, C, ψ) be an entwining structure, (H, A, C) the corresponding Doi-Koppinen structure, and write $F(H, A, C) = (A, C, \psi')$. Then

$$\psi'(a \otimes c) = a_{[0]} \otimes a_{[1]}c = \sum_{i=1}^{n} a_i \otimes [a_i^* \otimes a] \cdot c$$

$$= \sum_{i=1}^{n} a_i \otimes \langle a_i^*, a_{\psi}\rangle c^{\psi} = a_{\psi} \otimes c^{\psi} = \psi(a \otimes c)$$

If C is a left $H(A)$-module coalgebra, then we have a left-right Doi-Koppinen structure $(H(A), A, C)$, and an entwining structure $(A, C, \psi) = F(H(A), A, C)$. This entwining structure defines a left $H(A)$-module coalgebra structure on C, we denote this action temporarily by $\rightharpoonup$. This action equals the original one, since

$$[a^* \otimes a] \rightharpoonup c = \langle a^*, a_{\psi}\rangle c^{\psi} = \langle a^*, a_{[0]}\rangle a_{[1]} \cdot c$$

$$= \sum_{i=1}^{n}\langle a^*, a_i\rangle[a_i^* \otimes a] \cdot c = [a^* \otimes a] \cdot c$$

Adapting our arguments, we see that every entwining structure (A, C, ψ), with C finitely generated and projective, comes from an *alternative* Doi-Koppinen structure. This time, the involved bialgebra is $H' = H(C^*)^{\text{cop}}$. Observe that $H(C^*)^{\text{cop}} = T(C^* \otimes C)/I$, with I generated by elements of the form

$$\langle \varepsilon_C, c\rangle - \varepsilon_C \otimes c \quad \text{and} \quad c^* * d^* \otimes c - (c^* \otimes c_{(1)}) \otimes (d^* \otimes c_{(2)})$$

and

$$\Delta_{H'}([c^* \otimes c]) = \sum_{i=1}^{n} [c^* \otimes c_i] \otimes [c_i^* \otimes c] \quad \text{and} \quad \varepsilon_{H'}([c^* \otimes c]) = \langle c^*, c \rangle$$

From Proposition 20, it follows that C is a right H-comodule algebra, with

$$\rho^r(c) = \sum_{i=1}^{n} c_i \otimes [c_i^* \otimes c]$$

Given a left-right entwining structure (A, C, ψ), we define a left H'-action on A as follows:

$$[c^* \otimes c] \cdot a = \langle c^*, c^\psi \rangle a_\psi$$

and this makes A into a left H'-module algebra. (H', A, C) is a left-right alternative Doi-Koppinen structure. Further verifications are left to the reader. We summarize our results as follows:

Theorem 6. *Let C be a finitely generated projective coalgebra, and $H' = H(C^*)^{\mathrm{cop}}$. Then C is a right H'-comodule coalgebra. For a given k-algebra A, there is a bijective correspondence between left-right entwining structures (A, C, ψ) and left H'-module algebra structures on A. Consequently every entwining structure comes from an alternative Doi-Koppinen structure.*

We will now show that not every entwining structure arises from a Doi-Koppinen structure. Let k be a field. For a left-right entwining structure (A, C, ψ), $c \in C$ and $c^* \in C^*$, we consider the transformation

$$T_{c,c^*} : A \to A \; ; \; T_{c,c^*}(a) = \langle c^*, c^\psi \rangle a_\psi$$

If $(A, C, \psi) = F(H, A, C)$ arises from a Doi-Koppinen structure, then

$$T_{c,c^*}(a) = \langle c^*, a_{[1]}c \rangle a_{[0]}$$

and then every H-subcomodule of A is T_{c,c^*}-invariant. As every $a \in A$ is contained in a finite dimensional H-subcomodule of A (cf. [172, Theorem 2.1.3b]), the T_{c,c^*}-invariant subspace of A generated by a is finite dimensional.

Example 7. (Schauenburg [163]) Let $C = k \oplus kt$, with t primitive, and let A be the free algebra with generators X_i, where i ranges over the integers. We define $\psi : A \otimes C \to A \otimes C$ by

$$\psi(a \otimes 1) = a \otimes 1$$

for all $a \in k$, and

$$\psi(X_{i_1} X_{i_2} \cdots X_{i_n} \otimes t) = X_{i_1+1} X_{i_2+1} \cdots X_{i_n+1} \otimes t$$

A straightforward computation shows that ψ is entwining. Now take $c^* \in C^*$ such that $\langle c^*, t \rangle = 1$. Then

$$T_{t,c^*}(X_i) = X_{i+1}$$

and the T_{t,c^*}-invariant subspace of A generated by X_0 is infinite dimensional, and (A, C, ψ) cannot be derived from a Doi-Koppinen structure.

2.2 Doi-Koppinen modules and entwined modules

Entwined modules Let $(A, C, \psi) \in \mathbb{E}_\bullet^\bullet(k)$. An (A, C, ψ)-*entwined module* is a k-module with a right A-action and a right C-coaction such that

$$\rho^r(ma) = m_{[0]}a_\psi \otimes m_{[1]}^\psi \qquad (2.17)$$

The category of (A, C, ψ)-entwined modules and A-linear C-colinear maps is denoted by $\mathcal{M}(\psi)_A^C$. We also have left-left, left-right and right-left versions: For $(A, C, \varphi) \in {}_\bullet^\bullet\mathbb{E}(k)$, ${}_A^C\mathcal{M}(\varphi)$ consists of left A-modules and left C-comodules such that

$$\rho^l(am) = m_{[-1]}^\varphi \otimes a_\varphi m_{[0]} \qquad (2.18)$$

For $(A, C, \psi) \in {}_\bullet\mathbb{E}^\bullet(k)$, ${}_A\mathcal{M}^C(\psi)$ consists of left A-modules and right C-comodules such that

$$\rho^r(am) = a_\psi m_{[0]} \otimes m_{[1]}^\psi \qquad (2.19)$$

For $(A, C, \varphi) \in {}^\bullet\mathbb{E}_\bullet(k)$, ${}^C\mathcal{M}_A(\varphi)$ consists of right A-modules and left C-comodules such that

$$\rho^l(ma) = m_{[-1]}^\varphi \otimes m_{[0]}a_\varphi \qquad (2.20)$$

Examples 3. 1. For $(A, C, \psi) \in \mathbb{E}_\bullet^\bullet(k)$, $A \otimes C$ and $C \otimes A$ are right-right entwined modules. The structure is given by the formulas

$$(c \otimes a)b = c \otimes ab \qquad \rho^r(c \otimes a) = (c_{(1)} \otimes a_\psi) \otimes c_{(2)}^\psi$$
$$(a \otimes c)b = ab_\psi \otimes c^\psi \qquad \rho^r(a \otimes c) = a \otimes c_{(1)} \otimes c_{(2)}$$

$\psi : C \otimes A \to A \otimes C$ is a morphism in $\mathbb{E}_\bullet^\bullet(k)$. See also Examples 13 and 14

2. Let H be a bialgebra with twisted antipode $\overline{S}$, and consider the map $\psi : H \otimes H \to H \otimes H$ with

$$\psi(h \otimes k) = h_{(2)} \otimes h_{(3)}k\overline{S}(h_{(1)})$$

$(H, H, \psi) \in {}_\bullet\mathbb{E}^\bullet(k)$, and the objects of ${}_H\mathcal{M}(\psi)^H$ are k-modules with a left H-action and right H-coaction such that

$$\rho^r(hm) = h_{(2)}m_{[0]} \otimes h_{(3)}m_{[1]}\overline{S}(h_{(1)})$$

These modules are known under the name *Yetter-Drinfeld modules* . Yetter-Drinfeld modules will be investigated in Section 4.4 and Chapter 5.

3. For any bialgebra H, $(H, H, I_{H \otimes H}) \in {}_\bullet\mathbb{E}^\bullet(k)$. ${}_H\mathcal{M}(\psi)^H$ consists of k-modules with a left H-action and right H-coaction such that

$$\rho^r(hm) = hm_{[0]} \otimes m_{[1]}$$

i.e. the right H-coaction is left H-linear. In the situation where H is commutative and cocommutative, this type of modules has been considered first by Long in [118]; in the sequel, we will refer to them as *Long dimodules*. If H is commutative and cocommutative, then Long dimodules and Yetter-Drinfeld modules coincide. We will come back more extensively to Long dimodules in Section 4.5 and Chapter 7.

Doi-Koppinen Hopf modules Let $(H, A, C) \in \mathbb{DK}_\bullet^\bullet(k)$, and $(A, C, \psi) = F(H, A, C)$ the corresponding object in $\mathbb{E}_\bullet^\bullet(k)$. $\mathcal{M}(H)_A^C$ will be another notation for $\mathcal{M}(\psi)_A^C$. (2.17) takes the form

$$\rho^r(ma) = m_{[0]}a_{[0]} \otimes m_{[1]}a_{[1]} \tag{2.21}$$

The objects of $\mathcal{M}(H)_A^C$ are called *unified Hopf modules*, *Doi-Hopf modules* or *Doi-Koppinen Hopf modules*. If H has a twisted antipode, then ψ is bijective, and $(A, C, \psi^{-1}) \in {}_\bullet^\bullet\mathbb{E}(k)$. ${}_A^C\mathcal{M}(H)$ will be a new notation for ${}_A^C\mathcal{M}(\psi^{-1})$. This category consists of left A-modules and left C-comodules such that

$$\rho^l(am) = m_{[-1]}\overline{S}(a_{[1]}) \otimes a_{[0]}m_{[0]} \tag{2.22}$$

Similar constructions apply to left-right, right-left and left-left Doi-Koppinen structures.

Examples 4. 1. $(k, A, k) \in \mathbb{DK}_\bullet^\bullet(k)$, and the corresponding entwining structure is $(A, k, I_A) \in \mathbb{E}_\bullet^\bullet(k)$. The category of entwined modules is nothing else then the category of right A-modules.
2. $(k, k, C) \in \mathbb{DK}_\bullet^\bullet(k)$ corresponds to $(k, C, I_C) \in \mathbb{E}_\bullet^\bullet(k)$. An entwined module is now simply a right C-comodule.
3. Let H be a bialgebra. $(H, H, H) \in \mathbb{DK}_\bullet^\bullet(k)$ corresponds to $(H, C, \psi) \in \mathbb{E}_\bullet^\bullet(k)$, with

$$\psi(h \otimes k) = k_{(1)} \otimes hk_{(2)}$$

An entwined module is now a Hopf module in the sense of Sweedler [172].
4. If H is a bialgebra, and A is a right H-comodule algebra, then $(H, A, H) \in \mathbb{DK}_\bullet^\bullet(k)$ is a right-right Doi-Koppinen structure. The corresponding Doi-Koppinen modules are the well-known (A, H)-*relative Hopf modules* (see e.g. [67]).
5. In a similar way, if C is a right H-module coalgebra, then $(H, H, C) \in \mathbb{DK}_\bullet^\bullet(k)$ is a right-right Doi-Koppinen structure, and the corresponding Doi-Koppinen modules are the $[H, C]$)-Hopf modules studied in [67].
6. Let G be a group, and A a G-graded k-algebra. Then (kG, A, kG) is a Doi-Koppinen structure, and the corresponding category of Doi-Koppinen modules is the category of G-graded right A-modules.
7. Now let X be a right G-set. Then kX is a right kG-module coalgebra, and (kG, A, kX) is a Doi-Koppinen structure. $\mathcal{M}(kG)_A^{kX}$ consists of right A-modules graded by the G-set X (see [143] and [147] for a study of modules graded by G-sets).
8. Yetter-Drinfeld modules and Long dimodules are special cases of Doi-Hopf modules. This will be explained in Section 4.4 and 4.5.

Proposition 21. *For $(A, C, \psi) \in \mathbb{E}_\bullet^\bullet(k)$, the categories $\mathcal{M}(\psi)_A^C$, ${}_{A^{op}}\mathcal{M}(\psi \circ \tau)^C$, ${}^{C^{cop}}\mathcal{M}(\tau \circ \psi)_A$ and ${}_{A^{op}}^{C^{cop}}\mathcal{M}(\tau \circ \psi \circ \tau)$ are isomorphic. In particular, for $(H, A, C) \in \mathbb{DK}_\bullet^\bullet(k)$, the categories $\mathcal{M}(H)_A^C$, ${}_{A^{op}}\mathcal{M}(H^{op})^C$, ${}^{C^{cop}}\mathcal{M}(H^{cop})_A$ and ${}_{A^{op}}^{C^{cop}}\mathcal{M}(H^{opcop})$ are isomorphic.*

Proof. Everything is straightforward. For example, if $M \in \mathcal{M}(\psi)_A^C$, then the corresponding object in $_{A^{\mathrm{op}}}\mathcal{M}(\psi \circ \tau)^C$ is equal to M as a right C-comodule, but with left A^{op}-action given by

$$a^{\mathrm{op}} \cdot m = ma$$

All the other isomorphisms are defined in a similar way, and we leave further details to the reader.

2.3 Entwined modules and the smash product

Let A and B be k-algebras, and consider a map $R : B \otimes A \to A \otimes B$. We will use the following notation (summation understood):

$$R(b \otimes a) = a_R \otimes b_R = a_r \otimes b_r \qquad (2.23)$$

We put $A\#_R B = A \otimes B$ as a k-module, but with a new multiplication:

$$m_{A\#_R B} = (m_A \otimes m_B) \circ (I_A \otimes R \otimes I_B) \qquad (2.24)$$

or

$$(a\#b)(c\#d) = ac_R\#b_R d \qquad (2.25)$$

If this new multiplication makes $A\#_R B$ into an associative algebra with unit $1\#1$, then we call $A\#_R B$ a *smash product*, and (A, B, R) a *smash product structure* or a *factorization structure*.

Theorem 7. *([44]) (A, B, R) is a smash product structure if and only if*

$$R(b \otimes 1_A) = 1_A \otimes b \qquad (2.26)$$
$$R(1_B \otimes a) = a \otimes 1_B \qquad (2.27)$$
$$R(bd \otimes a) = a_{Rr} \otimes b_r d_R \qquad (2.28)$$
$$R(b \otimes ac) = a_R c_r \otimes b_{Rr} \qquad (2.29)$$

for all $a, c \in A$, $b, d \in B$.

Proof. Assume that (A, B, R) is a smash product structure. Then for all $b \in B$, we have
$$1_A\#b = (1_A\#b)(1_A\#1_B) = 1_R\#b_R$$
and (2.26) follows. (2.27) follows in a similar way. The multiplication is associative, so
$$(1\#b)\big((a\#1)(c\#1)\big) = \big((1\#b)(a\#1)\big)(c\#1)$$
and this implies (2.29). (2.28) follows from
$$\big((1\#b)(1\#d)\big)(a\#1) = (1\#b)\big((1\#d)(a\#1)\big)$$
the converse is left to the reader.

Remark 3. The smash product is related to the *factorization problem*. We say that a k-algebra X factorizes through k-algebras A and B if $X \cong A \otimes B$ as k-modules, and, after identifying X and $A \otimes B$, the maps

$$\iota_A : A \to A \otimes B = X \qquad \iota_A(a) = a \otimes 1_B$$
$$\iota_B : B \to A \otimes B = X \qquad \iota_B(b) = 1_A \otimes b$$

are algebra maps. It is not hard to show that there exists a bijective correspondence between the algebra structures on $A \otimes B$ for which ι_A and ι_B are algebra maps, and smash product structures of the form (A, B, R): given the multiplication m_X on X, we put

$$R(b \otimes a) = m_X(\iota_B(b) \otimes \iota_A(a))$$

Let $(A, B, R), (A', B', R')$ be smash product structures. A morphism $(A, B, R) \to (A', B', R')$ consists of a pair (α, β), where $\alpha : A \to A'$ and $\beta : B \to B'$ are algebra maps such that $(\alpha \otimes \beta) \circ R = R' \circ (\beta \otimes \alpha)$, or

$$\alpha(a_R) \otimes \beta(b_R) = \alpha(a)_{R'} \otimes \beta(b)_{R'}$$

for all $a \in A$ and $b \in B$. $\mathbb{S}(k)$ will denote the category of smash product structures over k.

Proposition 22. *If $(A, B, R) \in \mathbb{S}(k)$ is a smash product structure, then $(B^{op}, A^{op}, \tau \circ R \circ \tau)$ is also a smash product structure. Furthermore the switch map*

$$\tau : (A \#_R B)^{op} \to B^{op} \#_{\tau \circ R \circ \tau} A^{op}$$

is an algebra isomorphism.

Proof. The first statement is obvious. To prove the second one, we only need to show that τ is anti-multiplicative. Indeed,

$$\tau(c\#d)\tau(a\#b) = (d\#c)(b\#a) = d \cdot b_R \# c_R \cdot a$$
$$= b_R d \# R a c_R = \tau(a c_R \#_R b_R d)$$
$$= \tau((a\#b)(c\#d))$$

Proposition 23. *Let $(A, B, R) \in \mathbb{S}(k)$ be a smash product structure, and assume that R is invertible. Then $(B, A, S = R^{-1})$ is also a smash product structure, and*

$$R : B \#_S A \to A \#_R B$$

is an algebra isomorphism, with inverse S.

Proof. We write down conditions (2.26-2.29) as commutative diagrams. Change the direction of the morphisms involving R, and replace R by $R^{-1} = S$. We then obtain the diagram telling that (B, A, R^{-1}) is a smash product structure. We are left to prove that R is multiplicative. This works as follows: for all $b, d \in B$ and $a, c \in A$, we have

$$R((b\#a)(d\#c)) = R(bd_S\#a_Sc)$$
$$= (a_Sc)_R\#(bd_S)_R$$
$$(2.28) \quad = (a_Sc)_{R_1R_2}\#b_{R_2}d_{SR_1}$$
$$(2.29) \quad = (a_{SR_1}c_{R_3})_{R_2}\#b_{R_2}d_{SR_1R_3}$$
$$(S=R^{-1}) \quad = (ac_{R_3})_{R_2}\#b_{R_2}d_{R_3}$$
$$(2.29) \quad = a_{R_2}c_{R_3R_4}\#b_{R_2R_4}d_{R_3}$$
$$= (a_{R_2}\#b_{R_2})(c_{R_3}\#d_{R_3})$$
$$= R(b\#a)R(d\#c)$$

Theorem 8. *Let A be a k-algebra, and C a k-coalgebra which is finitely generated and projective as a k-module. Then there is a bijective correspondence between left-right entwining structures of the form (A,C,ψ) and smash product structures of the form (A,C^*,R). If R corresponds to ψ, then the categories $_A\mathcal{M}(\psi)^C$ and $_{A\#_RC^*}\mathcal{M}$ are isomorphic.*

Proof. Let $\{c_i, c_i^* \mid i = 1, \cdots, n\}$ be a dual basis for C. For an entwining structure $(A, C, \psi) \in {}_\bullet\mathbb{E}^\bullet(k)$, we define $f(\psi) = R : C^* \otimes A \to A \otimes C^*$ by

$$R(c^* \otimes a) = \sum_i \langle c^*, c_i^\psi \rangle a_\psi \otimes c_i^* \tag{2.30}$$

We claim that (A, C^*, R) is a smash product structure. For all $c^*, d^* \in C^*$ and $a, b \in A$, we have

$$a_{Rr} \otimes c_r^* * d_R^* = \sum_i \langle c^*, c_i^\psi \rangle (A_R)_\psi \otimes c_i^* * d_R^*$$
$$= \sum_{i,j} \langle c^*, c_i^\psi \rangle \langle d^*, c_j^\Psi \rangle a_{\Psi\psi} \otimes c_i^* * c_j^*$$
$$(1.5) \quad = \sum_i \langle c^*, (c_{i(1)})^\psi \rangle \langle d^*, (c_{i(2)})^\Psi \rangle a_{\Psi\psi} \otimes c_i^*$$
$$(2.3) \quad = \sum_i \langle c^*, (c_i^\psi)_{(1)} \rangle \langle d^*, (c_i^\psi)_{(2)} \rangle a_\psi \otimes c_i^*$$
$$= \sum_i \langle c^* * d^*, c_i^\alpha \rangle a_\psi \otimes c_i^*$$
$$= R(c^* * d^* \otimes a)$$

proving (2.28).

$$a_Rb_r \otimes (c^*)_{Rr} = \sum_i \langle c_R^*, c_i^\psi \rangle a_Rb_\psi \otimes c_i^*$$
$$= \sum_{i,j} \langle c_j^*, c_i^\psi \rangle \langle c^*, c_j^\Psi \rangle a_\Psi b_\psi \otimes c_i^*$$
$$= \sum_i \langle c^*, c_i^{\psi\Psi} \rangle a_\Psi b_\psi \otimes c_i^*$$

$$(2.7) \quad = \sum_i \langle c^*, c_i^\psi \rangle (ab)_\psi \otimes c_i^*$$

$$= R(c^* \otimes ab)$$

proving (2.29). (2.26) and (2.27) are left to the reader.
Conversely, if (A, C^*, R) is a smash product structure, then we define $g(R) = \psi : A \otimes C \to A \otimes C$ by

$$\psi(a \otimes c) = \sum_i \langle (c_i^*)_R, c \rangle a_R \otimes c_i \tag{2.31}$$

Then

$$a_{\Psi\psi} \otimes (c_{(1)})^\psi \otimes (c_{(2)})^\Psi = \sum_i \langle (c_i^*)_R, c_{(1)} \rangle a_{\Psi R} \otimes c_i \otimes (c_{(2)})^\Psi$$

$$= \sum_{i,j} \langle (c_i^*)_R, c_{(1)} \rangle \langle (c_j^*)_r, c_{(2)} \rangle a_{rR} \otimes c_i \otimes c_j$$

$$= \sum_{i,j} \langle (c_i^*)_R * (c_j^*)_r, c \rangle a_{rR} \otimes c_i \otimes c_j$$

$$(2.28) \quad = \sum_{i,j} \langle (c_i^* * c_j^*)_R, c \rangle a_R \otimes c_i \otimes c_j$$

$$(1.5) \quad = \sum_i \langle (c_i^*)_R, c \rangle a_R \otimes \Delta(c_i)$$

$$= a_\psi \otimes \delta(c^\psi)$$

proving (2.3). To prove (2.7):

$$a_\psi b_\Psi \otimes c^{\Psi\psi} = \langle (c_i^*)_R, c^\Psi \rangle a_R b_\Psi \otimes c_i$$

$$= \langle (c_i^*)_R, c_j \rangle \langle (c_j^*)_r, c \rangle a_R b_r \otimes c_i$$

$$= \left\langle \left(\langle (c_i^*)_R, c_j \rangle c_j^* \right)_r, c \right\rangle a_R b_r \otimes c_i$$

$$= \langle (c_i^*)_{Rr}, c \rangle a_R b_r \otimes c_i$$

$$(2.29) \quad = \langle (c_i^*)_R, c \rangle (ab)_R \otimes c_i$$

$$= \psi(ab \otimes c)$$

Next observe that

$$(g(f(\psi)))(a \otimes c) = \sum_i \langle (c_i^*)_R, c \rangle a_R \otimes c_i$$

$$= \sum_{i,j} \langle c_i^*, c_j^\psi \rangle \langle c_j^*, c \rangle a_\psi \otimes c_i$$

$$= \sum_j \langle c_j^*, c \rangle a_\psi \otimes c_j^\psi$$

$$= a_\psi \otimes c^\psi = \psi(a \otimes c)$$

and it follows that $(g \circ f)(\psi) = \psi$. In a similar way, we can prove that $(f \circ g)(R) = R$, and this finishes the proof of the first part of the Theorem. We will now define an isomorphism $F :\ _A\mathcal{M}(\psi)^C \to\ _{A\#_R C^*}\mathcal{M}$. For $M \in\ _A\mathcal{M}(\psi)^C$, we define $F(M) = M$ as a k-module, with left $A\#_R C^*$-action defined by

$$(a\#c^*) \cdot m = \langle c^*, m_{[1]} \rangle a \cdot m_{[0]} \tag{2.32}$$

It is clear that M is an $A\#_R C^*$-module, since

$$
\begin{aligned}
\left((a\#c^*)(b\#d^*) \right) \cdot m &= (ab_R \#(c_R^* * d^*)) \cdot m \\
&= \langle c_R^* * d^*, m_{[1]} \rangle ab_R m_{[0]} \\
&= \sum_i \langle c^*, c_i^\psi \rangle \langle c_i^*, m_{[1]} \rangle \langle d^*, m_{(2)} \rangle ab_\psi m_{[0]} \\
&= (a\#c^*) \cdot \left(\langle d^*, m_{[1]} \rangle bm_{[0]} \right) \\
&= (a\#c^*) \cdot \left((b\#d^*) \cdot m \right)
\end{aligned}
$$

Conversely, if M is a left $A\#_R C^*$-module, we define $G(M) \in\ _A\mathcal{M}(\psi)^C$: $G(M) = M$ as a k-module, with left A-action

$$am = (a\#\varepsilon_C) \cdot m$$

and right C-coaction

$$\rho^r(m) = \sum_i (1\#c_i^*) \cdot m \otimes c_i$$

Further details are left to the reader.

Theorem 9. *Let A be a k-algebra, and C a coalgebra which is finitely generated and projective as a k-module. Then there is a bijection between right-left entwining structures of the form (A, C, φ) and smash product structures of the form (C^*, A, S). In this situation we have an isomorphism of categories*

$$^C\mathcal{M}(\psi)_A \cong \mathcal{M}_{C^* \#_S A}$$

Proof. We use the left-right dictionary. If $(A, C, \varphi) \in\ ^\bullet\mathbb{E}_\bullet(k)$, then $(A^{\mathrm{op}}, C^{\mathrm{cop}}, \tau \circ \varphi \circ \tau) \in\ _\bullet\mathbb{E}^\bullet(k)$ (see Proposition 14). Using Theorem 8, we find $(A^{\mathrm{op}}, C^{\mathrm{cop}*} = C^{*\mathrm{op}}, R) \in \mathbb{S}(k)$. Finally Proposition 22 gives the corresponding smash product structure $(C^*, A, S = \tau \circ R \circ \tau)$. From (2.30) and (2.31), it follows that the correspondence between S and φ is given by the formulas

$$S(a \otimes c^*) = \sum_i \langle c^*, c_i^\varphi \rangle c_i^* \otimes a_\varphi \tag{2.33}$$

$$\varphi(c \otimes a) = \sum_i \langle (c_i^*)_R, c \rangle c_i \otimes a_R \tag{2.34}$$

Take an entwining structure $(A, C, \psi) \in {}_\bullet\mathbb{E}^\bullet(k)$. Assume that C is finitely generated and projective, and that ψ is invertible. Then we have a right-left entwining structure $(C, A, \varphi = \tau \circ \psi^{-1} \circ \tau) \in {}^\bullet\mathbb{E}_\bullet(k)$ (see Proposition 15). Let (A, C^*, R) and (C^*, A, S) be the two corresponding smash product structures from Theorems 8 and 9. One is then tempted to conjecture that $S = R^{-1}$, and therefore $A\#_R C^* \cong C^*\#_S A$, according to Proposition 23. Surprisingly, this is not true in general! a straightforward computation shows that $S = R^{-1}$ if and only if

$$c \otimes a = \sum_i \langle c_i^*, c^\varphi \rangle c_i^\psi \otimes a_{\psi\varphi} \tag{2.35}$$

$$c \otimes a = \sum_i \langle c_i^*, c^\psi \rangle c_i^\varphi \otimes a_{\varphi\psi} \tag{2.36}$$

for all $c \in C$ and $a \in A$. The condition $\varphi = \tau \circ \psi^{-1} \circ \tau$ amounts to

$$c \otimes a = \sum_i \langle c_i^*, c^\varphi \rangle c_i^\psi \otimes a_{\varphi\psi} \tag{2.37}$$

$$c \otimes a = \sum_i \langle c_i^*, c^\psi \rangle c_i^\varphi \otimes a_{\psi\varphi} \tag{2.38}$$

for all $c \in C$ and $a \in A$. We will make the difference clear in the Doi-Hopf case.

Examples 5. 1) Let A be a right H-comodule algebra, and B a right H-module algebra. Define $R : B \otimes A \to A \otimes B$ by

$$R(b \otimes a) = a_{[0]} \otimes ba_{[1]} \tag{2.39}$$

Then (A, B, R) is a smash product structure, and the multiplication on $A\#_R B$ is given by the formula

$$(a\#b)(c\#d) = ac_{[0]}\#(bc_{[1]})d \tag{2.40}$$

If H has a twisted antipode, then R is invertible, and R^{-1} is given by the formula

$$R^{-1}(b \otimes a) = b\overline{S}(a_{[1]}) \otimes a_{[0]} \tag{2.41}$$

2) In a similar way, if A is a left H-comodule algebra, and B is a left H-module algebra, then we have a smash product structure (B, A, R), with

$$R(a \otimes b) = a_{[-1]}b \otimes a_{[0]}$$

3) Let $(H, A, C) \in {}_\bullet a\mathbb{DK}^\bullet(k)$, i.e. A is a left H-module algebra, and C is a right H-comodule coalgebra. If C is finitely generated projective, then C^* is a left H-comodule algebra, cf. Proposition 7, so we find a smash product structure (A, C^*, R). Now (H, A, C) defines an entwining structure (see

Proposition 18), and Theorem 8 produces another smash product structure (A, C^*, R'). As one might expect, $R = R'$, since

$$R'(c \otimes a) = \sum_{i=1}^{n} \langle c^*, c_i^{\psi} \rangle a_{\psi} \otimes c_i^*$$

$$= \sum_{i=1}^{n} \langle c^*, c_{i[0]} \rangle c_{i[0]} a \otimes c_i^*$$

$$= c_{[-1]}^* a \otimes c_{[0]}^* = R(c \otimes a)$$

Let (A, B, R) be a smash product structure. Can we find a bialgebra H, an H-coaction on A and an H-action on B such that R is given by (2.39). We have discussed this question already for entwining structures. For smash product structures, the answer is the following:

Theorem 10. (Tambara [178]) *Let A be a finitely generated and projective algebra, and $H = H(A)$ as in Proposition 19. For every algebra B, we have a bijective correspondence between smash product structures of the form (A, B, R) and right H-module algebra structures on B.*
A similar result holds if B is finitely generated projective.

Proof. The proof is similar to the corresponding proofs for entwining structures (Theorems 5 and 6). We know that A is a right H-comodule algebra. Given R, we define a right H-action on B as follows:

$$b \cdot [a^* \otimes a] = \langle a^*, a_R \rangle b_R$$

We invite the reader to prove that this puts an H-module algebra structure on H. Conversely, if B is a right H-module algebra, then Example 5 1) tells us how to produce a smash product structure.

Example 8. If $(H, A, C) \in {}_{\bullet}\mathbb{DK}^{\bullet}(k)$, then C^* is a right H-module algebra, the right H-action on C^* is given by

$$\langle c^* {\leftharpoonup} h, c \rangle = \langle c^*, hc \rangle$$

and we obtain a smash product structure (A, C^*, R). We have a functor

$$F : {}_A\mathcal{M}(H)^C \to {}_{A\#_R C^*}\mathcal{M}$$

$F(M) = M$ as a k-module, with left $A\#_R C^*$-action

$$(a\#c^*) \cdot m = \langle c^*, m_{[1]} \rangle a m_{[0]}$$

If C is finitely generated and projective, then the map R coincides with the one from Theorem 8. First observe that

$$c^* {\leftharpoonup} h = \sum_i \langle c^*, hc_i \rangle c_i^*$$

To see this, apply both sides to an arbitrary $c \in C$. Thus, according to (2.30),

$$R(c^* \otimes a) = \sum_i \langle c^*, a_{[1]}c_i \rangle a_{[0]} \otimes c_i^*$$

$$= \sum_i a_{[0]} \otimes \langle c^* \leftharpoonup a_{[1]}, c_i \rangle c_i^*$$

$$= a_{[0]} \otimes (c^* \leftharpoonup a_{[1]}) \tag{2.42}$$

If C is projective, but not necessarily finitely generated, then F is fully faithful. Indeed, if $f : M \to N$ is a left $A \#_R C^*$-linear map between two Doi-Hopf modules M and N, then f is left A-linear and left C^*-linear, and therefore right C-colinear, by Proposition 3. Consequently, $_A\mathcal{M}(H)^C$ can be viewed as a full subcategory of $_{A\#_R C^*}\mathcal{M}$.

Example 9. Now assume that H has an antipode. To $(H, A, C) \in {}_\bullet\mathbb{DK}^\bullet(k)$, we can associate $(A, C, \psi) \in {}_\bullet\mathbb{E}^\bullet(k)$ and $(A, C, \varphi) \in {}^\bullet\mathbb{E}_\bullet(k)$ Recall that $\varphi : C \otimes A \to C \otimes A$ is given by

$$\varphi(c \otimes a) = S(a_{[1]})c \otimes a_{[0]}$$

We have associated smash product structures (A, C^*, R) and (C^*, A, S), with R given by (2.42), and S by

$$S(a \otimes c^*) = \big(c^* \leftharpoonup S(a_{[1]})\big) \otimes a_{[0]} \tag{2.43}$$

Even if C is not necessarily finitely generated and projective, (2.43) defines a smash product structure. In any case, we have a functor $F : {}^C\mathcal{M}(H)_A \to \mathcal{M}_{C^* \#_S A}$. $F(M) = M$ as a k-module, with action

$$m \cdot (c^* \# a) = \langle c^*, m_{[-1]} \rangle m_{[0]} a$$

If C is finitely generated and projective, then F is an isomorphism of categories.

If H has a twisted antipode, then the inverse of R exists and is given by (see (2.41))

$$R^{-1}(a \otimes c^*) = \big(c^* \leftharpoonup \overline{S}(a_{[1]})\big) \otimes a_{[0]} \tag{2.44}$$

If the antipode S of H is of order 2, then we can conclude from (2.43) and (2.44) that $S = R^{-1}$, and we have the following result.

Proposition 24. *Let $(H, A, C) \in {}_\bullet\mathbb{DK}(k)^\bullet$, and assume that H has an antipode of order 2. Let (A, C^*, R) and (C^*, A, S) be defined as in Example 9. Then the smash products $A\#_R C^*$ and $C^* \#_S A$ are isomorphic.*

Koppinen's smash product Let $(A, C, \psi) \in {}_\bullet\mathbb{E}^\bullet(k)$ be a left-right entwining structure. The *Koppinen smash* $\#_\psi(C, A)$ is equal to $\mathrm{Hom}(C, A)$ as a k-module, but with twisted multiplication

$$(f \bullet g)(c) = f(c_{(1)}^\psi)g(c_{(2)})_\psi$$

for all $f, g : C \to A$ and $c \in C$.

Proposition 25. *If (A, C, ψ) is a left-right entwining structure, then $\#_\psi(C, A)$ is an associative algebra, with unit $\eta_A \circ \varepsilon_C$.*

Proof. The proof of the associativity goes as follows:

$$
\begin{aligned}
((f \bullet g) \bullet h)(c) &= (f \bullet g)(c_{(1)}^\psi)h(c_{(2)})_\psi \\
&= f\Big(\big(c_{(1)}^\psi\big)_{(1)}^{\psi'}\Big)g\Big(\big(c_{(1)}^\psi\big)_{(2)}\Big)_{\psi'}h(c_{(2)})_\psi \\
(2.3) \quad &= f\Big(c_{(1)}^{\Psi\psi'}\Big)g\Big(c_{(2)}^\psi\Big)_{\psi'}h(c_{(3)})_{\psi\Psi} \\
(2.7) \quad &= f(c_{(1)}^\Psi)\Big(g(c_{(2)}^\psi)h(c_{(3)})_\psi\Big)_\Psi \\
&= f(c_{(1)}^\Psi)\big((g \bullet h)(c_{(2)})\big)_\Psi \\
&= (f \bullet (g \bullet h))(c)
\end{aligned}
$$

From (2.2) and (2.4), it follows easily that $\eta_A \circ \varepsilon_C$ is the unit element of $\#_\psi(C, A)$.

Proposition 26. C^* *and A are subalgebras of $\#_\psi(C, A)$, via*

$$
c^* \mapsto \eta_A \circ c^* \quad \text{and} \quad a \mapsto a\eta \circ \varepsilon_C
$$

Proof. Obvious.

Proposition 27. *For $(A, B, \psi) \in {}_\bullet\mathbb{E}^\bullet(k)$, we have a functor*

$$
F: \; {}_A\mathcal{M}(\psi)^C \to {}_{\#_\psi(C,A)}\mathcal{M}
$$

For an entwined module M, $F(M) = M$ as a k-module, with left $\#_\psi(C, A)$-action given by

$$
f \cdot m = f(m_{[1]})m_{[0]}
$$

Proof. We will prove that $F(M)$ is a $\#_\psi(C, A)$-module, leaving further details to the reader. For $f, g: \; C \to A$, we have

$$
\begin{aligned}
f \cdot (g \cdot m) &= f \cdot (g(m_{[1]})m_{[0]}) \\
&= f(m_{[1]}^\psi)g(m_{(2)})_\psi m_{[0]} \\
&= (f \bullet g)(m_{[1]})m_{[0]} \\
&= (f \bullet g) \cdot m
\end{aligned}
$$

Proposition 28. *Let $(A, B, \psi) \in {}_\bullet\mathbb{E}^\bullet(k)$, and assume that C is finitely generated and projective as a k-module. Let (A, C^*, R) be the corresponding smash product structure (cf. Theorem 8). Then we have an algebra isomorphism*

$$
s: \; A\#_R C^* \to \#_\psi(C, A)
$$

given by

$$
s(a\#c^*)(c) = \langle c^*, c \rangle a
$$

for all $a \in A$, $c \in C$ and $c^* \in C^*$.

Proof. It is well-known that s is a k-module isomorphism if C is finitely generated and projective. So we only need to show that s is an algebra map. Let $\{c_i, c_i^* \mid i = 1, \cdots, n\}$ be a finite dual basis of C. For all $a, b \in A$, $c \in C$ and $c^*, d^* \in C^*$, we have

$$
\begin{aligned}
s\big((a\#c^*)(b\#d^*)\big)(c) &= s\big(\langle c^*, c_i^\psi \rangle ab_\psi \# c_i^* * d^*\big)(c) \\
&= \langle c^*, c_i^\psi \rangle \langle c_i^* * d^*, c \rangle ab_\psi \\
&= \langle c^*, c_i^\psi \rangle \langle c_i^*, c_{(1)} \rangle \langle d^*, c_{(2)} \rangle ab_\psi \\
&= \langle c^*, c_{(1)}^\psi \rangle \langle d^*, c_{(2)} \rangle ab_\psi \\
&= (a\eta_A \circ c^*)(c_{(1)}^\psi)\big((b\eta_A \circ d^*)(c_{(2)})\big)_\psi \\
&= \big((a\eta_A \circ c^*) \bullet (b\eta_A \circ d^*)\big)(c) \\
&= \big(s(a\#c^*) \bullet s(b\#d^*)\big)(c)
\end{aligned}
$$

Example 10. Let $(H, A, C) \in {}_\bullet\mathbb{DK}^\bullet(k)$, and let (A, C, ψ) be the associated entwining structure. We will write $\#_H(C, A) = \#_\psi(C, A)$. The product on $\#_H(C, A)$ is given by the formula

$$
(f \bullet g)(c) = f\big(g(c_{(2)})_{(1)}c_{(1)}\big)g(c_{(2)})_{[0]}
$$

This multiplication appeared first in [111, Def. 2.2]. In the situation where $C = H$, it appears already in [68] and [110].

2.4 Entwined modules and the smash coproduct

The result in this Section may be viewed as the duals of the ones in Section 2.3. Let C and D be coalgebras, and consider a linear map $V : C \otimes D \to D \otimes C$. We will use the notation

$$
V(c \otimes d) = d^V \otimes c^V
$$

$C \bowtie^V D$ will be equal to $C \otimes D$ as a k-module, with comultiplication

$$
\Delta = (I_C \otimes V \otimes I_D) \circ (\Delta_C \otimes \Delta_D)
$$

or

$$
\Delta(c \bowtie d) = (c_{(1)} \bowtie d_{(1)}^V) \otimes (c_{(2)}^V \bowtie d_{(2)}) \tag{2.45}
$$

We call (C, D, V) a *smash coproduct structure* if $C \bowtie^V D$ is a coassociative coalgebra with counit $\varepsilon_C \bowtie \varepsilon_D$.

Theorem 11. (C, D, V) *is a smash coproduct structure if and only if the following conditions hold, for all $c \in C$ and $d \in D$.*

$$\varepsilon_C(c^V)d^V = \varepsilon_C(c)d \tag{2.46}$$
$$\varepsilon_D(d^V)c^V = \varepsilon_D(d)c \tag{2.47}$$
$$\Delta_D(d^V) \otimes c^V = d_{(1)}^V \otimes d_{(2)}^v \otimes c^{Vv} \tag{2.48}$$
$$d^V \otimes \Delta_C(c^V) = d^{Vv} \otimes c_{(1)}^v \otimes c_{(2)}^V \tag{2.49}$$

Proof. If $\varepsilon_C \ltimes \varepsilon_D$ is a counit, then for all $c \in C$ and $d \in D$,

$$c \ltimes d = ((\varepsilon_C \ltimes \varepsilon_D) \otimes I_{C \ltimes D})((c_{(1)} \ltimes d_{(1)}^V) \otimes (c_{(2)}^V \ltimes d_{(2)}))$$
$$= \varepsilon_D(d_{(1)}^V)c^V \ltimes d_{(2)}$$

Applying $I_C \ltimes \varepsilon_D$ to both sides, we find (2.47). (2.46) can be shown in a similar way.

If $C \ltimes D$ is coassociative, with counit $\varepsilon_C \ltimes \varepsilon_D$, then for all $c \in C$ and $d \in D$ we have

$$\left(c_{(1)} \ltimes \left(d_{(1)}^V\right)_{(1)}^v\right) \otimes \left(c_{(2)}^v \ltimes \left(d_{(1)}^V\right)_{(2)}\right) \otimes \left(c_{(3)}^V \otimes d_{(2)}\right)$$
$$= \left(c_{(1)} \ltimes d_{(1)}^V\right) \otimes \left((c_{(2)}^V)_{(1)} \ltimes d_{(2)}^v\right) \otimes \left((c_{(2)}^V)_{(2)}^v \ltimes d_{(3)}\right)$$

Applying ε_C to the first factor, and ε_D to the fourth and sixth factor, we find

$$d^{Vv} \otimes c_{(1)}^v \otimes c_{(2)}^V = d_{(1)}^V \otimes (c^V)_{(1)} \otimes \varepsilon_D(d_{(2)}^v)(c^V)_{(2)}^v$$
$$(2.47) \quad = d^V \otimes \Delta(c^V)$$

and this proves (2.49). Applying ε_C to the first and third factor, and ε_D to the sixth factor, we get

$$d_{(1)}^V \otimes d_{(2)}^v \otimes c^{Vv} = \varepsilon_C(c_{(1)}^v)(d^V)_{(1)}^v \otimes (d^V)_{(2)} \otimes c_{(3)}^V$$
$$(2.46) \quad = (d^V)_{(1)} \otimes (d^V)_{(2)} \otimes \varepsilon_C(c_{(1)})c_{(2)}^V$$
$$= \Delta(d^V) \otimes c^V$$

and (2.48) follows. The converse is left to the reader.

A morphism between to smash coproduct structures (C, D, V) and (C', D', V') consists of a pair of coalgebra maps $(\gamma : C \to C', \ \delta : D \to D')$ such that

$$(\delta \otimes \gamma) \circ V = V' \circ (\gamma \otimes \delta)$$

$\mathbb{CS}(k)$ will be the category of smash coproduct structures over the ground ring k. The proofs of the next two Proposition is similar to the proofs of the corresponding Propositions 22 and 23. Details are left to the reader.

Proposition 29. $(C, D, V) \in \mathbb{CS}(k)$ *if and only if* $(D^{\mathrm{cop}}, C^{\mathrm{cop}}, \tau \circ V \circ \tau) \in \mathbb{CS}(k)$. *In this situation, the switch map*

$$\tau : (C \ltimes^V D)^{\mathrm{cop}} \to D^{\mathrm{cop}} \ltimes^{\tau \circ V \circ \tau} C^{\mathrm{cop}}$$

is a coalgebra isomorphism.

Proposition 30. *Let $(C, D, V) \in \mathbb{CS}(k)$, and assume that V is invertible. Then $(D, C, V^{-1} = W) \in \mathbb{CS}(k)$, and*

$$V : C \ltimes^V D \to D \ltimes^W C$$

is a coalgebra isomorphism.

Theorem 12. *Let D be a coalgebra, and A a finitely generated projective algebra. We have a bijective correspondence between left-right entwining structures $(A, D, \psi) \in {}_{\bullet}\mathbb{E}^{\bullet}(k)$ and smash coproduct structures $(A^*, D, V) \in \mathbb{CS}(k)$. In this situation, we have an isomorphism of categories*

$$_A\mathcal{M}(\psi)^D \cong \mathcal{M}^{A^* \ltimes_V D}$$

Proof. Recall that the comultiplication on A^* is given by the formula

$$\Delta(a^*) = \sum_{i,j} \langle a^*, a_i a_j \rangle a_i^* \otimes a_j^* \tag{2.50}$$

where $\{a_i, a_i^* \mid i = 1, \cdots, n\}$ is a dual basis for A. Assume that $\psi : A \otimes D \to D \otimes A$ defines a left-right entwining structure, and define $V : A^* \otimes D \to D \otimes A^*$ by

$$V(a^* \otimes d) = \langle a^*, a_{i\psi} \rangle d^\psi \otimes a_i^* \tag{2.51}$$

for all $a^* \in A^*$ and $d \in D$. We will prove that $(A^*, D, V) \in \mathbb{CS}(k)$. To this end, we need to show that (2.46-2.49) hold. (2.46) and (2.47) are obvious. Let us check (2.48) first. For all $a^* \in A^*$ and $d \in D$, we have

$$d_{(1)}^V \otimes d_{(2)}^v \otimes (a^*)^{Vv} = \sum_i d_{(1)}^V \otimes \langle (a^*)^V, a_{i\psi} \rangle d_{(2)}^\psi \otimes a_i^*$$

$$= \sum_{i,j} \langle a^*, a_{j\Psi} \rangle d_{(1)}^\Psi \otimes \langle a_j^*, a_{i\psi} \rangle d_{(2)}^\psi \otimes a_i^* = \sum_i \langle a^*, a_{i\psi\Psi} d_{(1)}^\Psi \rangle \otimes d_{(2)}^\psi \otimes a_i^*$$

$$(2.3) \quad = \sum_i \sum_i \langle a^*, a_{i\psi} \rangle \Delta_D(d^\psi) \otimes a_i^* = \Delta_D(d^V) \otimes (A^*)^V$$

For fun, we will also verify (2.49). For all $a^* \in A^*$ and $d \in D$, we have

$$d^{Vv} \otimes a_{(1)}^{*v} \otimes a_{(2)}^{*V} = \sum_i \langle a_{(1)}*, a_{i\psi} \rangle d^{V\psi} \otimes a_i^* \otimes a_{(2)}^{*V}$$

$$= \sum_{i,j} \langle a_{(1)}*, a_{i\psi} \rangle \langle a_{(2)}*, a_{j\Psi} \rangle d^{\Psi\psi} \otimes a_i^* \otimes a_j^*$$

$$= \sum_{i,j} \langle a^*, a_{i\psi} a_{j\Psi} \rangle d^{\Psi\psi} \otimes a_i^* \otimes a_j^*$$

$$(2.7) \quad = \sum_{i,j} \langle a^*, (a_i a_j)_\psi \rangle d^\psi \otimes a_i^* \otimes a_j^*$$

and

$$d^V \otimes \Delta(a^{*V}) = \sum_i \langle a^*, a_{i\psi} \rangle d^\psi \otimes \Delta(a_i^*)$$

$$= \sum_{i,j,k} \langle a^*, a_{i\psi} \rangle \langle a_i^*, a_j a_k \rangle d^\psi \otimes a_j^* \otimes a_k^* = \sum_{j,k} \langle a^*, (a_j a_k)_\psi \rangle d^\psi \otimes a_j^* \otimes a_k^*$$

and (2.49) follows.

Conversely, if (A^*, D, V) is a smash coproduct structure, then we define $\psi : A \otimes D \to A \otimes D$ by

$$\psi(a \otimes d) = \langle a_i^{*V}, a \rangle a_i \otimes d^V \tag{2.52}$$

We leave it to the reader to show that (A, D, ψ) is a left-right entwining structure, and that (2.51) and (2.52) define a bijection between entwining structures (A, D, ψ) and smash coproduct structures (A^*, D, V). To complete the proof, we define a functor $F : {}_A\mathcal{M}(\psi)^D \to \mathcal{M}^{A^* \bowtie_V D}$ as follows: $F(M) = M$, with right $A^* \bowtie_V D$-coaction given by

$$\rho^r_{A^* \bowtie_V D}(m) = \sum_i a_i m_{[0]} \otimes (a_i^* \bowtie m_{[1]})$$

for all $m \in M$. A straightforward computation shows that $\rho^r_{A^* \bowtie_V D}$ is indeed a coaction. Conversely, given a right $A^* \bowtie_V D$-coaction on a k-module M. We define $G(M) = M$ with the following structure. For $m \in M$, write

$$\rho^r_{A^* \bowtie_V D}(m) = \sum_j m_j \otimes b_j^* \otimes d_j \in M \otimes A^* \otimes D$$

Then a left A-action and a right D-coaction are given by

$$a \cdot m = \sum_j \langle b_j^*, a \rangle \varepsilon(d_j) m_j \quad \text{and} \quad \rho^r(m) = \sum_j \langle b_j^*, 1 \rangle m_j \otimes d_j$$

Theorem 13. *Let C be a coalgebra, and A a finitely generated projective algebra. We have a bijective correspondence between right-left entwining structures $(A, C, \varphi) \in {}^\bullet\mathbb{E}_\bullet(k)$ and smash coproduct structures $(C, A^*, V) \in \mathbb{CS}(k)$. In this case, we have an isomorphism between the categories ${}^C\mathcal{M}(\psi)_A$ and ${}_{C \bowtie_V A^*}\mathcal{M}$.*

Proof. This follows from Theorem 12 if we apply the left-right dictionary: $(A, C, \varphi) \in {}^\bullet\mathbb{E}_\bullet(k)$ if and only if $(A^{\mathrm{op}}, C^{\mathrm{cop}}, \psi = \tau \circ \varphi \circ \tau) \in {}_\bullet\mathbb{E}^\bullet(k)$ (see Proposition 14). According to Theorem 12, these left-right entwining structures correspond to smash coproduct structures of the form $((A^{\mathrm{op}})^* = (A^*)^{\mathrm{cop}}, C^{\mathrm{cop}}, W)$. Finally $((A^*)^{\mathrm{cop}}, C^{\mathrm{cop}}, W) \in \mathbb{CS}(k)$ if and only if $(C, A^*, V = \tau \circ W \circ \tau) \in \mathbb{CS}$, see Proposition 29.

Example 11. Let D be a right H-comodule coalgebra, and C a right H-module coalgebra, and consider the map

$$V : C \otimes D \to D \otimes C, \quad V(c \otimes d) = d_{[0]} \otimes c d_{[1]}$$

We claim that $(C, D, V) \in \mathbb{CS}(k)$. Conditions (2.46) and (2.47) follow immediately from respectively (1.22) and (1.13). Furthermore

$$\Delta_D(d^V) \otimes c^V = d_{[0](1)} \otimes d_{[0](2)} \otimes cd_{[1]}$$

$$(1.21) \qquad = d_{(1)[0]} \otimes d_{(2)[0]} \otimes cd_{(1)[1]}d_{(2)[1]} = d_{(1)}^V \otimes d_{(2)}^v \otimes c^{Vv}$$

proving (2.48). (2.49) can be proved as follows:

$$d^{Vv} \otimes c_{(1)}^v \otimes c_{(2)}^V = (d_{[0]})^v \otimes c_{(1)}^v \otimes c_{(2)}d_{[1]}$$

$$= d_{[0]} \otimes c_{(1)}d_{[1]} \otimes c_{(2)}d_{[2]}$$

$$(1.13) \qquad = d_{[0]} \otimes \Delta_C(cd_{[1]}) = d^V \otimes \Delta_C(c^V)$$

If H has a twisted antipode $\overline{S}$, then V is invertible, with inverse

$$W(d \otimes c) = c\overline{S}(d_{[1]}) \otimes d_{[0]}$$

The comultiplication on $C \bowtie^V D$ is given by

$$\Delta(c \bowtie d) = \left(c_{(1)} \bowtie d_{(1)[0]}\right) \otimes \left(c_{(2)}d_{(1)[1]} \bowtie d_{(2)}\right)$$

In the particular situation where $C = H$, with right H-action given by right multiplication by elements in H, $C \bowtie^V D$ is isomorphic to the smash coproduct $D \bowtie H$ studied in [38] (at least in the case where H has an invertible antipode, see [38, Remark p.1654]).

Let (C, D, V) be a smash coproduct structure. Can it be obtained using Example 11? We have already discussed this question for entwining structures and smash product structures. As one might expect, the result for smash coproduct structures is similar, and uses the methods developed in [178].

Theorem 14. *Let D be a finitely generated projective coalgebra, and let $H' = H(D^*)^{\mathrm{cop}}$ (cf. Proposition 19). Then D is a right H'-comodule coalgebra. For any coalgebra C, there exists a bijective correspondence between right H'-module coalgebra structures on C and smash coproduct structures of the form (C, D, V).*

Proof. We know from Proposition 20 that D is a right H'-comodule coalgebra. If C is a right H'-module coalgebra, then Example 11 produces the required smash coproduct structure. Conversely, given $(C, D, V) \in \mathbb{CS}(k)$, we make C into a right H'-module coalgebra by putting

$$c \cdot [d^* \otimes d] = \langle d^*, d^V \rangle c^V$$

Further details are left to the reader.

Example 12. Now let C be a left H-comodule coalgebra, and D a left H-module coalgebra. We now have a smash coproduct structure (C, D, V), with

$$V(c \otimes d) = c_{[-1]}d \otimes c_{[0]} \qquad (2.53)$$

The comultiplication on $C \bowtie^V D$ is given by the formula

$$\Delta(c \bowtie d) = \left(c_{(1)} \bowtie c_{(2)[-1]}d_{(1)}\right) \otimes \left(c_{(2)[0]} \bowtie d_{(2)}\right)$$

If $D = H$, this is exactly the smash coproduct studied by Molnar [139], and this turns out to be the first version of the smash coproduct ever appearing in the literature.

Assume now that H is finitely generated and projective. Then C^* is a right H-comodule algebra, the right H-coaction is given by

$$\rho^r(c^*) = c_{[0]}^* \otimes c_{[1]}^* \in C \otimes H$$

if and only if

$$\langle c^*, c_{[0]} \rangle c_{[-1]} = \langle c_{[0]}^*, c \rangle c_{[1]}^* \qquad (2.54)$$

for all $c \in C$. Thus we obtain $(H, C^*, D) \in {}_\bullet\mathbb{DK}^\bullet(k)$, and $(C^*, D, \psi) \in {}_\bullet\mathbb{E}^\bullet(k)$. If C is finitely generated and projective, then ψ coincides with the map defined in the proof of Theorem 12. Indeed, if $\psi : C^* \otimes D \to C^* \otimes D$ is given by (2.52), then

$$
\begin{aligned}
\psi(c^* \otimes d) &= \langle c^*, c_i^V \rangle c_i^* \otimes d^V \\
(2.53) \qquad &= \langle c^*, c_{i[0]} \rangle c_i^* \otimes c_{i[-1]}d \\
(2.54) \qquad &= \langle c_{[0]}^*, c_i \rangle c_i^* \otimes c_{[1]}^*d \\
&= c_{[0]}^* \otimes c_{[1]}^*d
\end{aligned}
$$

as needed.

2.5 Adjoint functors for entwined modules

Let k be a commutative ring, and $(\alpha, \gamma) : (A, C, \psi) \to (A', C', \psi')$ a morphism in $\mathbb{E}_\bullet^\bullet(k)$. We will always assume that the coalgebras C and C' are flat as k-modules.

Lemma 8. *We have a functor* $F : \mathcal{M}(\psi)_A^C \to \mathcal{M}(\psi')_{A'}^{C'}$. *For any* $M \in \mathcal{M}(\psi)_A^C$, $F(M) = M \otimes_A A' \in \mathcal{M}(\psi')_{A'}^{C'}$, *where* A' *is viewed as a left A-module via* α. *The structure maps are given by the formulas*

$$(m \otimes a')b' = m \otimes a'b' \qquad (2.55)$$

$$\rho^r(m \otimes a') = (m_{[0]} \otimes a'_{\psi'}) \otimes \gamma(m_{[1]})^{\psi'} \qquad (2.56)$$

For a morphism $f : M \to N$ *in* $\mathcal{M}(\psi)_A^C$, $F(f) = f \otimes I_{A'}$.

Proof. It is clear that (2.55) is well-defined. Let us show that (2.55) is well-defined. For all $m \in M$, $a \in A$ and $a' \in A'$, we have

$$
\begin{aligned}
\rho^r(m \otimes \alpha(a)a') &= m_{[0]} \otimes (\alpha(a)a')_{\psi'} \otimes \gamma(m_{[1]})^{\psi'} \\
(2.1) \qquad &= m_{[0]} \otimes \alpha(a)_{\psi'} a'_{\Psi'} \otimes \gamma(m_{[1]})^{\psi'\Psi'} \\
(2.5) \qquad &= m_{[0]} \otimes \alpha(a_\psi)a'_{\Psi'} \otimes \gamma(m_{[1]}^\psi)^{\Psi'} \\
&= m_{[0]}a_\psi \otimes a'_{\Psi'} \otimes \gamma(m_{[1]}^\psi)^{\Psi'} \\
&= (ma)_{[0]} \otimes a'_{\Psi'} \otimes \gamma((ma)_{[1]})^{\Psi'} \\
&= \rho^r(ma \otimes a')
\end{aligned}
$$

Clearly (2.55-2.56) make $M \otimes_A A'$ into a right A'-module and a right C'-comodule. We still have to verify (2.17):

$$
\begin{aligned}
\rho^r((m \otimes a')b') &= \rho^r(m \otimes a'b') \\
&= m_{[0]} \otimes (a'b')_{\psi'} \otimes \gamma(m_{[1]})^{\psi'} \\
(2.1) \qquad &= m_{[0]} \otimes a'_{\psi'}b'_{\Psi'} \otimes \gamma(m_{[1]})^{\psi'\Psi'} \\
&= (m \otimes a')_{[0]}b'_{\Psi'} \otimes (m \otimes a')_{[1]}^{\Psi'}
\end{aligned}
$$

as needed.

Let $M' \in \mathcal{M}(\psi)_{A'}^{C'}$. Since C is assumed to be k-flat, it follows from Lemma 2 that the natural map

$$(M' \Box_{C'} C) \otimes C \to M' \Box_{C'}(C \otimes C)$$

mapping $(\sum_i m'_i \otimes c_i) \otimes c$ to $\sum_i m'_i \otimes (c_i \otimes c)$ is an isomorphism.

Lemma 9. *We have a functor $G : \mathcal{M}(\psi')_{A'}^{C'} \to \mathcal{M}(\psi)_A^C$. For any $M' \in \mathcal{M}(\psi')_{A'}^{C'}$, $G(M') = M' \Box_{C'} C \in \mathcal{M}(\psi)_A^C$, where C is viewed as a left C'-comodule via γ. The structure maps are given by the formulas*

$$\rho^r\left(\sum_i m'_i \otimes c_i\right) = \sum_i m'_i \otimes c_{i(1)} \otimes c_{i(2)} \tag{2.57}$$

$$\left(\sum_i m'_i \otimes c_i\right)a = \sum_i m'_i \alpha(a_\psi) \otimes c_i^\psi \tag{2.58}$$

For a morphism $f' : M' \to N'$ in $\mathcal{M}(\psi')_{A'}^{C'}$, $G(f') = f' \otimes I_C$.

Proof. It is easy to see that $\rho^r(\sum_i m'_i \otimes c_i) \in M' \Box_{C'}(C \otimes C)$, and from the flatness of C, we know that this can be viewed as an element of $(M' \Box_{C'} C) \otimes C$. Thus $G(M')$ is a right C-comodule.

Let us show that $G(M')$ is also a right A-module. We have to show that $(\sum_i m'_i \otimes c_i)a \in M' \Box_{C'} C$, or

$$\sum_i m'_{i(0)}\alpha(a_\psi)_{\Psi'}\otimes(m'_{i(1)})^{\Psi'}\otimes c_i^\psi = \sum_i m'_i\alpha(a_\psi)\otimes\gamma((c_i^\psi)_{(1)})\otimes(c_i^\psi)_{(2)} \quad (2.59)$$

We know that $\sum_i m'_i \otimes c_i \in M'\Box_{C'}C$, or

$$\sum_i m'_{i(0)} \otimes m'_{i(1)} \otimes c_i = \sum_i m'_i \otimes \gamma(c_{i(1)}) \otimes c_{i(2)} \quad (2.60)$$

Using (2.3), we find that the right hand side of (2.59) is equal to

$$\sum_i m'_i\alpha(a_{\psi\Psi}) \otimes \gamma\big((c_{i(1)})^\Psi\big) \otimes (c_{i(2)})^\psi$$

$$(2.5) \quad = \sum_i m'_i\alpha(a_\psi)_{\Psi'} \otimes \gamma(c_{i(1)})^{\Psi'} \otimes (c_{i(2)})^\psi$$

$$(2.60) \quad = \sum_i m'_{i(0)}\alpha(a_\psi)_{\Psi'} \otimes \big(m'_{i(1)}\big)^{\Psi'} \otimes (c_i)^\psi$$

which is exactly the left hand side of (2.59).

Finally, we check that (2.17) holds. For any $m = \sum_i m'_i \otimes c_i \in M'\Box_{C'}C$, we have

$$\rho^r(ma) = \rho^r\big(\sum_i m'_i\alpha(a_\psi) \otimes c_i^\psi$$

$$= \sum_i m'_i\alpha(a_\psi) \otimes \big(c_i^\psi\big)_{(1)} \otimes \big(c_i^\psi\big)_{(2)}$$

$$(2.3) \quad = \sum_i m'_i\alpha(a_{\psi\Psi}) \otimes \big(c_{i(1)}\big)^\Psi \otimes \big(c_{i(2)}\big)^\psi$$

$$= (m'_i \otimes c_{i(1)})a_\psi \otimes \big(c_{i(2)}\big)^\psi$$

$$= m_{[0]}a_\psi \otimes m_{[0]}^\psi$$

Theorem 15. *Let* $(\alpha,\gamma) : (A,C,\psi) \to (A',C',\psi')$ *be a morphism in* $\mathbb{E}^\bullet_\bullet(k)$, *and let* F *and* G *be the functors defined in Lemmas 8 and 9. Then* (F,G) *is an adjoint pair of functors.*

Proof. The unit $\eta : 1_C \to GF$ and counit $\varepsilon : FG \to 1_{C'}$ are given by the following formulas, for all $M \in \mathcal{M}(\psi)_A^C$ and $M' \in \mathcal{M}(\psi')_{A'}^{C'}$:

$$\eta_M : M \to GF(M) \quad \eta_M(m) = (m_{[0]} \otimes 1_{A'}) \otimes m_{[1]} \quad (2.61)$$

$$\varepsilon_{M'} : FG(M') \to M' \quad \varepsilon_{M'}\big(\sum_i(m'_i \otimes c_i) \otimes a'\big) = \sum_i \varepsilon_C(c_i)m'_i a' \quad (2.62)$$

We leave it to the reader to verify that η_M and $\varepsilon_{M'}$ are well-defined, and that they define natural transformations. In order to have an adjoint pair of functors, we need the commutativity of the following diagrams:

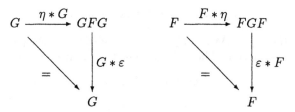

where $*$ is the Godement product. This means

$$G(\varepsilon_{M'}) \circ \eta_{G(M')} = I_{G(M')} \tag{2.63}$$

$$\varepsilon_{F(M)} \circ F(\eta_M) = I_{F(M)} \tag{2.64}$$

for all $M \in \mathcal{M}(\psi)_A^C$ and $M' \in \mathcal{M}(\psi')_{A'}^{C'}$. These two conditions are easily verified.

Example 13. Let (A, C, ψ) be a right-right entwining structure, and consider the morphism

$$(\eta_A, I_C) : (k, C, I_C) \to (A, C, \psi)$$

As we have observed earlier, $\mathcal{M}(I_C)_k^C \cong \mathcal{M}^C$ is just the category of C-comodules. In this situation, the functor

$$G : \mathcal{M}(\psi)_A^C \to \mathcal{M}^C$$

is nothing else then the functor forgetting the right A-action. Its left adjoint F is given by

$$F(M) = M \otimes A$$

with structure maps

$$(m \otimes a)b = m \otimes ab \quad ; \quad \rho^r(m \otimes a) = m_{[0]} \otimes a_\psi \otimes m_{[1]}^\psi \tag{2.65}$$

In particular, $F(C) = C \otimes A$ is an entwined module, with structure

$$(c \otimes a)b = c \otimes ab \quad ; \quad \rho^r(c \otimes a) = c_{(1)} \otimes a_\psi \otimes c_{(2)}^\psi \tag{2.66}$$

In the special case where (A, C, ψ) corresponds to a Doi-Koppinen datum, we obtain a functor $F : \mathcal{M}^C \to \mathcal{M}(\psi)_A^C$ left adjoint to the functor forgetting the A-action. Now the structure on $C \otimes A$ is given by the formulas

$$(c \otimes a)b = c \otimes ab \quad ; \quad \rho^r(c \otimes a) = c_{(1)} \otimes a_{[0]} \otimes c_{(2)}a_{[1]} \tag{2.67}$$

Observe that this garantees that $\mathcal{M}(\psi)_A^C$ always contains at least one object, namely $C \otimes A$.

Example 14. Again, let (A, C, ψ) be a right-right entwining structure, but now consider the morphism

$$(I_A, \varepsilon_C): \ (A, C, \psi) \to (A, k, I_A)$$

Recall that $\mathcal{M}(I_A)_A^k \cong \mathcal{M}_A$, and now $F: \ \mathcal{M}(\psi)_A^C \to \mathcal{M}_A$ is the functor forgetting the C-coaction. The right adjoint G of F is given by

$$G(M) = M \otimes C$$

with

$$(m \otimes c)a = ma_\psi \otimes c^\psi \quad ; \quad \rho^r(m \otimes c) = m \otimes c_{(1)} \otimes c_{(2)} \qquad (2.68)$$

In particular, $G(A) = A \otimes C \in \mathcal{M}(\psi)_A^C$, with action and coaction given by

$$(b \otimes c)a = ba_\psi \otimes c^\psi \quad ; \quad \rho^r(b \otimes c) = b \otimes c_{(1)} \otimes c_{(2)} \qquad (2.69)$$

If (A, C, ψ) corresponds to a Doi-Koppinen datum (H, A, C), then takes the form

$$(b \otimes c)a = ba_{[0]} \otimes ca_{(1)} \quad ; \quad \rho^r(b \otimes c) = b \otimes c_{(1)} \otimes c_{(2)} \qquad (2.70)$$

We now have two objects in $\mathcal{M}(\psi)_A^C$, namely $C \otimes A$ and $A \otimes C$. In the next Lemma, we show that there is a morphism from one to the other, and that they are often isomorphic.

Proposition 31. *Let (A, C, ψ) be a right-right entwining structure. Then $\psi: \ C \otimes A \to A \otimes C$ is a morphism in $\mathcal{M}(\psi)_A^C$. If ψ is invertible, then $C \otimes A$ and $A \otimes C$ are isomorphic objects in $\mathcal{M}(\psi)_A^C$.*

Proof. We have to show that ψ is right A-linear and right C-colinear. ψ is right A-linear, since

$$\psi(c \otimes ab) = (ab)_\psi \otimes c^\psi$$
$$= a_\psi b_\Psi \otimes c^{\psi\Psi}$$
$$= \psi(c \otimes a)b$$

ψ is right C-colinear, since

$$(\psi \otimes I_C)\rho^r(c \otimes a) = (\psi \otimes I_C)(c_{(1)} \otimes a_\psi \otimes c_{(2)}^\psi)$$
$$= a_{\psi\Psi} \otimes (c_{(1)})^\Psi \otimes (c_{(2)})^\psi$$
$$= a_\psi \otimes (c^\psi)_{(1)} \otimes (c^\psi)_{(2)}$$
$$= \rho^r(\psi(c \otimes a))$$

2.6 Two-sided entwined modules

Consider a right-right entwining structure $(A, C, \psi) \in \mathbb{E}_\bullet^\bullet(k)$ and a left-left entwining structure $(B, D, \varphi) \in {}_\bullet^\bullet\mathbb{E}(k)$. A *two-sided entwined module* is a k-module M having the structure of a left-left (B, D, φ)-module and right-right (A, C, ψ)-module such that the following additional compatibility conditions hold:

1. M is a (B, A)-bimodule;
2. M is a (D, C)-bicomodule;
3. the right A-action is left D-colinear:

$$\rho^l(ma) = m_{[-1]} \otimes m_{[0]}a$$

4. the left B-action is right C-colinear:

$$\rho^r(bm) = bm_{[0]} \otimes m_{[1]}$$

The category of two-sided entwined modules will be denoted by ${}_B^D\mathcal{M}(\varphi, \psi)_A^C$. In particular, we will be interested in the situation where $A = B$, $C = D$ and $\varphi = \psi^{-1}$. Then $A \otimes C$ and $C \otimes A$ are isomorphic objects in $\mathcal{M}(\psi)_A^C$, and the left-left version of the same result (use the left-right dictionary!) tells us that they are also isomorphic objects in ${}_A^C\mathcal{M}(\psi^{-1})$. The structure maps are the following: on $C \otimes A$:

$$(c \otimes a)b = c \otimes ab$$
$$\rho^r(c \otimes a) = c_{(1)} \otimes a_\psi \otimes c_{(2)}^\psi$$
$$b(c \otimes a) = c^\varphi \otimes b_\varphi a$$
$$\rho^l(c \otimes a) = c_{(1)} \otimes c_{(2)} \otimes a$$

On $A \otimes C$:

$$(a \otimes c)b = ab_\psi \otimes c^\psi$$
$$\rho^r(a \otimes c) = a \otimes c_{(1)} \otimes c_{(2)}$$
$$b(a \otimes c) = ba \otimes c$$
$$\rho^l(a \otimes c) = c_{(1)}^\varphi \otimes a_\varphi \otimes c_{(2)}$$

We leave it to the reader to write down these structure maps in the situation where $(H, A, C) \in \mathbb{DK}_\bullet^\bullet(k)$, and H has a twisted antipode $\overline{S}$. We have shown the following result.

Proposition 32. *Let $(A, C, \psi) \in \mathbb{E}^{*\bullet}_\bullet(k)$, and $\varphi = \psi^{-1}$. Then*

$$\psi : \ C \otimes A \to A \otimes C$$

is an isomorphism in ${}_A^C\mathcal{M}(\psi^{-1}, \psi)_A^C$.

Proposition 33. *Let $(\alpha, \gamma) : \ (A, C, \psi) \to (A', C', \psi')$ be a morphism in $\mathbb{E}_\bullet^\bullet(k)$, and $(B, D, \varphi) \in {}_\bullet^\bullet\mathbb{E}(k)$. Then we have a pair of adjoint functors*

$$F : \ {}_B^D\mathcal{M}(\varphi, \psi)_A^C \to {}_B^D\mathcal{M}(\varphi, \psi')_{A'}^{C'}; \quad G : \ {}_B^D\mathcal{M}(\varphi, \psi')_{A'}^{C'} \to {}_B^D\mathcal{M}(\varphi, \psi)_A^C$$

Proof. We define $F(M) = M \otimes_A A'$ with right structure as defined in Section 2.5. The left structure is the one induced by the left structure on M:

$$b(m \otimes a') = bm \otimes a'$$
$$\rho^l(m \otimes a') = m_{[-1]} \otimes m_{[0]} \otimes a'$$

In a similar way, $G(M') = M' \square_{C'} C$ with right structure as in Section 2.5, and left structure

$$b(\sum_i m'_i \otimes c_i) = bm'_i \otimes c_i$$

$$\rho^l(\sum_i m'_i \otimes c_i) = \sum_i m'_{i(-1)} \otimes m'_{i(0)} \otimes c_i$$

Assume that ψ is invertible. We have seen that $C \otimes A \cong A \otimes C \in {}^C_A\mathcal{M}(\psi^{-1}, \psi)^C_A$, and therefore

$$F(C \otimes A) = (C \otimes A) \otimes_A A' \cong C \otimes A' \cong F(A \otimes C)$$
$$= (A \otimes C) \otimes_A A' \in {}^C_A\mathcal{M}(\psi^{-1}, \psi')^{C'}_{A'}$$

and

$$G(F(C \otimes A)) = ((C \otimes A) \otimes_A A')\square_{C'}C \cong (C \otimes A')\square_{C'}C$$
$$\cong G(F(A \otimes C)) = ((A \otimes C) \otimes_A A') \in {}^C_A\mathcal{M}(\psi^{-1}, \psi)^C_A$$

For later use, we give the structure maps:

$$(c \otimes a')b' = c \otimes a'b' \tag{2.71}$$
$$\rho^{lr}(c \otimes a') = c_{(1)} \otimes (c_{(2)} \otimes a'_{\psi'}) \otimes \gamma(c_{(3)})^{\psi'} \tag{2.72}$$
$$b(c \otimes a') = c^{\varphi} \otimes \alpha(b_{\varphi})a' \tag{2.73}$$

On $((C \otimes A) \otimes_A A')\square_{C'}C \cong (C \otimes A')\square_{C'}C$, we have

$$\sum_i ((c_i \otimes a'_i) \otimes d_i)b = \sum_i (c_i \otimes a'_i\alpha(b_{\psi})) \otimes d_i^{\psi} \tag{2.74}$$

$$\rho^{lr}\left(\sum_i ((c_i \otimes a'_i) \otimes d_i)\right) = \sum_i c_{i(1)} \otimes (c_{i(2)} \otimes a'_i) \otimes d_{i(1)} \otimes d_{i(2)} \tag{2.75}$$

$$b\left(\sum_i (c_i \otimes a'_i) \otimes d_i)\right) = \sum_i (c_i^{\varphi} \otimes \alpha(b_{\varphi})a'_i) \otimes d_i \tag{2.76}$$

On $(A \otimes C) \otimes_A A'$, we have

$$b((a \otimes c) \otimes a')b' = (ba \otimes c) \otimes a' \tag{2.77}$$
$$\rho^r((a \otimes c) \otimes a') = (a \otimes c_{(1)}) \otimes a'_{\psi'} \otimes \gamma(c_{(2)})^{\psi'} \tag{2.78}$$
$$\rho^l((a \otimes c) \otimes a') = c^{\varphi}_{(1)} \otimes (a_{\varphi} \otimes c_{(2)}) \otimes a' \tag{2.79}$$

On $((A \otimes C) \otimes_A A') \square_{C'} C$, we have

$$b\Big(\sum_i ((a_i \otimes c_i) \otimes a_i') \otimes d_i\Big)e = \sum_i ((ba_i \otimes c_i) \otimes a_i' \alpha(e_\psi)) \otimes d_i^\psi \quad (2.80)$$

$$\rho^r \Big(\sum_i ((a_i \otimes c_i) \otimes a_i') \otimes d_i\Big) = \sum_i ((a_i \otimes c_i) \otimes a_i') \otimes d_{i(1)} \otimes d_{i(2)} \quad (2.81)$$

$$\rho^l \Big(\sum_i ((a_i \otimes c_i) \otimes a_i') \otimes d_i\Big) = \sum_i c_{i(1)}^\varphi \otimes ((a_{i\varphi} \otimes c_{i(2)}) \otimes a_i') \otimes d_i \quad (2.82)$$

If ψ is not invertible, then we have no left A-action on $C \otimes A$, since we need ψ^{-1} to describe this action. But we still have a left C-coaction, making $C \otimes A$ into an object of $_k^C \mathcal{M}(I_C, \psi)_A^C$. consequently $G(F(C \otimes A)) \in {_k^C}\mathcal{M}(I_C, \psi)_A^C$. In a similar way, we have that

$$A \otimes C, \ G(F(A \otimes C)) \in {_A^k}\mathcal{M}(I_A, \psi)_A^C$$

2.7 Entwined modules and comodules over a coring

Let A be a ring (with unit). An A-*coring* C is an (A, A)-bimodule together with two (A, A)-bimodule maps

$$\Delta_C : C \to C \otimes_A C \text{ and } \varepsilon_C : C \to A$$

such that the usual coassociativity and counit properties hold, i.e.

$$(\Delta_C \otimes_A I_C) \circ \Delta_C = (I_C \otimes_A \Delta_C) \circ \Delta_C \quad (2.83)$$

$$(\varepsilon_C \otimes_A I_C) \circ \Delta_C = (I_C \otimes_A \varepsilon_C) \circ \Delta_C = I_C \quad (2.84)$$

Corings were introduced by Sweedler, see [173]. A right C-comodule is a right A-module M together with a right A-module map $\rho^r : M \to M \otimes_A C$ such that

$$(\rho^r \otimes_A I_C) \circ \rho^r = (I_M \otimes_A \Delta_C) \circ \rho^r \quad (2.85)$$

$$(I_M \otimes_A \varepsilon_C) \circ \rho^r = I_M \quad (2.86)$$

In a similar way, we can define left C-comodules and (C, C)-bicomodules. If $R = k$ is a commutative ring, then an R-coring is nothing else then a k-coalgebra. We will use the Sweedler-Heyneman notation for corings and comodules over corings:

$$\Delta_C(c) = c_{(1)} \otimes_A c_{(2)} \ ; \ \rho^r(m) = m_{[0]} \otimes_A m_{[1]}$$

etc. A map $f : M \to N$ between (right) C-comodules is called a C-comodule map if f is a right A-module map, and

$$\rho^r(f(m)) = f(m_{[0]}) \otimes_A m_{[1]}$$

for all $m \in M$. $\mathcal{M}^C$ is the category of right C-comodules and C-comodule maps. In a similar way, we introduce the categories

$$^C\mathcal{M}, \ ^C\mathcal{M}^C, \ _A\mathcal{M}^C$$

For example, $_A\mathcal{M}^C$ is the category of right C-comodules that are also (A, A)-bimodules such that the right C-comodule map is left A-linear.

Takeuchi observed recently that entwined modules can be considered as co-modules over a certain coring; this has been investigated further by Brzeziński [26]. Let A be a k-algebra, and C a k-coalgebra. We say that an A-coring $\mathcal{C}$ *factorizes* through A and C if $\mathcal{C} \cong A \otimes C$ as k-modules and (identifying the elements of $\mathcal{C}$ and $A \otimes C$):

$$a(b \otimes c) = ab \otimes c \tag{2.87}$$

$$\Delta_{\mathcal{C}}(a \otimes c) = (a \otimes c_{(1)}) \otimes_A (1 \otimes c_{(2)}) \tag{2.88}$$

$$\varepsilon_{\mathcal{C}}(a \otimes c) = \varepsilon_C(c)a \tag{2.89}$$

for all $a, b \in A$ and $c \in C$.

Theorem 16. *Let k be a commutative ring, A a k-algebra, and C a k-coalgebra. There exists a bijective correspondence between the A-coring structures on $\mathcal{C} = A \otimes C$ satisfying (2.87-2.89) and right-right entwining structures of the form (A, C, ψ).*

Proof. Assume first that $\mathcal{C} = A \otimes C$ satisfies (2.87-2.89), and define

$$\psi : \ C \otimes A \to A \otimes C \ ; \ \ \psi(c \otimes a) = (1 \otimes c)a$$

We claim that (A, C, ψ) is a right-right entwining structure. Indeed, for all $c \in C$ and $a, b \in A$, we have

$$\psi(c \otimes ab) = (1 \otimes c)(ab) = ((1 \otimes c)a)b$$
$$= (a_\psi \otimes c^\psi)b = a_\psi(1 \otimes c^\psi)b$$
$$= a_\psi(b_\Psi \otimes c^{\psi\Psi}) = a_\psi b_\Psi \otimes c^{\psi\Psi}$$
$$\psi(c \otimes 1) = (1 \otimes c)1 = 1 \otimes c$$

proving (2.1) and (2.2) In $(A \otimes C) \otimes_A (A \otimes C) \cong A \otimes C \otimes A$, we have

$$\Delta_{\mathcal{C}}(a_\psi \otimes c^\psi) = a_\psi \otimes \Delta_{\mathcal{C}}(c^\psi)$$

and

$$\Delta_{\mathcal{C}}(a_\psi \otimes c^\psi) = \Delta_{\mathcal{C}}((1 \otimes c)a) = \Delta_{\mathcal{C}}(1 \otimes c)a$$
$$= (1 \otimes c_{(1)}) \otimes_A (1 \otimes c_{(2)})a = (1 \otimes c_{(1)}) \otimes_A (a_\psi \otimes c_{(2)}^\psi)$$
$$= (1 \otimes c_{(1)})a_\psi \otimes c_{(2)}^\psi = a_{\psi\Psi} \otimes c_{(1)}^\Psi \otimes c_{(2)}^\psi$$

and (2.3) follows. Finally

$$\varepsilon_C(c^\psi)a_\psi = \varepsilon_C((1 \otimes c)a) = \varepsilon_C(1 \otimes c)a = \varepsilon_C(c)a$$

proving (2.4). Conversely, let (A, C, ψ) be a right-right entwining structure. Being an object of $^k_A\mathcal{M}(I_A, \psi)^C_A$, $C = A \otimes C$ is an A-bimodule. Δ_C and ε_C are defined by (2.88) and (2.89), and an immediate computation shows that (2.83) and (2.84) are satisfied.

Theorem 17. *Let $C = A \otimes C$ be a coring factorized through A and C, and assume that (A, C, ψ) is the corresponding right-right entwining structure. Then we have an isomorphism of categories*

$$\mathcal{M}^C \cong \mathcal{M}(\psi)^C_A$$

Proof. We have a functor $F : \mathcal{M}^C \to \mathcal{M}(\psi)^C_A$, with $F(M) = M$ as a k-module. The right A-action on $F(M)$ is the same as the one on M, and the right C-coaction on $F(M)$ is the C-comodule map on M followed by the isomorphism $M \otimes_A (A \otimes C) \cong M \otimes C$. We easily compute that

$$\rho^r_{F(M)}(ma) = \rho^r_M(ma) = \rho^r_M(m)a$$
$$= \sum m_{[0]} \otimes_A (1 \otimes m_{[1]})a = \sum m_{[0]} \otimes_A a_\psi \otimes m^\psi_{[1]}$$
$$= \sum m_{[0]}a_\psi \otimes m^\psi_{[1]}$$

as needed. We leave it to the reader to construct the inverse of F.

We have a left-handed version of Theorems 16 and 17. We say that a coring C factorizes through C and A if $C \cong C \otimes A$ as k-modules, and (after identifying C and $C \otimes A$):

$$(c \otimes b)a = c \otimes ba$$
$$\Delta_C(c \otimes a) = (c_{(1)} \otimes 1) \otimes_A (c_{(2)} \otimes a)$$
$$\varepsilon_C(c \otimes a) = \varepsilon_C(c)a$$

for all $a, b \in A$ and $c \in C$.

Theorem 18. *We have a bijective correspondence between A-coring structures on $C = C \otimes A$ and left-left entwining structures of the form (A, C, φ), and we have an isomorphism of categories*

$$^C\mathcal{M} \cong {}^C_A\mathcal{M}(\varphi)$$

Proof. Similar to the proof of Theorems 16 and 17; we only mention that φ is given by the formula

$$a(c \otimes 1) = \varphi(a \otimes c)$$

Proposition 34. *Consider $(A, C, \psi) \in \mathbb{E}_\bullet^\bullet(k)$, with ψ invertible, and put $\psi^{-1} = \varphi$. Then $\psi : C \otimes A \to A \otimes C$ is an isomorphism of corings. Consequently we have an isomorphism of categories*

$$\,_A^C\mathcal{M}(\psi^{-1}, \psi)_A^C \cong \,^C\mathcal{M}^C$$

Proof. A straightforward computation: for all $a, b \in A$ and $c \in C$, we have

$$\psi(a(c \otimes b)) = \psi(c^\varphi \otimes a_\varphi b) = (a_\varphi b)_\psi \otimes c^{\varphi\psi}$$
$$= a_{\varphi\psi} b_\Psi \otimes c^{\varphi\psi\Psi} = ab_\Psi \otimes c^\Psi$$
$$= a\psi(c \otimes b)$$
$$\psi((c \otimes b)a) = (ba)_\psi \otimes c^\psi = b_\psi a_\Psi \otimes c^{\psi\Psi}$$
$$= (b_\psi \otimes c^\psi)a = \psi(c \otimes b)a$$
$$\Delta_{A \otimes C}(\psi(c \otimes a)) = \left(a_\psi \otimes (c^\psi)_{(1)} \right) \otimes_A \left(1 \otimes (c^\psi)_{(2)} \right)$$
$$= \left(a_{\psi\Psi} \otimes c_{(1)}^\Psi \right) \otimes_A \left(1 \otimes c_{(2)}^\psi \right)$$
$$= (1 \otimes c_{(1)}) a_\psi \otimes_A (1 \otimes c_{(2)}^\psi)$$
$$= (1 \otimes c_{(1)}) \otimes_A (a_\psi \otimes c_{(2)}^\psi)$$
$$= (\psi \otimes \psi) \left((c_{(1)} \otimes 1) \otimes_A (c_{(2)} \otimes a) \right)$$
$$= (\psi \otimes \psi) \Delta_{C \otimes A}(c \otimes a)$$

Proposition 35. *Let $\mathcal{C}$ be an A-coring, and write $R = \,_A\text{Hom}(\mathcal{C}, A)$. The multiplication rule*

$$(f \cdot f')(c) = f'\left(c_{(1)} f(c_{(2)}) \right) \tag{2.90}$$

(for $f, f' \in R$ and $c \in C$) makes R into a ring with unit ε_C. The map $\iota : A \to R$, $\iota(a)(c) = \varepsilon_C(c)a$ is a ring homomorphism. Consequently R can be viewed as an A-bimodule:

$$(afb)(c) = f(ca)b \tag{2.91}$$

Proof. It is easy to see that $f \cdot f'$ is left and right A-linear. The multiplication is associative since

$$(f \cdot (f' \cdot f''))(c) = (f' \cdot f'')(c_{(1)} f(c_{(2)}))$$
$$= f'' \left((c_{(1)} f(c_{(2)}))_{(1)} f'((c_{(1)} f(c_{(2)}))_{(1)}) \right)$$
$$= f'' \left(c_{(1)} f'(c_{(2)} f(c_{(3)})) \right)$$
$$= ((f \cdot f') \cdot f'')(c)$$

It is easy to see that ε_C is a left and right unit. Finally

$$(af)(c) = (\iota(a) \cdot f)(c) = f(c_{(1)}\varepsilon_C(c_{(2)})a) = f(ca)$$

and

$$(fb)(c) = (f \cdot \iota(b))(c) = \iota(b)(c_{(1)}f(c_{(2)})) = f(c)b$$

The ring $R = {}_A\mathrm{Hom}(C, A)$ is due to Sweedler [173]. In the case where $C = A \otimes C$ factorizes through A and C, the multiplication on R is related to the Koppinen smash product:

Proposition 36. Let (A, C, ψ) be a right-right entwining structure, and $C = A \otimes C$. Then the canonical isomorphism of k-modules

$$R = {}_A\mathrm{Hom}(A \otimes C, A) \cong \#_{\psi \circ \tau}(C, A^{\mathrm{op}})$$

is an isomorphism of k-algebras.

Proof. For $f, g \in R$, we have

$$(f \cdot g)(1 \otimes c) = g((1 \otimes c_{(1)})f(1 \otimes c_{(2)}))$$
$$= g(f(1 \otimes c_{(2)})_\psi \otimes c_{(1)}^\psi)$$
$$= f(1 \otimes c_{(2)})_\psi g(1 \otimes c_{(1)}^\psi)$$

and this is exactly the multiplication on the Koppinen smash product $\#_{\psi \circ \tau}(C, A^{\mathrm{op}})$ $((A^{\mathrm{op}}, C, \psi \circ \tau \in {}_\bullet\mathbb{E}(k)^\bullet)$.

Proposition 37. *Let (A, C, ψ) be a right-right entwining structure, $C = A \otimes C$, and assume that C is finitely generated and projective as a k-module. Then $(A, C^{\mathrm{cop}}, \tau \circ \psi) \in {}^\bullet\mathbb{E}(k)_\bullet$, and we can consider $C^{*\mathrm{op}}\#_R A$. We have an isomorphism of k-algebras*

$$C^{*\mathrm{op}}\#_R A \cong R = {}_A\mathrm{Hom}(A \otimes C, A)$$

Proof. Let $\{c_i, c_i^*, \mid i = 1, \cdots, m\}$ be a dual basis for C. Take $f = c^*\#a$, $g = d^*\#b \in R \cong C^{*\mathrm{op}}\#_R A$. In $C^{*\mathrm{op}}\#_R A$, we have, using (2.33)

$$(c^*\#a)(d^*\#b) = d_R^* * c^*\#a_R b = \sum_i \langle d^*, c_i^\psi \rangle c_i^* * c^*\#a_\psi b$$

and

$$((c^*\#a)(d^*\#b))(c) = \sum_i \langle d^*, c_i^\psi \rangle \langle c_i^*, c_{(1)} \rangle \langle c^*, c_{(2)} \rangle a_\psi b$$

$$= \langle c^*, c_{(2)} \rangle a_\psi \langle d^*, c_{(1)}^\psi \rangle b = f(c_{(2)})_\psi g(c_{(1)}^\psi) = (f \cdot g)(c)$$

proving our result.

Lemma 10. *Consider an A-coring C, which is finitely generated and projective as a left A-module, and let $\{c_k, f_k \mid k = 1, \cdots, m\}$ be a finite dual basis for C. Then*

$$\sum_k f_k \otimes_A \Delta_C(c_k) = \sum_{k,l} f_k \cdot f_l \otimes_A c_l \otimes_A c_k \qquad (2.92)$$

Proof. The fact that $\{c_k, f_k \mid k = 1, \cdots, m\}$ is a dual basis means

$$c = \sum_k f_k(c)c_k \quad \text{and} \quad f = \sum_k f_k f(c_k) \qquad (2.93)$$

for all $c \in C$ and $f \in R$. Now

$$\Delta_C(c) = \sum_k f_k(c)c_{k(1)} \otimes_A c_{k(2)}$$

and

$$\Delta_C(c) = c_{(1)} \otimes_A c_{(2)} = \sum_k c_{(1)} \otimes_A f_k(c_{(2)})c_k$$

$$= \sum_k c_{(1)}f_k(c_{(2)}) \otimes_A c_k = \sum_{k,l} f_l(c_{(1)}f_k(c_{(2)}))c_l \otimes_A c_k$$

$$= \sum_{k,l} (f_k \cdot f_l)(c)c_l \otimes_A c_k$$

and the result follows.

Proposition 38. *Let C be an A-coring. We have a functor*

$$F: \mathcal{M}^C \to \mathcal{M}_R$$

F is an isomorphism if C is finitely generated and projective as a left A-module.

Proof. We put $F(M) = M$ with

$$mf = m_{[0]}f(m_{[1]})$$

for all $f \in R$ and $m \in M$. $F(M)$ is a right R-module, since

$$m\varepsilon_C = m_{[0]}\varepsilon_C(m_{[1]}) = m$$

and

$$(mf)f' = (m_{[0]}f(m_{[1]}))_{[0]}f'\left((m_{[0]}f(m_{[1]}))_{[1]}\right)$$

$$= m_{[0]}f'(m_{[1]}f(m_{[2]})) = m(f \cdot f')$$

Now let C be finitely generated and projective as a left A-module; take $M \in \mathcal{M}_R$, and put $G(M) = M$, with C-comodule structure

$$\rho^r(m) = \sum_k m f_k \otimes_A c_k \tag{2.94}$$

ρ^r defines a right C-comodule structure on M since

$$\sum m_{[0]} \varepsilon_C(m_{[1]}) = \sum_k m f_k \varepsilon_C(c_k) = m \varepsilon_C = m$$

and

$$(\rho^r \otimes_A I_C)(\rho^r(m)) = \sum_{k,l} m(f_k \cdot f_l) \otimes_A c_l \otimes_A c_k$$

$$((2.93)) \qquad = \sum_k m f_k \otimes_A \Delta_C(c_k) = (I_M \otimes_A \Delta_C)(\rho^r(m))$$

It is not difficult to show that F and G are functors, and that they are each others inverses: for $M \in \mathcal{M}_R$, the structure on $FG(M)$ is given by

$$m \bullet f = \sum_k m f_k f(c_k) = mf$$

while for $M \in \mathcal{M}^C$, the structure on $GF(M)$ is given by

$$\rho^r_{GF(M)}(m) = \sum_k m_{[0]} f_k(m_{[1]}) \otimes_A c_k = m_{[0]} \otimes m_{[1]}$$

Corollary 1. *Let C be an A-coring which is finitely generated and projective as a left A-module. Then $R = {}_A\mathrm{Hom}(C, A) \in {}_A\mathcal{M}^C$. The (A, A)-bimodule structure is given by (2.91):*

$$(afb)(c) = f(ca)b$$

and the right C-comodule structure is given by

$$\rho^r(f) = f_{[0]} \otimes_A f_{[1]} = \sum_k f \cdot f_k \otimes_A c_k \tag{2.95}$$

This means that

$$f_{[0]}(c)f_{[1]} = c_{(1)}f(c_{(2)}) \tag{2.96}$$

Proof. $R \in \mathcal{M}_R \cong \mathcal{M}^C$. (2.95) is then obtained using (2.94). (2.96) follows immediately.

Corollary 2. *Let (A, C, ψ) be a right-right entwining structure, and assume that C is finitely generated and projective as a left A-module. Then $C^* \otimes A \in {}_A\mathcal{M}(I_C, \psi)^C_A$. The structure maps are*

$$(c^* \otimes a)b = c^* \otimes ab \tag{2.97}$$

$$b(c^* \otimes a) = \sum_i \langle c^*, d_i^\psi \rangle c_i^* \otimes b_\psi a \tag{2.98}$$

$$\rho^r(c^* \otimes a) = \sum_i c_i^* * c^* \otimes a_\psi \otimes d_i^\psi \tag{2.99}$$

Proof. In Corollary 1, we take $\mathcal{C} = A \otimes C$, $R = {}_A\mathrm{Hom}(A \otimes C, A) \cong \#_R(C, A) \cong C^* \otimes A$, and we translate the structure on R into the structure on $C^* \otimes A$. Let us do the left A-module structure in detail: for all $a, b \in A$, $c \in C$, and $c^* \in C^*$, we have

$$
\begin{aligned}
\big(b(c^* \otimes a)\big)(c) &= \Big(\iota(b) \cdot (c^* \otimes a)\Big)(c) \\
&= \iota(b)(c_{(2)})_\psi(c^* \otimes a)(c_{(1)}^\psi) = \varepsilon_C(c_{(2)})b_\psi \langle c^*, c_{(1)}^\psi \rangle a \\
&= \langle c^*, c^\psi \rangle b_\psi a = \langle c^*, c_i^\psi \rangle \langle c_i^*, c \rangle b_\psi a
\end{aligned}
$$

and (2.98) follows.

In Proposition 5, we have seen that the category of comodules over a flat k-coalgebra is a Grothendieck category. Brzeziński informed us that the same proof can be used to prove the following result:

Theorem 19. *Let C be an A-coring. $\mathcal{M}^C$ is a Grothendieck category and the forgetful functor $\mathcal{M}^C \to \mathcal{M}_A$ is exact if and only if C is flat as a left A-module. In particular, if (A, C, ψ) is a right-right entwining structure, and C is flat as a k-module, then $\mathcal{M}(\psi)_A^C$ is a Grothendieck category.*

2.8 Monoidal categories

Let A and B be algebras and coalgebras (but not necessarily bialgebras), and assume that we have $(A, B, R) \in \mathbb{S}(k)$ and $(A, B, V) \in \mathbb{CS}(k)$. Then we have an algebra structure (the smash product) and a coalgebra structure (the smash coproduct) on $A \otimes B$; if these structures make $A \otimes B$ into a bialgebra, then we denote this bialgebra by $A_V \bowtie_R B$. The smash biproduct has been investigated in [44] and, more extensively, in [18], [19] and [17]. In this Section, we restrict attention to the situation where A and B are bialgebras, and either R or V is the switch map. In these situation, we keep the old notation:

$$A_\tau \bowtie_R B = A \#_R B \quad \text{and} \quad C_V \bowtie_\tau D = C \bowtie^V D$$

Let us first investigate the smash product.

Proposition 39. *Let A and B be bialgebras, and $(A, B, R) \in \mathbb{S}(k)$. $A \#_R B$ with comultiplication Δ and counit ε given by*

$$\Delta(a \# b) = (a_{(1)} \# b_{(1)}) \otimes (a_{(2)} \# b_{(2)}) \quad \text{and} \quad \varepsilon(a \# b) = \varepsilon_A(a)\varepsilon_B(b)$$

is a bialgebra if and only if

$$a_{(1)R} \otimes b_{(1)R} \otimes a_{(2)r} \otimes b_{(2)r} = a_{R(1)} \otimes b_{R(1)} \otimes a_{R(2)} \otimes b_{R(2)} \qquad (2.100)$$

$$\varepsilon_A(a_R)\varepsilon_B(b_R) = \varepsilon_A(a)\varepsilon_B(b) \qquad (2.101)$$

for all $a \in A$ and $b \in B$.

Proof. A direct consequence of Proposition 1: (2.100) is equivalent to (1.6), (2.101) is equivalent to (1.7), and (1.8-1.9) are automatically fulfilled.

We have a similar property for the smash coproduct:

Proposition 40. *Let C and D be bialgebras, and $(C, D, V) \in \mathbb{CS}(k)$. $C \ltimes^V D$ with multiplication $m = (m_C \otimes m_D) \circ (I_C \otimes \tau \otimes I_D)$ and unit $1 = 1 \ltimes 1$ is a bialgebra if and only if*

$$(df)^V \otimes (ce)^V = d^V f^v \otimes c^V e^v \qquad (2.102)$$
$$V(1 \otimes 1) = 1 \otimes 1 \qquad (2.103)$$

for all $c, e \in C$ and $d, f \in D$.

Example 15. (cf. [185]) Let H be a Hopf algebra, and C an H-bicomodule coalgebra. Consider the map $V : C \otimes H \to H \otimes C$ given by

$$V(c \otimes h) = c_{[-1]} h S(c_{[1]}) \otimes c_{[0]}$$

Then $(C, H, V) \in \mathbb{CS}(k)$; the smash coproduct is denoted $C \times_r H$, and is called the *right twisted smash coproduct*. Now assume that C is also a bialgebra. Then $C \times_r H$ is a bialgebra if and only if the following equations hold, for all $c, d \in C$ and $h, k \in H$:

$$\left(c_{[-1]} h S(c_{[1]})\right)\left(d_{[-1]} k S(d_{[1]})\right) \otimes c_{[0]} d_{[0]}$$
$$= (cd)_{[-1]} (hk) S((cd)_{[1]}) \otimes (cd)_{[0]} \qquad (2.104)$$
$$1_{[-1]} S(1_{[1]}) \otimes 1_{[0]} = 1_H \otimes 1_C \qquad (2.105)$$

We will now present similar results for entwining structures; we will closely follow [54], where the Doi-Hopf case is discussed. The basic idea is the following: if H is a bialgebra, then the categories $_H\mathcal{M}$ and $\mathcal{M}^H$ are monoidal categories: for two H-modules M and N, we find the following H-module structure on $M \otimes N$:

$$h \cdot (m \otimes n) = h_{(1)} m \otimes h_{(2)} n \qquad (2.106)$$

If M and N are H-comodules, then H coacts on $M \otimes N$ as follows:

$$\rho^r (m \otimes n) = m_{[0]} \otimes n_{[0]} \otimes m_{[1]} n_{[1]} \qquad (2.107)$$

In both cases, the unit object is k with respectively the trivial action and coaction. In fact, the two previous Propositions tell us how to make the category of modules over the smash product (resp. the category of comodules over the smash coproduct) into a monoidal category. We refer to [123, Chapter 7] for the definition of a monoidal category. Basically, a monoidal category is a category $\mathcal{C}$ together with a functor

$$\otimes : \mathcal{C} \times \mathcal{C} \to \mathcal{C}$$

and a unit object $k \in \mathcal{C}$ such that $\otimes$ is coherently associative, that is, we have isomorphisms

$$C \otimes (D \otimes E) \cong (C \otimes D) \otimes E$$

and

$$C \otimes k \cong k \otimes C \otimes C$$

for all $C, D, E \in \mathcal{C}$, satisfying certain coherence conditions. In our setting, the associativity condition will always be the natural one, and there will be no need to worry about the coherence conditions.

Now let (A, C, ψ) be a left-right entwining structure, and assume that A and C are both bialgebras. For $M, N \in {}_A\mathcal{M}^C$, we put the following left A-action and right C-coaction on $M \otimes N$:

$$a(m \otimes n) = a_{(1)}m \otimes a_{(2)}n \tag{2.108}$$
$$\rho^r(m \otimes n) = m_{[0]} \otimes n_{[0]} \otimes m_{[1]}n_{[1]} \tag{2.109}$$

With this structure, $M \otimes N$ is a left A-module and a right C-comodule. $M \otimes N$ is an entwined module if and only if

$$\rho^r(a(m \otimes n)) = a_\psi(m_{[0]} \otimes n_{[0]}) \otimes (m_{[1]}n_{[1]})^\psi$$

or

$$a_{(1)\psi}m_{[0]} \otimes a_{(2)\Psi}n_{[0]} \otimes m_{[1]}^\psi n_{[1]}^\Psi = a_{\psi(1)}m_{[0]} \otimes a_{\psi(2)}n_{[0]} \otimes (m_{[1]}n_{[1]})^\psi \tag{2.110}$$

for all $m \in M$, $n \in N$ and $a \in A$. It clearly suffices that

$$\Delta_A(a_\psi) \otimes (cd)^\psi = a_{(1)\psi} \otimes a_{(2)\Psi} \otimes c^\psi d^\Psi \tag{2.111}$$

for all $a \in A$ and $c, d \in C$. (2.111) is also necessary: we take $M = N = A \otimes C$ (with $a(b \otimes c) = ab \otimes c$, $\rho^r(b \otimes c) = b_\psi \otimes c_{(1)} \otimes c_{(2)}^\psi$), $m = 1_A \otimes c$, $n = 1_A \otimes d$. Then (2.110) amounts to

$$a_{\psi(1)} \otimes c_{(1)} \otimes a_{\psi(2)} \otimes d_{(1)} \otimes (c_{(2)}d_{(2)})^\psi = a_{(1)\psi} \otimes c_{(1)} \otimes a_{(2)\Psi} \otimes d_{(1)}c_{(2)}^\psi d_{(2)}^\Psi$$

and (2.111) follows after we apply ε_C to the second and the fourth factor.
k is a left A-module, using the trivial A-action, and a right C-comodule, using the right C-coaction:

$$a \cdot x = \varepsilon_A(a)x \quad \text{and} \quad \rho^r(x) = x1_C \tag{2.112}$$

for all $x \in k$ and $a \in A$. Here we identified $k \otimes C$ and C. With this structure, k is an entwined module if and only if

$$\varepsilon_A(a)1_C = \varepsilon_A(a_\psi)1_C^\psi \tag{2.113}$$

for all $a \in A$.

Theorem 20. *Let (A, C, ψ) be a left-right entwining structure, and assume that A and C are bialgebras. For $M, N \in {}_A\mathcal{M}(\psi)^C$, we define an A-action and a C-coaction on $M \otimes N$ using (2.108-2.109); k is a left A-module and a right C-comodule using (2.112).*
Then $({}_A\mathcal{M}(\psi)^C, \otimes, k)$ is a monoidal category if and only if (2.111) and (2.113) hold for all $a \in A$ and $c, d \in C$. In this case, we say that (A, C, ψ) is a monoidal entwining structure.

Proof. We leave it to the reader to show that the natural isomorphisms

$$(M \otimes N) \otimes P \cong M \otimes (N \otimes P) \quad \text{and} \quad M \otimes k \cong k \otimes M \cong M$$

of k-modules are also isomorphisms of entwined modules.

If (A, C, ψ) is a monoidal entwining structure, then A and C can be made into objects of ${}_A\mathcal{M}(\psi)^C$:

Proposition 41. *Let (A, C, ψ) be a monoidal entwining structure. On A and C, we consider the following left A-action and right C-coaction:*

$$b \cdot a = ba \quad \text{and} \quad \rho^r(a) = \psi(1_C \otimes a) = a_\psi \otimes 1_C^\psi$$

$$a \cdot c = \varepsilon_A(a_\psi)c^\psi \quad \text{and} \quad \rho^r(c) = c_{(1)} \otimes c_{(2)}$$

Then A and C are entwined modules.

Proof. We will show that $A \in {}_A\mathcal{M}(\psi)^C$, and leave the other statement to the reader. First, A is a right C-comodule, since

$$(I_A \otimes \varepsilon_C)\rho^r(a) = \varepsilon_C(1_C^\psi)a_\psi = \varepsilon_C(1_C)a = a$$

and

$$(I_A \otimes \Delta_C)\rho^r(a) = a_\psi \otimes \Delta(1_C^\psi) = a_{\psi\Psi} \otimes 1_C^\Psi \otimes 1_C^\psi = (\rho^r \otimes I_C)\rho^r(a)$$

$A \in {}_A\mathcal{M}(\psi)^C$ since

$$a_\psi b_{[0]} \otimes b_{[1]}^\psi = a_\psi b_\Psi \otimes 1_C^{\Psi\psi} = (ab)_\psi \otimes 1_C^\psi = \rho^r(ab)$$

Examples 6. 1. Let H be a Hopf algebra, and C a left H-module bialgebra. This means that C is a bialgebra, and that H acts on C in such a way that C is an H-module algebra and an H-module coalgebra. Now let A be a bialgebra and a right H-comodule algebra such that the following compatibility relation holds, for all $a \in A$:

$$a_{[0](1)} \otimes a_{[0](2)} \otimes a_{[1](1)} \otimes a_{[1](2)} = a_{(1)[0]} \otimes a_{(2)[0]} \otimes a_{(1)[1]} \otimes a_{(2)[1]} \quad (2.114)$$

We know that (H, A, C) is a left-right Doi-Hopf datum, and we have a corresponding left-right entwining structure (A, C, ψ). It is straightforward to check that (A, C, ψ) is monomial.

In particular, let H be cocommutative, and let $A = H$, with right H-coaction given by right comultiplication. Then (2.114) holds, and we have a monomial entwining structure.

As a special case, let $H = k$, and A and C bialgebras. Then $(A, C, I_{A \otimes C}$ is a monomial entwining structure.

2. Our first example can be dualized: let A be a right H-comodule bialgebra, and C a bialgebra and a left H-module coalgebra such that

$$(h \cdot c)(h' \cdot c') = (hh') \cdot (cc') \tag{2.115}$$

Then we obtain a left-right entwining structure (A, C, ψ) which is monomial.

Remark 4. Assume that C is finitely generated and projective. In Theorem 9, we have seen that we have a bijective correspondence between $(A, C, \psi) \in {}_\bullet\mathbb{E}^\bullet(k)$ and $(A, C^*, R) \in \mathbb{S}(k)$. It can be shown that (A, C, ψ) is monoidal if and only if $A \#_R C^*$ is a bialgebra as in Proposition 39. Similar observations apply to the smash coproduct: if A is finitely generated projective, then we have a one-to-one correspondence between $(A, C, \psi) \in {}_\bullet\mathbb{E}^\bullet(k)$ and $(A^*, C, V) \in \mathbb{CS}(k)$ (cf. Theorem 12); (A, C, ψ) is monoidal if and only if $A^* \bowtie^V C$ is a bialgebra as in Proposition 40.

A morphism

$$(\alpha, \gamma): (A, C, \psi) \to (A', C', \psi')$$

is called *monoidal* if α and γ are bialgebra maps.

Proposition 42. *If $(\alpha, \gamma): (A, C, \psi) \to (A', C', \psi')$ is monoidal, then the induction functor $F: {}_A\mathcal{M}(\psi)^C \to {}_{A'}\mathcal{M}(\psi')^{C'}$ is a monoidal functor. Its adjoint G is also monoidal, if we view it as a functor between the corresponding opposite categories.*

Proof. Take $M, N \in {}_A\mathcal{M}(\psi)^C$, and define a map

$$f = f_{M,N}: \ F(M \otimes N) = A' \otimes_A (M \otimes N)$$
$$\to F(M) \otimes F(N) = (A' \otimes_A M) \otimes (A' \otimes_A N)$$

by

$$f(a' \otimes_A (m \otimes n)) = (a'_{(1)} \otimes_A m) \otimes (a'_{(2)} \otimes_A n)$$

It is easily verified that f is a well-defined left A'-linear map; let us check that f is also right C'-colinear. For all $a' \in A'$, $m \in M$ and $n \in N$, we have

$$(f \otimes I_{C'})\rho^r(a' \otimes (m \otimes n))$$
$$= \left(a'_{\psi'(1)} \otimes_A m_{[0]}\right) \otimes \left(a'_{\psi'(2)} \otimes_A n_{[0]}\right) \otimes \left(\gamma(m_{[1]}n_{[1]})\right)^{\psi'}$$
$$(2.111) \quad = \left(a'_{(1)\psi'} \otimes_A m_{[0]}\right) \otimes \left(a'_{(2)\Psi'} \otimes_A n_{[0]}\right) \otimes \left(\gamma(m_{[1]})^{\psi'}\gamma(n_{[1]})^{\Psi'}\right)$$
$$= \rho^l\big(f(a' \otimes (m \otimes n))\big)$$

We also have a map $g : F(k) = A' \otimes_A k \to k$, given by $g(a' \otimes_A x) = \varepsilon_{A'}(a')$. It is straightforward to verify that the maps $f_{M,N}$ and g satisfy all the necessary coherence axioms for having a monoidal functor (cf. [123, Sec. XI.2]).

The second part of the proof is formally dual to the proof of the first part. For two entwined modules $M', N' \in {}_{A'}\mathcal{M}(\psi')^{C'}$, we have a map $f' = f'_{M',N'} :$
$G(M') \otimes G(N') = (M' \square_{C'} C) \otimes (N' \square_{C'} C) \to G(M' \otimes N') = ((M' \otimes N') \square_{C'} C)$
given by

$$f'\left(\sum_i (m_i' \otimes c_i) \otimes (n_i' \otimes d_i) \right) = \sum_i (m_i' \otimes n_i') \otimes c_i d_i$$

The map $g' : k \to G(k) = k \square_{C'} c$ is defined by $g(x) = x \otimes 1_C$. We leave it to the reader to check that the $f'_{M',N'}$ and g' are well-defined morphisms in ${}_A\mathcal{M}(\psi)^C$, and that they satisfy the axioms for having a monoidal functor.

We now consider the situation where $A = A'$ and $\alpha = I_A$. We keep the same notation as above: $F : {}_A\mathcal{M}(\psi)^C \to {}_A\mathcal{M}(\psi')^{C'}$ is the induction functor, and G is its adjoint. Observe that $F(M) = M$ as a left A-module, with C'-coaction $\rho_{C'}(m) = m_{[0]} \otimes \gamma(m_{[1]})$. Our next result is a generalization of [152, Theorem 7.1 (1) and (3)] and [54, Theorem 3.1].

Theorem 21. Let $(I_A, \gamma) : (A, C, \psi) \to (A, C', \psi')$ be a monoidal morphism of monoidal entwining structures. Then

$$G(C') = C \tag{2.116}$$

Let $M \in {}_A\mathcal{M}(\psi)^C$ be flat as a k-module, and take $N \in {}_A\mathcal{M}(\psi')^{C'}$. If C is a Hopf algebra, then

$$M \otimes G(N) \cong G(F(M) \otimes N) \quad \text{in } {}_A\mathcal{M}(\psi)^C \tag{2.117}$$

If C has a twisted antipode $\overline{S}$, then

$$G(N) \otimes M \cong G(N \otimes F(M)) \quad \text{in } {}_A\mathcal{M}(\psi)^C \tag{2.118}$$

Proof. We know that $\varepsilon_{C'} \otimes I_C : C' \square_C C \to C$ is an isomorphism; the inverse map is $(\gamma \otimes I_C) \circ \Delta_C$. It is clear that $\varepsilon_{C'} \otimes I_C$ is A-linear and C-colinear, and this proves (2.116).
We now define a map

$$\Gamma : M \otimes G(N) = M \otimes (N \square_{C'} C) \to G(F(M) \otimes N) = (F(M) \otimes N) \square_{C'} C$$

by

$$\Gamma(m \otimes \sum_i (n_i \otimes c_i)) = \sum_i (m_{[0]} \otimes n_i) \otimes m_{[1]} c_i$$

1) Γ is well-defined. We have to show that

$$\Gamma(m \otimes \sum_i (n_i \otimes c_i)) \in (F(M) \otimes N)\square_{C'} C$$

This may be seen as follows: for all $m \in M$ and $\sum_i n_i \otimes c_i \in N\square_{C'}C$, we have that

$$(\rho_{F(M)\otimes N} \otimes I_C) \sum_i \left((m_{[0]} \otimes n_i) \otimes m_{[1]}c_i \right)$$

$$= \sum_i (m_{[0]} \otimes n_{i[0]}) \otimes \gamma(m_{[1]})n_{i[1]} \otimes m_{[2]}c_i$$

$$= \sum_i (m_{[0]} \otimes n_i) \otimes \gamma(m_{[1]})\gamma(n_{i[1]}) \otimes m_{[2]}c_{i[2]}$$

$$= \sum_i (I_{F(M)\otimes N} \otimes \rho_{C'})\left((m_{[0]} \otimes n_i) \otimes m_{[1]}c_i \right)$$

2) Γ is A-linear. For all $a \in A$, $m \in M$ and $\sum_i n_i \otimes c_i \in N\square_{C'}C$, we have that

$$\Gamma\left(a(m \otimes \sum_i (n_i \otimes c_i)) \right) = \sum_i \Gamma\left(a_{(1)}m \otimes (a_{(2)\psi}n_i \otimes c_i\psi) \right)$$

$$= \sum_i (a_{(1)\Psi}m_{[0]} \otimes a_{(2)\psi}n_i) \otimes m_{[1]}^{\Psi}c_i^{\psi}$$

$$(2.111) \quad = \sum_i (a_{\psi(1)}m_{[0]} \otimes a_{\psi(2)}n_i) \otimes (m_{[1]}c_i)^{\psi}$$

$$= \sum_i a_{\psi}(m_{[0]} \otimes n_i) \otimes (m_{[1]}c_i)^{\psi}$$

$$= a\Gamma(m \otimes \sum_i (n_i \otimes c_i))$$

3) Γ is C-colinear. For all $m \in M$ and $n_i \otimes c_i \in N\square_{C'}C$, we have that

$$\rho(\Gamma(m \otimes \sum_i (n_i \otimes c_i))) = \sum_i \rho((m_{[0]} \otimes n_i) \otimes m_{[1]}c_i)$$

$$= \sum_i (m_{[0]} \otimes n_i) \otimes m_{[1]}c_{i1} \otimes m_{[2]}c_{i2}$$

$$= \sum_i (\Gamma \otimes I_C)(m_{[0]} \otimes (n_i \otimes c_{i1})) \otimes m_{[1]}c_{i2}$$

$$= (\Gamma \otimes I_C)(\rho m \otimes \sum_i (n_i \otimes c_i)))$$

Assume now that C has an antipode, and define

$$\Psi : (F(M) \otimes N)\square_{C'}C \to M \otimes (N\square_{C'}C)$$

by

$$\Psi((m_i \otimes n_i) \otimes c_i) = m_{i[0]} \otimes (n_i \otimes S(m_{i[1]})c_i)$$

We have to show that Ψ is well-defined. M is flat, so $M \otimes (N \square_{C'} C)$ is the equalizer of the maps

$$\begin{cases} I_M \otimes I_N \otimes \rho_C \\ I_M \otimes \rho_N \otimes I_C \end{cases} : M \otimes N \otimes C \rightrightarrows M \otimes N \otimes C' \otimes C$$

Now take $\sum_i (m_i \otimes n_i) \otimes c_i \in (F(M) \otimes N) \square_{C'} C$. Then

$$(m_{i[0]} \otimes n_{i[0]}) \otimes \gamma(m_{i[1]})n_{i[1]} \otimes c_i = (m_i \otimes n_i) \otimes \gamma(c_{i(1)}) \otimes c_{i(2)} \qquad (2.119)$$

Now

$$(I_M \otimes I_N \otimes \rho_C)(\sum_i m_{i[0]} \otimes (n_i \otimes S(m_{i[1]})c_i))$$

$$= \sum_i m_{i[0]} \otimes n_i \otimes \gamma(S(m_{i[2]})c_{i(1)}) \otimes S(m_{i[1]})c_{i(2)}$$

and

$$(I_M \otimes \rho_N \otimes I_C)(\sum_i m_{i[0]} \otimes (n_i \otimes S(m_{i[1]})c_i)) = \sum_i m_{i[0]} \otimes n_{i[0]} \otimes n_{i[1]} \otimes S(m_{i[1]})c_i$$

Apply $(I_M \otimes \gamma \otimes I_C) \circ (I_H \otimes (\Delta_C \circ S_C)) \circ \rho_M$ to the first factor of (2.119) Then we obtain that

$$\sum_i m_{i[0]} \otimes \gamma(S(m_{i[2]})) \otimes S(m_{i[1]}) \otimes n_{i[0]} \otimes \gamma(m_{i[3]})n_{i[1]} \otimes c_i$$

$$= \sum_i m_{i[0]} \otimes \gamma(S(m_{i[2]})) \otimes S(m_{i[1]}) \otimes n_i \otimes \gamma(c_{i(1)}) \otimes c_{i(2)}$$

Multiplying the second and the fifth factor, and also the third and the sixth factor, we obtain that

$$\sum_i m_{i[0]} \otimes n_{i[0]} \otimes n_{i[1]} \otimes S(m_{i[1]})c_i$$

$$= \sum_i m_{i[0]} \otimes n_i \otimes \gamma(S(m_{i[2]}))\gamma(c_{i(1)}) \otimes S(m_{i[1]})c_{i(2)}$$

or

$$(I_M \otimes \rho_N \otimes I_C)(\Psi(\sum_i (m_i \otimes n_i) \otimes c_i)) = (I_M \otimes I_N \otimes \rho_C)(\Psi(\sum_i (m_i \otimes n_i) \otimes c_i))$$

Let us finally point out that Γ and Ψ are each others inverses.

$$(\Gamma \circ \Psi)(\sum_i (m_i \otimes n_i) \otimes c_i) = \Gamma(\sum_i m_{i[0]} \otimes (n_i \otimes S(m_{i[1]})c_i))$$

$$= \sum_i m_{i[0]} \otimes n_i \otimes m_{i[1]}S(m_{i[2]})c_i))$$

$$= \sum_i (m_i \otimes n_i) \otimes c_i$$

In a similar way, we have for all $m \in M$ and $\sum_i n_i \otimes c_i \in N \square_{C'} C$ that

$$(\Psi \circ \Gamma)(m \otimes \sum_i (n_i \otimes c_i)) = \sum_i \Psi((m_{[0]} \otimes n_i) \otimes m_{[1]} c_i$$

$$= \sum_i m_{[0]} \otimes n_i \otimes S(m_{[1]}) m_{[2]} c_i$$

$$= \sum_i m \otimes (n_i \otimes c_i)$$

This finishes the proof of (2.117). The proof of (2.118) is similar; we restrict here to giving the connecting isomorphisms

$$\Gamma' : G(N) \otimes M \to G(N \otimes F(M))$$
$$(n_i \otimes c_i) \otimes m \mapsto (n_i \otimes m_{[0]}) \otimes c_i m_{[1]}$$
$$\Psi' : G(N \otimes F(M)) \to G(N) \otimes M$$
$$(n_i \otimes m_i) \otimes c_i \mapsto (n_i \otimes c_i \overline{S}(m_{i[1]}) \otimes m_{i[0]}$$

Corollary 3. *Let (A, C, ψ) be a monoidal entwining structure, and let $F = \bullet : {}_A\mathcal{M}(\psi)^C \to {}_A\mathcal{M}$ be the functor forgetting the C-coaction. For any flat entwined module M, we have an isomorphism*

$$M \otimes C \cong \underline{M} \otimes C \qquad (2.120)$$

in ${}_A\mathcal{M}(\psi)^C$. If k is a field, then ${}_A\mathcal{M}(\psi)^C$ has enough injective objects, and any injective object in ${}_A\mathcal{M}(\psi)^C$ is a direct summand of an object of the form $I \otimes C$, where I is an injective A-module.

Proof. We apply Theorem 21 to the monoidal morphism

$$(I_A, \varepsilon_C) : (A, C, \psi) \to (A, k, I_A)$$

Now

$$M \otimes C \cong M \otimes G(k) \cong G(F(M) \otimes k) = G(F(M)) = F(M) \otimes C = \underline{M} \otimes C$$

If k is a field, then every k-module is flat, and (2.120) holds for every entwined module M. $\underline{M}$ embeds into an injective object I of ${}_A\mathcal{M}$. F is exact, so its right adjoint $G = \bullet \otimes C$ preserves injective objects. Finally, we have the following monomorphisms in ${}_A\mathcal{M}(\psi)^C$:

$$M \xrightarrow{I_M \otimes \eta_C} M \otimes C \cong \underline{M} \otimes C \to I \otimes C = G(I)$$

If M is injective, then $M \otimes I \otimes C$ splits, so M is a direct summand of $I \otimes C$. ∎

Corollary 4. *Let (A, C, ψ) be a monomial entwining structure over a field k. Then the category of entwined modules ${}_A\mathcal{M}(\psi)^C$ is a Grothendieck category having enough injective objects.*

For completeness sake, let us mention the dual version of Theorem 21. We omit the proof, as it can be derived from the proof of Theorem 21 using duality arguments.

Theorem 22. *Let $(\alpha, I_C) : (A, C, \psi) \to (A', C, \psi')$ be a monoidal morphism of monoidal entwining structures. Then*

$$F(A) = A' \tag{2.121}$$

If A' is a Hopf algebra, then

$$F(M) \otimes N \cong F(M \otimes G(N)) \tag{2.122}$$

for all $M \in {}_A\mathcal{M}(H)^C$ and $N \in {}_{A'}\mathcal{M}(H)^C$. If A' has a twisted antipode, then

$$N \otimes F(M) \cong F(G(N) \otimes M) \tag{2.123}$$

Remark 5. Assume that (A, C, ψ) is a monoidal entwining structure. It is possible to examine when the category of entwined modules is braided monoidal. This was done first in the case of Doi-Hopf modules, and it was shown that braiding on the category of Yetter-Drinfeld modules can be considered as a special case (see [54]). Recently, Hobst and Pareigis [95] discussed the entwining case.

3 Frobenius and separable functors for entwined modules

We begin this Chapter with the introduction of separable functors and Frobenius functors. We discuss general properties, such as Maschke's Theorem and Rafael's Theorem, and the relationship between the two notions. The terminology comes from separable algebras and Frobenius algebras, as introduced in Section 1.3: an algebra is separable, resp. Frobenius, if and only if the restriction of scalars functor is separable, resp. Frobenius. We apply our results to Hopf algebras, recovering the classical result that a finite dimensional Hopf algebra is Frobenius, and the Larson-Sweedler version of Maschke's Theorem. The techniques can be easily applied to decide when forgetfull functors from the category of entwined modules, or from the category of modules over a smash product is separable or Frobenius. This leads to several new versions of Maschke's Theorem.

3.1 Separable functors and Frobenius functors

Frobenius functors Let $F : \mathcal{C} \to \mathcal{D}$ be a covariant functor. If there exists a functor $G : \mathcal{D} \to \mathcal{C}$ which is at the same time a right and left adjoint of F, then we call F a *Frobenius functor*, and we say that (F, G) is a *Frobenius pair* for $\mathcal{C}$ and $\mathcal{D}$. Observe that this notion is symmetric in F and G: if (F, G) is a Frobenius pair, then (G, F) is also a Frobenius pair. This concept was first introduced by Morita [141], and the terminology is inspired by the fact that, for a ring homomorphism $R \to S$, the restriction of scalars functor is Frobenius if and only if S/R is Frobenius in the classical sense (see [105], [142]). We will make this clear in Section 3.1. Frobenius functors regained popularity recently (see [49] and [58]). We collect the following properties of Frobenius functors; we keep the notation introduced at the end of the previous Section. Our first result is nothing else then a restatement of one of the equivalent definitions of a pair of adjoint functors; nevertheless, it will be a key result in the further development of our theory.

Theorem 23. *Let G be a right adjoint of $F : \mathcal{C} \to \mathcal{D}$. Then (F, G) is a Frobenius pair if and only if there exist natural transformations $\nu \in V = \underline{\mathrm{Nat}}(GF, 1_{\mathcal{C}})$ and $\zeta \in W = \underline{\mathrm{Nat}}(1_{\mathcal{D}}, FG)$ such that*

$$F(\nu_C) \circ \zeta_{F(C)} = I_{F(C)} \tag{3.1}$$

$$\nu_{G(D)} \circ G(\zeta_D) = I_{G(D)} \tag{3.2}$$

for all $C \in \mathcal{C}$ and $D \in \mathcal{D}$

From Theorem 2, we immediately obtain the folowing:

Proposition 43. *Let (F, G) be a Frobenius pair. Then F and G preserve limits and colimits. If $\mathcal{C}$ and $\mathcal{D}$ are abelian, then F and G are exact and preserve injective and projective objects.*

The following result is straightforward, and we omit the proof.

Proposition 44. *The composition of two Frobenius functors is Frobenius. The direct sum of two Frobenius functors between abelian categories is Frobenius.*

Assume that we know that (F, G) is an adjoint pair. We know from Proposition 11 that we have natural isomorphisms

$$\alpha : \underline{\mathrm{Nat}}(GF, 1_{\mathcal{C}}) \xrightarrow{\cong} \underline{\mathrm{Nat}}(\mathrm{Hom}_{\mathcal{D}}(F, F), \mathrm{Hom}_{\mathcal{C}}(\bullet, \bullet))$$

$$\beta : \underline{\mathrm{Nat}}(1_{\mathcal{D}}, FG) \xrightarrow{\cong} \underline{\mathrm{Nat}}(\mathrm{Hom}_{\mathcal{C}}(G, G), \mathrm{Hom}_{\mathcal{D}}(\bullet, \bullet))$$

Assume that (F, G) is Frobenius, and let $\nu : GF \to 1_{\mathcal{C}}$ and $\zeta : 1_{\mathcal{D}} \to FG$ be the counit and unit of the adjunction (G, F). Put

$$\alpha(\nu) = \mathrm{Tr}_F : \mathrm{Hom}_{\mathcal{D}}(F, F) \to \mathrm{Hom}_{\mathcal{C}}(\bullet, \bullet))$$

$$\beta(\zeta) = \mathrm{Tr}_G : \mathrm{Hom}_{\mathcal{C}}(G, G) \to \mathrm{Hom}_{\mathcal{D}}(\bullet, \bullet)$$

be the natural transformations corresponding to ν and ζ. Tr_F and Tr_G were introduced by Madar and Marcus [124], and are called the *transfer maps* associated to (F, G). They can be used to characterize Frobenius pairs:

Proposition 45. *Let (F, G) be an adjoint pair of functors. (F, G) is a Frobenius pair if and only if there exist natural transformations*

$$\mathrm{Tr}_F : \mathrm{Hom}_{\mathcal{D}}(F, F) \to \mathrm{Hom}_{\mathcal{C}}(\bullet, \bullet)$$

$$\mathrm{Tr}_G : \mathrm{Hom}_{\mathcal{C}}(G, G) \to \mathrm{Hom}_{\mathcal{D}}(\bullet, \bullet)$$

such that

$$F(\mathrm{Tr}_F(\varepsilon_{F(C)})) \circ \mathrm{Tr}_G(\eta_{GF(C)}) = I_{F(C)} \tag{3.3}$$

$$\mathrm{Tr}_F(\varepsilon_{FG(D)}) \circ G(\mathrm{Tr}_G(\eta_{G(D)})) = I_{G(D)} \tag{3.4}$$

for all $C \in \mathcal{C}$ and $D \in \mathcal{D}$.

Proof. Let (F, G) be a Frobenius pair, and let $\text{Tr}_F = \alpha(\tau)$, $\text{Tr}_G = \beta(\xi)$. (3.3) and (3.4) then follow immediately from (3.1) and (3.2). Conversely, if (F, G) is an adjoint pair, and Tr_F and Tr_G satisfy (3.3) and (3.4), then $\tau = \alpha^{-1}(\text{Tr}_F)$ and $\nu = \beta^{-1}(\text{Tr}_G)$ satisfy (3.1) and (3.2).

Let (F, G) be an adjoint pair of functors. (F, G) is called a *Frobenius pair of the second kind* if their exist category equivalences $\alpha : \mathcal{C} \to \mathcal{C}$ and $\beta : \mathcal{D} \to \mathcal{D}$ such that G is a left adjoint of $\beta \circ F \circ \alpha$. For a more detailed study of Frobenius pairs of the second kind, we refer to [40].

Separable functors and Maschke's Theorem Let $F : \mathcal{C} \to \mathcal{D}$ be a covariant functor. F induces a natural transformation

$$\mathcal{F} : \text{Hom}_{\mathcal{C}}(\bullet, \bullet) \to \text{Hom}_{\mathcal{D}}(F(\bullet), F(\bullet)) \; ; \; \mathcal{F}_{C,C'}(f) = F(f)$$

We say that F is a *separable functor* if $\mathcal{F}$ splits, i. e. we have a natural transformation

$$\mathcal{P} : \text{Hom}_{\mathcal{D}}(F(\bullet), F(\bullet)) \to \text{Hom}_{\mathcal{C}}(\bullet, \bullet)$$

such that

$$\mathcal{P} \circ \mathcal{F} = 1_{\text{Hom}_{\mathcal{C}}(\bullet, \bullet)}$$

the identity natural transformation on $\text{Hom}_{\mathcal{C}}(\bullet, \bullet)$. Separable functors were introduced in [145], were the definition can be found in the following more explicit form: F is separable if and only if for all $C, C' \in \mathcal{C}$, we have a map

$$\mathcal{P}_{C,C'} : \text{Hom}_{\mathcal{D}}(F(C), F(C')) \to \text{Hom}_{\mathcal{C}}(C, C')$$

such that the following conditions hold:
SF1 for any $f : C \to C'$ in $\mathcal{C}$, $\mathcal{P}_{C,C'}(F(f)) = f$.
SF2 If we have morphisms $f : C \to C'$ and $f_1 : C_1 \to C_1'$ in $\mathcal{C}$ and $g : F(C) \to F(C_1)$, $g' : F(C') \to F(C_1')$ in $\mathcal{D}$ such that the diagram

$$
\begin{array}{ccc}
F(C) & \xrightarrow{\ g\ } & F(C_1) \\
{\scriptstyle F(f)}\big\downarrow & & \big\downarrow{\scriptstyle F(f_1)} \\
F(C') & \xrightarrow{\ g'\ } & F(C_1')
\end{array}
$$

commutes in $\mathcal{D}$, then the diagram

$$
\begin{array}{ccc}
C & \xrightarrow{\ \mathcal{P}_{C,C_1}(g)\ } & C_1 \\
{\scriptstyle f}\big\downarrow & & \big\downarrow{\scriptstyle f_1} \\
C' & \xrightarrow{\ \mathcal{P}_{C',C_1'}(g)\ } & C_1'
\end{array}
$$

commutes in $\mathcal{C}$. The terminology comes from the fact that, for a ring homomorphism $R \to S$, the restriction of scalars functor is separable if and only if S/R is separable. We will make this clear in Section 3.1. If F is separable, then for all $C, C' \in \mathcal{C}$, the map $\mathcal{F}_{C,C'} : \mathrm{Hom}_\mathcal{C}(C, C') \to \mathrm{Hom}_\mathcal{D}(F(C), F(C'))$ is injective, since it has a left inverse, and it follows that F is a faithful functor. Faithful functors are functors for which all $\mathcal{F}_{C,C'}$ have a left inverse; if the left inverse can be chosen to be natural in C and C', then the functor is separable. From the categorical point of view, a better name is perhaps *naturally faithful functor*, but we will stick to the terminology introduced in [145], and commonly used in the literature afterwards.

Now we give some general properties of separable functors.

Proposition 46. *Consider functors $F : \mathcal{C} \to \mathcal{D}$ and $G : \mathcal{D} \to \mathcal{E}$.*

1. *If F and G are separable, then $G \circ F$ is also separable;*

2. *if $G \circ F$ is separable, then F is separable.*

Proof. 1. is obvious; let us show 2. Consider the natural transformation

$$\mathcal{G} : \mathrm{Hom}_\mathcal{D}(\bullet, \bullet) \to \mathrm{Hom}_\mathcal{E}(G(\bullet), G(\bullet))$$

induced by G, and

$$\mathcal{Q} : \mathrm{Hom}_\mathcal{E}(GF(\bullet), GF(\bullet)) \to \mathrm{Hom}_\mathcal{C}(\bullet, \bullet)$$

coming from the fact that $G \circ F$ is separable. $\mathcal{P} = \mathcal{Q} \circ \mathcal{G}$ satisfies **SF1** and **SF2**.

Separable functors are important because they satisfy (a functorial version of) Maschke's Theorem: an exact sequence that is split after we apply the functor F is itself split.

Proposition 47. *Let $F : \mathcal{C} \to \mathcal{D}$ be a separable functor, and $f : C \to C'$ a morphism in $\mathcal{C}$. If $F(f)$ has a left (or right or two-sided) inverse g in $\mathcal{D}$, then f has a left (or right or two-sided) inverse in $\mathcal{C}$, namely $\mathcal{P}_{C,C'}(g)$.*

Proof. Let g be a left inverse of $F(f)$. In $\mathcal{D}$, we have a commutative diagram

$$
\begin{array}{ccc}
F(C) & \xrightarrow{F(I_C)} & F(C) \\
{\scriptstyle F(f)} \downarrow & & \downarrow {\scriptstyle F(I_C)} \\
F(C') & \xrightarrow{\quad g \quad} & F(C)
\end{array}
$$

and, using **SF1** and **SF2**, we find the following commutative diagram in $\mathcal{C}$:

$$C \xrightarrow{\mathcal{P}_{C,C_1}(F(I_C)) = I_C} C$$

$$f \downarrow \qquad \qquad \downarrow I_C$$

$$C' \xrightarrow{\mathcal{P}_{C',C}(g)} C$$

so $\mathcal{P}_{C',C}(g) \circ f = I_C$.

Corollary 5. (Maschke's Theorem for separable functors) *Let* F : $C \to \mathcal{D}$ *be a separable functor, and consider a short exact sequence*

$$0 \longrightarrow C' \longrightarrow C \longrightarrow C'' \longrightarrow 0 \qquad (3.5)$$

in C. *If the sequence*

$$0 \longrightarrow F(C') \longrightarrow F(C) \longrightarrow F(C'') \longrightarrow 0$$

is split in $\mathcal{D}$, *then (3.5) is also split.*

Proposition 48. *Let* F : $C \to \mathcal{D}$ *be a separable functor between two categories* C *and* $\mathcal{D}$.

1. *If* F *preserves epimorphisms, then* F *reflects projective objects;*
2. *if* F *preserves monomorphisms, then* F *reflects injective objects.*

Proof. We will prove the first statement. Assume that $M \in C$ is such that $F(M)$ is projective, and take an epimorphism $g : C \to C'$, and a morphism $f : M \to P'$ in C. $F(g)$ is also an epimorphism, so there exists $u : F(M) \to F(C)$ in $\mathcal{D}$ such that $F(f) = F(g) \circ u$. In $\mathcal{D}$ we have the following commutative diagram

$$F(M) \xrightarrow{u} F(C)$$

$$F(f) \downarrow \qquad \qquad \downarrow F(g)$$

$$F(C') \xrightarrow{F(I_{C'})} F(C')$$

F is separable, so we have the following commutative diagram in C

$$M \xrightarrow{\mathcal{P}_{M,C}(u)} C$$

$$f \downarrow \qquad \qquad \downarrow g$$

$$C' \xrightarrow{I_{C'}} C'$$

so $f = g \circ \mathcal{P}_{M,C}(u)$, and it follows that M is projective.

Our next result is Rafael's Theorem, giving necessary and sufficient conditions for a functor having an adjoint to be separable (see [158] and [159]). Rafael's Theorem is a key result in the development of our further theory.

In the sequel, $G : \mathcal{D} \to \mathcal{C}$ will be a right adjoint of $F : \mathcal{C} \to \mathcal{D}$. In the proof of Theorem 15 we have explained that this is equivalent to the existence of natural transformations $\eta : 1_{\mathcal{C}} \to GF$ and $\varepsilon : FG \to 1_{\mathcal{D}}$ such that

$$G(\varepsilon_N) \circ \eta_{G(N)} = I_{G(N)} \tag{3.6}$$

$$\varepsilon_{F(M)} \circ F(\eta_M) = I_{F(M)} \tag{3.7}$$

for all $M \in \mathcal{C}$ and $N \in \mathcal{D}$.

Theorem 24. (Rafael) *Assume that the functor $F : \mathcal{C} \to \mathcal{D}$ has a right adjoint G.*

1. *F is separable if and only if the unit $\eta : 1_{\mathcal{C}} \to GF$ of the adjunction splits; by this we mean that there exists a natural transformation $\nu : GF \to 1_{\mathcal{C}}$ such that $\nu \circ \eta = 1_{1_{\mathcal{C}}}$, the identity natural transformation on $1_{\mathcal{C}}$;*
2. *G is separable if and only if the counit $\varepsilon : FG \to 1_{\mathcal{D}}$ of the adjunction cosplits; by this we mean that there exists a natural transformation $\zeta : 1_{\mathcal{D}} \to FG$ such that $\varepsilon \circ \zeta = 1_{1_{\mathcal{D}}}$, the identity natural transformation on $1_{\mathcal{D}}$.*

Proof. We prove the first statement, the second one then follows by duality arguments. Let $\theta : \operatorname{Hom}_{\mathcal{B}}(F, F) \to \operatorname{Hom}_{\mathcal{A}}(\bullet, \bullet)$ be a natural transformation. Let $\nu = \alpha^{-1}(\theta)$, using Proposition 11. The naturality of θ entails that

$$\nu_C \circ \eta_C = \theta_{GF(C),C}(\varepsilon_{F(C)}) \circ \eta_C$$
$$= \theta_{C,C}(\varepsilon_{F(C)} \circ F(\eta_C))$$
$$= \theta_{C,C}(I_{F(C)}) = \theta_{C,C}(\operatorname{Res}_F(I_C))$$

Assume that F is separable, and take $\theta = \phi$. Then we find $\nu : GF \to 1_{\mathcal{A}}$ such that

$$\nu_C \circ \eta_C = I_C$$

i.e. $\nu \circ \eta$ is the identity natural transformation.

Conversely, let $\nu : GF \to 1_{\mathcal{A}}$ be natural, and put $\phi = \alpha(\nu)$. Then

$$\phi_{C,C'}(F(f)) = \nu_{C'} \circ GF(f) \circ \eta_C = f \circ \nu_C \circ \eta_C$$

Clearly if $\nu \circ \eta$ is the identity natural transformation, then ϕ splits Res_F, and F is separable.

In Rafael's Theorem, we assume that (F, G) is an adjoint pair, and then deduce an easier criterion for F or G to be separable. If we know that (F, G) is Frobenius, then we can find an even easier necessary and sufficient condition for the separability of F or G. From Proposition 10, we obtain immediately the following:

Corollary 6. *Let (F, G) be a Frobenius pair of functors, then we have isomorphisms*

$$\underline{\text{Nat}}(F, F) \cong \underline{\text{Nat}}(G, G) \cong \underline{\text{Nat}}(1_C, GF) \cong \underline{\text{Nat}}(FG, 1_D)$$
$$\cong \underline{\text{Nat}}(GF, 1_C) \cong \underline{\text{Nat}}(1_D, FG)$$

For a Frobenius pair (F, G), we will write $\nu : GF \to 1_C$ and $\zeta : 1_D \to FG$ for the counit and unit of the adjunction (G, F). For all $C \in C$ and $D \in D$, we then have

$$F(\nu_C) \circ \zeta_{F(C)} = I_{F(C)} \quad \text{and} \quad \nu_{G(D)} \circ G(\zeta_D) = I_{G(D)} \tag{3.8}$$

We can now state and prove the "Frobenius version" of Rafael's Theorem:

Proposition 49. *[46] Let (F, G) be a Frobenius pair, and let η, ε, ν and ζ be as above. The following statements are equivalent:*

1. *F is separable;*
2. *$\exists \alpha \in \underline{\text{Nat}}(F, F) : \nu_C \circ G(\alpha_C) \circ \eta_C = I_C$ for all $C \in C$;*
3. *$\exists \beta \in \underline{\text{Nat}}(G, G) : \nu_C \circ \beta_{F(C)} \circ \eta_C = I_C$ for all $C \in C$.*

We have a similar characterization for the separability of G: the statements

1. *G is separable;*
2. *$\exists \alpha \in \underline{\text{Nat}}(F, F) : \varepsilon_D \circ \alpha_{G(D)} \circ \zeta_D = I_D$ for all $D \in D$;*
3. *$\exists \beta \in \underline{\text{Nat}}(G, G) : \varepsilon_D \circ F(\beta_D) \circ \zeta_D = I_D$ for all $D \in D$.*

are equivalent.

Proof. Assume that F is separable. By Rafael's Theorem, there exists $\tilde{\nu} \in \underline{\text{Nat}}(GF, 1_C)$ such that $\tilde{\nu}_C \circ \eta_C = I_C$ for all $C \in C$. Let $\alpha : F \to F$ be the corresponding natural transformation of Corollary 6, i.e. $\alpha_C = F(\tilde{\nu}_C) \circ \zeta_{F(C)}$, and $\tilde{\nu}_C = \nu_C \circ G(\alpha_C)$, and the first implication of the Proposition follows. The converse follows trivially from Rafael's Theorem. All the other equivalences can be proved in a similar way.

Let us finish this Section fixing some notation for pairs of adjoint functors. Let $F : C \to D$ be a functor having a right adjoint G. As above, we let

$$\eta : 1_C \to GF \quad \text{and} \quad \varepsilon : FG \to 1_D$$

be the unit and counit of the adjunction. We will write

$$V = \underline{\text{Nat}}(GF, 1_C) \quad \text{and} \quad W = \underline{\text{Nat}}(1_D, FG)$$

V and W are classes, but in many particular examples they turn out to be sets. According to Theorems 23 and 24, the members of V and W decide whether F and G are Frobenius or separable functors.

Remark also that there exists an associative multiplication on V and W. For $\nu, \nu' \in V$, and $\zeta, \zeta' \in W$, we put

$$\nu' \cdot \nu = \nu \circ \eta \circ \nu' \quad \text{and} \quad \zeta' \cdot \zeta = \zeta \circ \varepsilon \circ \zeta' \tag{3.9}$$

If F is separable, then there exists a right unit for the multiplication on V; if G is separable, then there exists a left unit for the multiplication on W. These one-sided units are idempotents and are called *separability idempotents*.

Relative injective and projective objects Let $F : \mathcal{C} \to \mathcal{D}$ be a covariant functor with right adjoint G. An object $M \in \mathcal{C}$ is called *relative injective* if the following condition is satisfied: if $i : \mathcal{C} \to C'$ in $\mathcal{C}$ is such that $F(i) : F(C) \to F(C')$ has a left inverse p in $\mathcal{D}$, then for every $f : C \to M$ in $\mathcal{C}$, there exists a morphism $g' : C' \to M$ such that $f = g \circ i$, i.e. the diagram

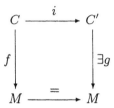

commutes. In a similar way, we define *relative projective objects* in $\mathcal{D}$.

Proposition 50. *As above, let (F, G) be an adjoint pair of functors. $M \in \mathcal{C}$ is relative injective if and only if the unit map $\eta_M : M \to GF(M)$ has a left inverse ν_M. $N \in \mathcal{D}$ is relative projective if and only if $\varepsilon_N : FG(N) \to N$ has a right inverse.*

Proof. First assume that η_M has a section ν_M in $\mathcal{C}$. Let $i : C \to C'$, $p : F(C') \to F(C)$ and $f : C \to M$ be as above. We define $g : C' \to M$ as the composition

$$C' \xrightarrow{\eta_{C'}} GF(C') \xrightarrow{G(p)} GF(C) \xrightarrow{GF(f)} GF(M) \xrightarrow{\nu_M} M$$

In order to prove that $g \circ i = f$, it suffices to prove that

$$GF(f) \circ G(p) \circ \eta_{C'} \circ i = \eta_M \circ f \tag{3.10}$$

since $\nu_M \circ \eta_M = I_M$. We have a commutative diagram

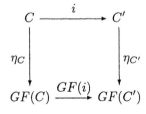

hence

$$G(p) \circ \eta_{C'} \circ i = G(p) \circ GF(i) \circ \eta_C = G(p \circ F(i)) \circ \eta_C = \eta_C$$

so (3.10) is equivalent to

$$GF(f) \circ \eta_C = \eta_M \circ f$$

and this follows from the fact that η is a natural transformation.

Conversely assume that M is relative injective, and put $i = \eta_M : M \to GF(M)$. $F(i) = F(\eta_M)$ has a section, namely $\varepsilon_{F(M)}$, since (F, G) is an adjoint pair of functors, so there exists $g : GF(M) \to M$ in C such that $g \circ i = g \circ \eta_M = I_M$.

The statement about projective objects is similar.

Comparing Proposition 50 and Theorem 24, we obtain the following result:

Corollary 7. *Let (F, G) be an an adjoint pair of functors between C and $\mathcal{D}$. If F is separable, then every object of C is relative injective, and if G is separable, then every object of $\mathcal{D}$ is relative projective.*

Remark 6. In [47], a functor $F : C \to \mathcal{D}$ is called a *Maschke functor* if every object of C is relative injective. It is called a *dual Maschke functor* if every object is relative projective. It can be shown that a separable functor is always Maschke and dual Maschke. The converse is not true in general. We refer to the forthcoming [47] for details.

Separable functors of the second kind Let $F : C \to \mathcal{D}$ and $H : C \to \mathcal{E}$ be covariant functors. We then have functors

$$\mathrm{Hom}_C(\bullet, \bullet), \ \mathrm{Hom}_{\mathcal{D}}(F, F), \ \mathrm{Hom}_{\mathcal{E}}(H, H) : C^{\mathrm{op}} \times C \to \underline{\mathrm{Sets}}$$

and natural transformations

$$\mathcal{F} : \mathrm{Hom}_C(\bullet, \bullet) \to \mathrm{Hom}_{\mathcal{D}}(F, F) \ ; \ \mathcal{H} : \mathrm{Hom}_C(\bullet, \bullet) \to \mathrm{Hom}_{\mathcal{E}}(H, H)$$

given by

$$\mathcal{F}_{C,C'}(f) = F(f) \ ; \ \mathcal{H}_{C,C'}(f) = H(f)$$

for $f : C \to C'$ in C.

Definition 4. The functor F is called H-separable if there exists a natural transformation

$$\mathcal{P} : \mathrm{Hom}_{\mathcal{D}}(F, F) \to \mathrm{Hom}_{\mathcal{E}}(H, H)$$

such that

$$\mathcal{P} \circ \mathcal{F} = \mathcal{H} \tag{3.11}$$

that is, $\mathcal{H}$ factors through $\mathcal{F}$ as a natural transformation, and we have a commutative diagram

$$
\begin{array}{ccc}
\mathrm{Hom}_C(\bullet, \bullet) & \overset{\mathcal{F}}{\longrightarrow} & \mathrm{Hom}_{\mathcal{D}}(F, F) \\
\Big\downarrow{\scriptstyle \mathcal{H}} & \swarrow{\scriptstyle \mathcal{P}} & \\
\mathrm{Hom}_{\mathcal{E}}(H, H) & &
\end{array}
$$

An H-separable functor is also called a separable functor of the second kind; for a detailed study, we refer the reader to the forthcoming paper [47]. Obviously, a 1_C-separable functor is the same thing as a separable functor. The basic properties are very similar. Adapting the proof of Propositions 46 and 47, Corollary 5 and Theorem 24, we obtain the following results (see [47] for more detail).

Proposition 51. *Consider functors*

$$\mathcal{C} \xrightarrow{F} \mathcal{D} \xrightarrow{F_1} \mathcal{D}_1 \text{ and } \mathcal{C} \xrightarrow{H} \mathcal{E}$$

1. *If $F_1 \circ F$ is H-separable, then F is H-separable.*
2. *If F is H-separable, and F_1 is separable, then $F_1 \circ F$ is H-separable.*

Proposition 52. *Let F be an H-separable functor. If $f : C \to C'$ in $\mathcal{C}$ is such that $F(f)$ has a left, right, or two-sided inverse in $\mathcal{D}$, then $H(f)$ has a left, right, or two-sided inverse in $\mathcal{E}$.*

Corollary 8. (Maschke's Theorem for H-separable functors) *Let $\mathcal{C}$, $\mathcal{D}$ and $\mathcal{E}$ be abelian categories, and assume that $F : \mathcal{C} \to \mathcal{D}$ is H-separable. An exact sequence in $\mathcal{C}$ that becomes split after we apply the functor F, also becomes split after we apply the functor H.*

Theorem 25. (Rafael's Theorem for H-separability) *Let $G : \mathcal{D} \to \mathcal{C}$ be a right adjoint of $F : \mathcal{C} \to \mathcal{D}$, and consider functors $H : \mathcal{C} \to \mathcal{E}$ and $K : \mathcal{D} \to \mathcal{E}$.*

1. *F is H-separable if and only if there exists a natural transformation $\nu : HGF \to H$ such that*

$$\nu_C \circ H(\eta_C) = I_{H(C)} \tag{3.12}$$

for any $C \in \mathcal{C}$.
2. *G is K-separable if and only if there exists a natural transformation $\zeta : K \to KFG$ such that*

$$K(\varepsilon_D) \circ \zeta_D = I_{K(D)} \tag{3.13}$$

for any $D \in \mathcal{D}$.

Fully faithful functors Let (F, G) be a pair of adjoint functors between the categories $\mathcal{C}$ and $\mathcal{D}$. We have seen that

$$V = \underline{\mathrm{Nat}}(GF, 1_{\mathcal{C}}) \text{ and } W = \underline{\mathrm{Nat}}(1_{\mathcal{D}}, FG)$$

decide whether F and G are separable or Frobenius (see Theorems 23 and 24). It is well-known that V and W can also be used to determine when F and G are fully faithful, and when they establish an equivalence of categories.

Theorem 26. *Let G be a right adjoint of $F : \mathcal{C} \to \mathcal{D}$.*

1. F is fully faithful if and only if there exists $\nu \in V$ such that $\nu_C = \eta_C^{-1}$, for all $C \in \mathcal{C}$;
2. G is fully faithful if and only if there exists $\zeta \in W$ such that $\zeta_D = \varepsilon_D^{-1}$, for all $D \in \mathcal{D}$.

Proof. We will prove 1.; the proof of 2. is similar. First assume that F is fully faithful. For each $C \in \mathcal{C}$, we consider $\varepsilon_{F(C)} : FGF(C) \to F(C)$. F is full, so $\varepsilon_{F(C)} = F(\nu_C)$, for some $\nu_C : GF(C) \to C$. Now

$$F(\nu_C \circ \eta_C) = F(\nu_C) \circ F(\eta_C) = \varepsilon_{F(C)} \circ F(\eta_C) = I_{F(C)} = F(I_C)$$

and it follows that $\nu_C \circ \eta_C = I_C$, since F is faithful.
In a similar way, we compute that

$$\varepsilon_{F(C)} \circ F(\eta_C \circ \nu_C) = F(\nu_C) \circ F(\eta_C) \circ F(\nu_C) = F(\nu_C)$$
$$= F(\nu_C) \circ F(I_{GF(C)}) = \varepsilon_{F(C)} \circ F(I_{GF(C)}) \qquad (3.14)$$

(F, G) is an adjoint pair, and this implies that the map

$$\theta_{C,D} : \mathrm{Hom}_{\mathcal{C}}(C, G(D)) \to \mathrm{Hom}_{\mathcal{D}}(F(C), D), \quad \theta_{C,D}(f) = \varepsilon_D \circ F(f)$$

is an isomorphism. (3.14) tells us that $\theta_{GF(C),F(C)}$ takes the same values at $\eta_C \circ \nu_C$ and $I_{GF(C)}$, hence $\eta_C \circ \nu_C = I_{GF(C)}$. The naturality of ν now follows from the naturality of η.
Conversely, if η_C is an isomorphism for all $C \in \mathcal{C}$, then we have isomorphisms

$$\mathrm{Hom}_{\mathcal{C}}(C, C') \cong \mathrm{Hom}_{\mathcal{C}}(C, GF(C')) \cong \mathrm{Hom}_{\mathcal{D}}(F(C), F(C'))$$

The composition of these two isomorphisms sends $f : C \to C'$ to

$$\theta_{C,F(C')}(\eta_{C'} \circ f) = \varepsilon_{F(C')} \circ F(\eta_{C'}) \circ F(f) = I_{F(C')} \circ F(f) = F(f)$$

and we have shown that F is fully faithful.

By definition, an adjoint pair of functors (F, G) establishes a *category equivalence* if F and G are fully faithful.

Corollary 9. *Let (F, G) be a category equivalence. Then (F, G) is a Frobenius pair, and F and G are separable.*

3.2 Restriction of scalars and the smash product

Let $i : R \to S$ be a ring homomorphism, and let $F = \bullet \otimes_R S : \mathcal{M}_R \to \mathcal{M}_S$ be the induction functor. The restriction of scalars functor $G : \mathcal{M}_S \to \mathcal{M}_R$ is a right adjoint of F; the unit and counit of the adjunction are described as follows, for all $M \in \mathcal{M}_R$ and $N \in \mathcal{M}_S$:

$$\eta_M : M \to M \otimes_R S \; ; \; \eta_M(m) = m \otimes 1$$

$$\varepsilon_N : N \otimes_R S \to N \; ; \; \varepsilon_N(n \otimes s) = ns$$

We will now use the results of the previous Sections to decide when F and G are Frobenius or separable. To this end, we give some explicit descriptions of V and W. Given $\nu : GF \to 1_{\mathcal{M}_R}$ in V, it is not hard to prove that $\bar{\nu} = \nu_R : S \to R$ is a morphism in $_R\mathcal{M}_R$. Conversely, given an R-bimodule map $\bar{\nu} : S \to R$, we can construct a natural transformation $\nu \in V$; ν_M is given by

$$\nu_M(m \otimes s) = m\bar{\nu}(s)$$

Thus we have

$$V \cong V_1 = {_R}\mathrm{Hom}_R(S, R) \tag{3.15}$$

Now let $\zeta : 1_{\mathcal{M}_S} \to FG$ be a natural transformation in W. By definition, $\zeta_S : S \to S \otimes_R S$ is a right S-bimodule map. It is also a left S-bimodule map: for $s \in S$, we consider the map $f_s : S \to S$, $f_s(t) = st$. f_s is a morphism in $\mathcal{M}_S$, so that the naturality of ζ gives us a commutative diagram

$$
\begin{array}{ccc}
S & \xrightarrow{\zeta_S} & S \otimes_R S \\
\downarrow{\scriptstyle f_s} & & \downarrow{\scriptstyle f_s \otimes I_S} \\
S & \xrightarrow{\zeta_S} & S \otimes_R S
\end{array}
$$

from which it follows that $\nu_S(st) = s\nu_S(t)$. Now $\zeta_S(1) = e = e^1 \otimes e^2 = \zeta_S(1) \in S \otimes_R S$ satisfies the following:

$$se^1 \otimes e^2 = \nu_S(s) = e^1 \otimes e^2 s \tag{3.16}$$

for all $s \in S$. Conversely if e satisfies (3.16), then we can recover ζ:

$$\zeta_N : N \to N \otimes_R S \; ; \; \zeta_N(n) = ne^1 \otimes e^2$$

In the sequel, we will omit the summation symbol, and write $e = e^1 \otimes e^2$, where it is understood implicitly that we have a summation. So we have

$$W \cong {_S}\mathrm{Hom}_S(S, S \otimes_R S) \cong W_1$$
$$= \{e = e^1 \otimes e^2 \in S \otimes_R S \mid se^1 \otimes e^2 = e^1 \otimes e^2 s, \text{ for all } s \in S\} \tag{3.17}$$

Theorem 27. *Let $i : R \to S$ be a ring homomorphism, F the induction functor, and G the restriction of scalars functor.*

1. *F is separable if and only if there exists a conditional expectation, that is $\bar{\nu} \in V_1$ such that $\bar{\nu}(1) = 1$. This means that S/R is a split extension in the sense of [153].*

2. G is separable if and only if there exists a separability idempotent, that is $e \in W_1$ such that $e^1 e^2 = 1$. In the case where R is commutative, this means that S/R is a separable extension in the sense of [66] and [109], as discussed in Section 1.3.

3. (F, G) is a Frobenius pair if and only if there exist $\bar{\nu} \in V_1$ and $e \in W_1$ such that

$$\bar{\nu}(e^1) e^2 = e^1 \bar{\nu}(e^2) = 1 \tag{3.18}$$

In the literature, $(\bar{\nu}, e)$ is called a Frobenius pair.

Remarks 2. 1. Observe that the results in Theorem 27 are left-right symmetric. For example, $\mathcal{M}_S \to \mathcal{M}_R$ is separable if and only if $_S\mathcal{M} \to {_R}\mathcal{M}$ is separable. 2. explains the terminology for separable functors.

2. The monoid structure on W translates into a monoid structure on W_1. The multiplication on W_1 is given by the formula

$$e \cdot f = e^1 f^1 \otimes f^2 e^2$$

We also have an addition on W_1, which makes W_1 into a ring (usually without unit).

In Proposition 12, we have seen that a projective separable algebra over a commutative ring is finitely generated. For Frobenius algebras, we have the following result.

Corollary 10. *We use the same notation as in Theorem 27. If (F, G) is a Frobenius pair, then S is finitely generated and projective as a (left or right) R-module.*

Proof. For all $s \in S$, we have $s = se^1 \bar{\nu}(e^2) = e^1 \bar{\nu}(e^2 s)$, hence $\{e^1, \bar{\nu}(e^2 \bullet)\}$ is a dual basis for S as a right R-module. In the same way, $\{e^2, \bar{\nu}(\bullet e^1)\}$ is a dual basis for S as a left R-module.

Using different descriptions of V and W, we find other criteria for F and G to be separable or Frobenius. Let $\mathrm{Hom}_R(S, R)$ be the set of right R-module homomorphisms from S to R. $\mathrm{Hom}_R(S, R)$ is an (R, S)-bimodule:

$$(rfs)(t) = rf(st) \tag{3.19}$$

for all $f \in \mathrm{Hom}_R(S, R)$, $r \in R$ and $s, t \in S$.

Proposition 53. *Let $i : R \to S$ be a ring homomorphism and use the notation introduced above. Then*

$$V = \underline{\mathrm{Nat}}(GF, 1_C) \cong V_2 = {_R}\mathrm{Hom}_S(S, \mathrm{Hom}_R(S, R))$$

Proof. We define $\alpha_1 : V_1 \to V_2$ as follows: for $\bar{\nu} \in V_1$, let $\alpha_1(\bar{\nu}) = \bar{\phi} : S \to \mathrm{Hom}_R(S, R)$ be given by

$$\bar{\phi}(s)(t) = \bar{\nu}(st)$$

Given $\bar{\phi} \in V_2$, we put

$$\alpha^{-1}(\bar{\phi}) = \bar{\phi}(1)$$

We invite the reader to verify that α_1 and α_1^{-1} are well-defined and that they are each others inverses.

Proposition 54. *Let $i : R \to S$ be a ring homomorphism and assume that S is finitely generated and projective as a right R-module. Using the notation introduced above, we have that*

$$W = \underline{\mathrm{Nat}}(1_{\mathcal{D}}, FG) \cong W_2 = {}_R\mathrm{Hom}_S(\mathrm{Hom}_R(S, R), S)$$

Proof. Let $\{s_i, \sigma_i \mid i = 1, \cdots, m\}$ be a finite dual basis of S as a right R-module. For all $s \in S$ and $f \in \mathrm{Hom}_R(S, R)$, we have

$$s = \sum_i s_i \sigma_i(s) \quad \text{and} \quad f = \sum_i f(s_i)\sigma_i$$

We define $\beta_1 : W_1 \to W_2$ as follows: $\beta_1(e) = \phi$, with

$$\phi(f) = f(e^1)e^2$$

for all $f \in \mathrm{Hom}_R(S, R)$. Let us show that ϕ is left R-linear and right S-linear:

$$\phi(fs) = f(se^1)e^2 = f(e^1)e^2s = \phi(f)s$$

$$\phi(rf) = rf(e^1)e^2 = r\phi(f)$$

Conversely, for $\phi \in W_2$, we put

$$\beta_1^{-1}(\phi) = e = \sum_i s_i \otimes \phi(\sigma_i)$$

$e \in W_1$, since

$$\sum_i s_i \otimes \phi(\sigma_i)s = \sum_i s_i \otimes \phi(\sigma_i s) = \sum_{i,j} s_i \otimes \phi(\sigma_i(ss_j)\sigma_j)$$

$$= \sum_{i,j} s_i \sigma_i(ss_j) \otimes \phi(\sigma_j) = \sum_j ss_j \otimes \phi(\sigma_j)$$

Finally, β_1 and β_1^{-1} are each others inverses:

$$\beta_1(\beta_1^{-1}(\phi))(f) = \beta_1\left(\sum_i s_i \otimes \phi(\sigma_i)\right)(f)$$

$$= \sum_i f(s_i)\phi(\sigma_i) = \sum_i \phi(f(s_i)\sigma_i) = \phi(f)$$

$$\beta_1^{-1}(\beta_1(e)) = \sum_i s_i \otimes \beta_1(e)(\sigma_i) = \sum_i s_i \otimes \sigma_i(e^1)e^2 = \sum_i s_i \sigma_i(e^1) \otimes e^2 = e$$

Theorem 28. *Let $i : R \to S$ be a ring homomorphism. We use the notation introduced above.*

1. $F : \mathcal{M}_R \to \mathcal{M}_S$ *is separable if and only if there exists $\overline{\phi} \in V_2$ such that $\overline{\phi}(1)(1) = 1$.*
2. *Assume that S is finitely generated and projective as a right R-module. Then G is separable if and only if there exists $\phi \in W_2$ such that*

$$\sum_i s_i \phi(\sigma_i) = 1$$

3. (F, G) *is a Frobenius pair if and only if S is finitely generated and projective as a right R-module and $\mathrm{Hom}_R(S, R)$ and S are isomorphic as (R, S)-bimodules. This means that S/R is Frobenius in the sense of [105]. In the case where $R = k$ is a field, we recover Definition 3.*

Proof. The result is a translation of Theorem 27 in terms of V_2 and W_2, using Proposition 12 (for 2.) and Corollary 10 (for 3.). Let us prove one implication of 3. Assume that (F, G) is a Frobenius pair. From Corollary 10, we know that S is finitely generated and projective. Let $\nu \in V_1$ and $e \in W_1$ be as in part 3. of Theorem 27, and take $\overline{\phi} = \alpha_1(\overline{\nu}) \in V_2$, $\phi = \beta_1(e) \in W_2$. For all $f \in \mathrm{Hom}_R(S, R)$ and $s \in S$, we have

$$(\overline{\phi} \circ \phi)(f)(s) = \overline{\nu}(\phi(f)s) = \overline{\nu}(f(e^1)e^2 s)$$
$$= f(e^1)\overline{\nu}(e^2 s) = f(se^1)\overline{\nu}(e^2) = f(se^1\overline{\nu}(e^2)) = f(s)$$

and

$$(\phi \circ \overline{\phi})(s) = \overline{\phi}(s)(e_1)e_2 = \overline{\nu}(se^1)e^2 = \overline{\nu}(e^1)e^2 s = s$$

We will now give alternative interpretations of V_2 and W_2. Consider the functor

$$F' : \mathcal{M}_R \to \mathcal{M}_S, \quad F'(M) = \mathrm{Hom}_R(S, M)$$

where the right S-action on $\mathrm{Hom}_R(S, M)$ is the following

$$(f \cdot s)(t) = f(st)$$

for all $f \in \mathrm{Hom}_R(S, M)$ and $s, t \in S$. It is easy to check that F' is a right adjoint of the restriction of scalars functor G, so that S/R is Frobenius if and only F and F' are isomorphic functors. In fact we have the following

Proposition 55. *Let $i : R \to S$ be a ring homomorphism, and assume that S is finitely generated and projective as a right R-module. Then*

$$W' = \underline{\mathrm{Nat}}(F', F) \cong W_2 = {}_R\mathrm{Hom}_S(\mathrm{Hom}_R(S, R), S)$$

$$V' = \underline{\mathrm{Nat}}(F, F') \cong V_2 = {}_R\mathrm{Hom}_S(S, \mathrm{Hom}_R(S, R))$$

Proof. Take $\Phi : F' \to F$. Then $\Phi_R = \phi \in W_2$. Conversely, for $\phi \in W_2$, we consider $e = \beta_1^{-1}(\phi) \in W_1$, and we define $\Phi : F' \to F$ by

$$\Phi_M(f) = f(e^1) \otimes e^2 \in M \otimes_R S = F(M) \tag{3.20}$$

for all $M \in \mathcal{M}_R$ and $f \in F'(M) = \operatorname{Hom}_R(S, M)$.

Now if $\overline{\Phi} : F' \to F$ is in V', then $\overline{\Phi}_R = \overline{\phi} \in V_2$. Given $\overline{\phi} \in V_2$, we take $\overline{\nu} = \alpha_1^{-1}(\overline{\phi}) \in V_1$, and we define $\overline{\Phi} \in V'$ by

$$\overline{\Phi}_M(m \otimes s)(t) = m\overline{\nu}(st) = \nu_M(m \otimes st)$$

(F, G) is a Frobenius pair if and only if we can find $\phi \in W_2$ and $\overline{\phi} \in V_2$ that are each others inverses, or equivalently, the corresponding $\Phi \in W'$ and $\overline{\Phi} \in V'$ are each others inverses. One direction is easy, the other one can be seen as follows:

$$\begin{aligned}
(\Phi_M \circ \overline{\Phi}_M)(m \otimes s) &= \overline{\Phi}_M(m \otimes s)(e^1) \otimes e^2 \\
&= m\overline{\nu}(se^1) \otimes e^2 = m \otimes s \\
((\overline{\Phi}_M \circ \Phi_M)(f))(s) &= \overline{\Phi}_M(f(e^1) \otimes e^2)(s) = f(e^1)\overline{\nu}(e^2 s) \\
&= f(se^1\overline{\nu}(e^2)) = f(s)
\end{aligned}$$

This provides an alternative explanation for the fact that (F, G) is a Frobenius pair if and only if F and F' are isomorphic functors.

As before, let $i : R \to S$ be a ring homomorphism, and F and G the induction and restriction of scalars functor. Also assume that we know that S/R is Frobenius. Arguments that are similar to the proof of e.g. (3.15) show that

$$\underline{\operatorname{Nat}}(F, F) \cong {}_R\operatorname{Hom}_S(S, S) \cong C_R(S) \tag{3.21}$$

To a natural transformation $\alpha : F \to F$, there corresponds $\alpha_R \in \operatorname{Hom}_S(S, S)$, and $x = \alpha_R(1) \in C_R(S)$. From Proposition 49, we conclude that the separability of the functors F and G (i.e. the fact whether S/R is split or separable) can be decided using some $x \in C_R(S)$. The following result could be deduced from Proposition 49, but a direct proof can also be given.

Proposition 56. *Assume that S/R is Frobenius, and let $(\overline{\nu}, e)$ be a Frobenius pair, as in Theorem 27.*

1. *S/R is separable if and only if there exists some $x \in C_R(S)$ such that $e^1 x e^2 = 1$.*
2. *S/R is split if and only if there exists some $x \in C_R(S)$ such that $\overline{\nu}(x) = 1$.*

Proof. 1. For any $x \in C_R(S)$, $e^1 \otimes x e^2$ is Casimir, and one implication follows. Conversely, if $f = f^1 \otimes f^2$ is a separability idempotent, then $x = \overline{\nu}(f^1)f^2 \in C_R(S)$, and

$$e^1 \otimes_R xe^2 = e^1 \otimes_R \overline{\nu}(f^1)f^2 e^2 = e^1 \otimes_R \overline{\nu}(e^2 f^1)f^2$$
$$= f^1 e^1 \otimes_R \overline{\nu}(e^2)f^2 = f^1 e^1 \overline{\nu}(e^2) \otimes_R f^2$$
$$= f^1 \otimes_R f^2 = f$$

and the first statement follows.

2. One direction is obvious: $\overline{\mu}$ defined by

$$\overline{\mu}(s) = \overline{\nu}(xs)$$

is a conditional expectation. Conversely, assume that $\overline{\mu}$ is a conditional expectation. Then $x = \overline{\mu}(e^1)e^2 \in C_R(S)$, and

$$\overline{\nu}(xs) = \overline{\nu}(\overline{\mu}(e^1)e^2 s) = \overline{\nu}(\overline{\mu}(se^1)e^2)$$
$$= \overline{\mu}(se^1)\overline{\nu}(e^2) = \overline{\mu}(se^1 \overline{\nu}(e^2)) = \overline{\mu}(s)$$

and the result follows.

An immediate consequence is the following result, generalizing [5, Theorem 3.4].

Corollary 11. *Let $i : R \to S$ be a morphism of commutative rings, and assume that S/R is Frobenius, with Frobenius pair $(\overline{\nu}, e)$. Then S/R is separable if and only if $e^1 e^2$ is invertible in S. In this case, the separability idempotent is unique.*

Proof. According to Proposition 56, there exists $x \in S$ such that $e^1 x e^2 = (e^1 e^2)x = 1$.

Abrams [5] calls $e^1 e^2$ the *characteristic element*.

Application to the smash product Let (A, B, R) be a smash product structure (over a commutative ring k), and take $R = A$, $S = B\#_R A$. For $\overline{\nu} \in V_1 = {}_R\mathrm{Hom}_R(S, R)$, define $\kappa : B \to A$ by

$$\kappa(b) = \overline{\nu}(b\#1)$$

Then $\overline{\nu}$ can be recovered form κ, since $\overline{\nu}(b\#a) = \kappa(b)a$. Furthermore

$$a\kappa(b) = a\overline{\nu}(b\#1) = \overline{\nu}(b_R\#a_R) = \kappa(b_R)a_R$$

and we find that

$$V \cong V_1 \cong V_3 = \{\kappa : B \to A \mid a\kappa(b) = \kappa(b_R)a_R\} \tag{3.22}$$

Now we simplify the description of $W \cong W_1 \subset (B\#_R A) \otimes_A (B\#_R A)$. To this end, we observe that we have a k-module isomorphism

$$\gamma : (B\#_R A) \otimes_A (B\#_R A) \to B \otimes B \otimes A$$

defined by

$$\gamma((b\#a) \otimes (d\#c)) = b \otimes d_R \otimes a_R c$$
$$\gamma^{-1}(b \otimes d \otimes c) = (b\#1) \otimes (d\#c)$$

Now let $W_3 = \gamma(W_1) \subset B \otimes B \otimes A$. Take $e = b^1 \otimes b^2 \otimes a^2 \in B \otimes B \otimes A$ (summation implicitly understood). $e \in W_3$ if and only if (3.16) holds, for all $s = b\#1$ and $s = 1\#a$ with $b \in B$ and $a \in A$, if and only if

$$bb^1 \otimes b^2 \otimes a^2 = b^1 \otimes b^2 b_R \otimes (a^2)_R \tag{3.23}$$
$$(b^1)_R \otimes (b^2)_r \otimes a_{Rr} a^2 = b^1 \otimes b^2 \otimes a^2 a \tag{3.24}$$

for all $a \in A$, $b \in B$. We find

$$W \cong W_1 \cong W_3 = \{e = b^1 \otimes b^2 \otimes a^2 \in B \otimes B \otimes A \mid (3.23) \text{ and } (3.24) \text{ hold}\} \tag{3.25}$$

Using these descriptions of V and W, we find immediately that Theorem 27 takes the following form.

Theorem 29. *Let (B, A, R) be a factorization structure over a commutative ring k.*

1. *$B\#_R A/A$ is separable (i.e. the restriction of scalars functor $G : \mathcal{M}_{B\#_R A} \to \mathcal{M}_A$ is separable) if and only if there exists $e = b^1 \otimes b^2 \otimes a^2 \in W_3$ such that*
$$b^1 b^2 \otimes a^2 = 1_B \otimes 1_A \in B \otimes A \tag{3.26}$$

2. *$B\#_R A/A$ is split (i.e. the induction functor $F : \mathcal{M}_A \to \mathcal{M}_{B\#_R A}$ is separable) if and only if there exists $\kappa \in V_3$ such that*
$$\kappa(1_B) = 1_A \tag{3.27}$$

3. *$B\#_R A/A$ is Frobenius (i.e. (F, G) is Frobenius pair) if and only if there exist $\kappa \in V_3$, $e \in W_3$ such that*
$$(b^2)_R \otimes \kappa(b^1)_R a^2 = b^1 \otimes \kappa(b^2) a^2 = 1_B \otimes 1_A \tag{3.28}$$

In the same style, we can reformulate Theorem 28. In our situation,

$$\operatorname{Hom}_R(S, R) = \operatorname{Hom}_A(B\#_R A, A) \cong \operatorname{Hom}(B, A)$$

The $(A, B\#_R A)$-bimodule structure on $\operatorname{Hom}(B, A)$ is the following (cf. (3.19)):

$$(cf(b\#a))(d) = cf(db)a$$

for all $a, c \in C$ and $b, d \in B$. From Proposition 53, we deduce that

$$V \cong V_2 \cong V_4 = {}_A\operatorname{Hom}_{B\#_R A}(B\#_R A, \operatorname{Hom}(B, A)) \tag{3.29}$$

If B is finitely generated and projective as a k-module, then we find using Proposition 54

$$W \cong W_2 \cong W_4 = {}_A\mathrm{Hom}_{B\#_RA}(\mathrm{Hom}(B, A), B\#_RA) \qquad (3.30)$$

Theorem 28 now takes the following form:

Theorem 30. *Let (B, A, R) be a factorization structure over a commutative ring k, and assume that B is finitely generated and projective as a k-module. Let $\{b_i, b_i^* \mid i = 1, \cdots, m\}$ be a finite dual basis for B.*

1. *$B\#_RA/A$ is separable if and only if there exists an $(A, B\#_RA)$-bimodule map $\phi:\ \mathrm{Hom}(B, A) \cong B^* \otimes A \to B\#_RA$ such that*

$$\sum_i (b_i\#1)\phi(b_i^*\#1) = 1_B \otimes 1_A$$

2. *$B\#_RA/A$ is split if and only if there exists an $(A, B\#_RA)$-bimodule map $\overline{\phi}:\ B\#_RA \to \mathrm{Hom}(B, A)$ such that*

$$\overline{\phi}(1_B\#1_A)(1_B) = 1_A$$

3. *$B\#_RA/A$ is Frobenius if and only if $B^*\otimes A$ and $B\#_RA$ are isomorphic as $(A, B\#_RA)$-bimodules. This is also equivalent to the existence of $\kappa \in V_3$, $e = b^1 \otimes b^2 \otimes a^2 \in W_3$ such that the maps*

$$\phi:\ \mathrm{Hom}(B, A) \to B\#_RA\ ;\quad \phi(f) = f(b^1)b^2\#a^2$$

and

$$\overline{\phi}:\ B\#_RA \to \mathrm{Hom}(B, A)\ ;\quad \overline{\phi}(b\#a)(d) = \kappa(bd_R)a_R$$

are each others inverses.

The same method can be applied to the extension $B\#_RA/B$. There are two ways to proceed: as above, but applying the left-handed version of Theorem 28 (using the left-right symmetry of the notion of separable and Frobenius extension). Another possibility is the use of "op"-arguments: if (B, A, R) is a factorization structure, then

$$\tilde{R}:\ B^{\mathrm{op}} \otimes A^{\mathrm{op}} \to A^{\mathrm{op}} \otimes B^{\mathrm{op}}$$

makes $(A^{\mathrm{op}}, B^{\mathrm{op}}, \tilde{R})$ into a factorization structure, and it is not hard to see that we have an algebra isomorphism

$$(A^{\mathrm{op}}\#_{\tilde{R}}B^{\mathrm{op}})^{\mathrm{op}} \cong B\#_RA$$

Using the left-right symmetry again, we find that $B\#_RA/B$ is Frobenius if and only if $(A^{\mathrm{op}}\#_{\tilde{R}}B^{\mathrm{op}})^{\mathrm{op}}/B^{\mathrm{op}}$ is Frobenius if and only if $(A^{\mathrm{op}}\#_{\tilde{R}}B^{\mathrm{op}})/B^{\mathrm{op}}$ is Frobenius, and we can apply Theorems 29 and 30. We invite the reader to write down explicit results.

Now let (B, A, R) be a factorization structure, with R invertible. We write $\overline{R}$ for the inverse of R. We also assume that A is a bialgebra. Our next aim is to compare the properties of the extensions $A \to B \#_R A$ and $k \to B$. Let $F : k \to B$ and $F^\# : A \to B \#_R A$ be the respective induction functors. We use similar notation for the restriction of scalars functors, and for the corresponding modules consisting of natural transformations, for example

$$V = \underline{\mathrm{Nat}}(GF, 1_{\mathcal{M}_k}) \quad \text{and} \quad V^\# = \underline{\mathrm{Nat}}(G^\# F^\#, 1_{\mathcal{M}_A})$$

Proposition 57. *Let (B, A, R) be a factorization structure; assume that R is invertible and that A is a bialgebra. With notation as above, we have k-module homomorphisms*

$$\gamma : V_3^\# \to V_1 \quad ; \quad \gamma(\kappa) = \varepsilon_A \circ \kappa$$

$$\delta : W_3^\# \to W_1 \quad : \quad \delta(b^1 \otimes b^2 \otimes a) = b^1 \otimes \varepsilon(a_{\overline{R}}^2) b_{\overline{R}}^2$$

Proof. The first property is obvious, since $V_1 = B^*$. We have to show that δ is well-defined. From (3.23) and (2.28) (applied to the factorization structure $(A, B, \overline{R})$), we find

$$bb^1 \otimes a_{\overline{R}}^2 \otimes b_{\overline{R}}^2 = b^1 \otimes a_{R\overline{R}}^2 \otimes (b^2 b_R)_{\overline{R}} = b^1 \otimes a_{R\overline{R}\overline{r}}^2 \otimes b_{\overline{r}}^2 b_{R\overline{S}}$$

Applyig ε_A to the second factor, we see that

$$bb^1 \otimes \varepsilon(a_{\overline{R}}^2) b_{\overline{R}}^2 = b^1 \otimes \varepsilon(a_{\overline{R}}^2) b_{\overline{R}}^2 b$$

so $\delta(b^1 \otimes b^2 \otimes a) \in W_1$, as needed.

Corollary 12. *Let (B, A, R) be a factorization structure, with R invertible and A a bialgebra. If $A \#_R B / A$ is Frobenius, then B/k is Frobenius.*

Proof. Take $e = b^1 \otimes b^2 \otimes a \in W_3$ and $\kappa \in V_3$ satisfying (3.28). It suffices to show that $\delta(e)$ and $\gamma(\kappa) = \overline{\nu}$ satisfy (3.18). Applying $\overline{R}$ to (3.28), and using (2.27) and (2.29) (applied to $(B, A, \overline{R})$), we find

$$1_B \otimes 1_A = b_{R\overline{R}}^2 \otimes (\kappa(b^1)_R a^2)_{\overline{R}} = b_{R\overline{R}\overline{r}}^2 \otimes \kappa(b^1)_{R\overline{R}} a_{\overline{r}}^2 = b_r^2 \otimes \kappa(b^1) a_{\overline{r}}^2$$

Applying ε_A to the second factor, we find

$$\overline{\nu}(b^1) \varepsilon(a_{\overline{r}}^2) b_{\overline{r}}^2 = 1_B$$

proving one equality from (3.18). (3.18) tells us also that

$$b^1 \otimes \kappa(b^2) a^2 = 1_B \otimes 1_A$$

or, using (3.22)

$$b^1 \otimes a^2_{\overline{R}} \kappa(b^2_{\overline{R}}) = 1_B \otimes 1_A$$

and, applying ε_A to the second factor:

$$b^1 \varepsilon(a^2_{\overline{R}}) \overline{\nu}(b^2_{\overline{R}}) = 1_B$$

proving the second equality from (3.18).

For later use, we now will describe the natural transformation $\Phi : F \to F'$ from Proposition 55, in the case where $S = B \# _R A$, and $R = A$. We will assume that R is invertible, and that B is finitely generated and projective as a k-module. First, we give alternative descriptions of F, $F' : \mathcal{M}_A \to \mathcal{M}_{B \# _R A}$. Using the fact that R is invertible, we find

$$F(M) = M \otimes_A (B \# _R A) \cong M \otimes B$$

where the right $B \# _R A$-action on $M \otimes B$ is the following

$$(m \otimes b') \triangleleft (b \# a) = ma_{\overline{R}} \otimes (b'b)_{\overline{R}} \tag{3.31}$$

Also

$$F'(M) = \mathrm{Hom}_A(B \# _R A, M) \cong \mathrm{Hom}(B, M) \cong B^* \otimes M$$

with right $B \# _R A$-action given by

$$(b^* \otimes m) \triangleleft (b \# a) = \sum_i \langle b^*, bb_{iR} \rangle b_i^* \otimes ma_R \tag{3.32}$$

where $\{b_i, b_i^* \mid i = 1, \cdots, n\}$ is a dual basis for B, as usual. Now (3.20) tells us, for all $M \in \mathcal{M}_A$, that

$$\Phi_M : B^* \otimes M \to M \otimes B$$

is given by

$$\Phi_M(b^* \otimes m) = \langle b^*, b^1 \rangle ma^2_{\overline{R}} \otimes b^2_{\overline{R}} \tag{3.33}$$

where $e = b^1 \otimes b^2 \otimes a^2 \in W_3$. From the comments following Proposition 55, we conclude that $B \# _R A / A$ is Frobenius if and only if $e \in W_3$ can be chosen in such a way that Φ_M is an isomorphism for every right A-module M.

Application to Hopf algebras We now look more closely at the situation where H is a Hopf algebra over a commutative ring k. Then the additional structure on H allows us to simplify the conditions in Theorem 27. The results we present here go back to [117] (if k is a field), and [151] (if k is a commutative ring).

First recall that $t \in H$ is called a *left* (resp. *right*) *integral* in H if

$$ht = \varepsilon(h)t \quad \text{resp.} \quad th = \varepsilon(h)t$$

for all $h \in H$. $\int_H^l$ (resp. $\int_H^r$) denote the k-modules consisting respectively of left and right integrals in H. In a similar way, we introduce left and right integrals in H^* (or on H). These are functionals $\varphi \in H^*$ that have to verify respectively

$$h^* * \varphi = \langle h^*, 1 \rangle \varphi \quad \text{resp.} \quad \varphi * h^* = \langle h^*, 1 \rangle \varphi$$

for all $h^* \in H^*$. The k-modules consisting of left and right integrals in H^* are denoted by $\int_{H^*}^l$ and $\int_{H^*}^r$. Let us first show that there is a close relation between integrals and the elements $e \in W_1$ (cf. (3.17)).

Proposition 58. *Let H be a Hopf algebra. We have the following maps*

$$p: W_1 \to \int_H^l \quad ; \quad p(e) = e^1 \varepsilon(e^2)$$

$$p': W_1 \to \int_H^r \quad ; \quad p'(e) = \varepsilon(e^1) e^2$$

$$i: \int_H^l \to W_1 \quad ; \quad i(t) = t_{(1)} \otimes S(t_{(2)})$$

$$i': \int_H^r \to W_1 \quad ; \quad i'(t) = S(t_{(1)}) \otimes t_{(2)}$$

satisfying

$$(p \circ i)(t) = t \quad ; \quad (p' \circ i')(t) = t$$

for every left (resp. right) integral t.

Proof. We will show that $i(t) \in W_1$ if t is a left integral, and leave all the other assertions to the reader.

$$
\begin{aligned}
ht_{(1)} \otimes S(t_{(2)}) &= h_{(1)} t_{(1)} \otimes S(t_{(2)}) S(h_{(2)}) h_{(3)} \\
&= (h_{(1)} t)_{(1)} \otimes S((h_{(1)} t)_{(2)}) h_{(2)} \\
&= (\varepsilon(h_{(1)}) t)_{(1)} \otimes S((\varepsilon(h_{(1)}) t)_{(2)}) h_{(2)} \\
&= t_{(1)} \otimes S(t_{(2)}) h
\end{aligned}
$$

Corollary 13. *A Hopf algebra H is separable if and only if there exists a (left or right) integral $t \in H$ such that $\varepsilon(t) = 1$.*

Proof. An immediate consequence of Theorem 27: if t is a left integral with $\varepsilon(t) = 1$, then $e = i(t) \in W_1$ satisfies $e^1 e^2 = t_{(1)} S(t_{(2)}) = \varepsilon(t) = 1$. The converse is similar: if $e^1 e^2 = 1$, then $\varepsilon(p(e)) = \varepsilon(e^1 e^2) = 1$.

Corollary 14. *A Hopf algebra H over a field k is finite dimensional semisimple if and only if there exists a (left or right) integral $t \in H$ such that $\varepsilon(t) = 1$.*

Proof. One direction follows immediately from Corollary 13 and Proposition 13. Conversely, if H is semisimple, then $H = I \oplus \text{Ker}(\varepsilon)$ for some two-sided ideal I of H. We claim that $I \subset \int_H^l$: For $z \in I$, and $h \in H$, we have $(h - \varepsilon(h)) \in \text{Ker}(\varepsilon)$

$$hz = (h - \varepsilon(h))z + \varepsilon(h)z = 0 + hz$$

are two decompositions of hz in $\text{Ker}(\varepsilon) \oplus I$, so $\varepsilon(h)z = hz$, and z is a left integral. Choose $z \neq 0$ in I (this is possible since I is one-dimensional). $\varepsilon(z) \neq 0$ since $z \notin \text{Ker}(\varepsilon)$, and $t = z/\varepsilon(z)$ is a left integral with $\varepsilon(t) = 1$.

We now want to characterize Hopf algebras that are Frobenius. This problem is closely connected to the Fundamental Theorem for Hopf modules:

Proposition 59. (Fundamental Theorem for Hopf modules) *Let H be a Hopf algebra, and $M \in \mathcal{M}(H)_H^H$ a right-right Hopf module. Then we have an isomorphism*

$$\alpha : M^{coH} \otimes H \to M$$

of right-right Hopf modules. α and α^{-1} are given by the following formulas:

$$\alpha(m' \otimes h) = m'h \quad \text{and} \quad \alpha^{-1}(m) = m_{[0]}S(m_{[1]}) \otimes m_{[2]} \qquad (3.34)$$

for all $m' \in M^{coH}$, $m \in M$ and $h \in H$.

An immediate application is the following:

Proposition 60. *Let H be a finitely generated projective Hopf algebra. H^* is a left H^*-module (by multiplication), and therefore a right H-comodule. It is also a right H-module, we let*

$$\langle h^* \leftharpoonup h, k \rangle = \langle h^*, kS(h) \rangle$$

for all $h^* \in H^*$, and $h, k \in H$. With these structure maps, H^* is a right-right Hopf module, and $(H^*)^{coH} \cong \int_{H^*}^l$. Consequently we have an isomorphism

$$\alpha : \int_{H^*}^l \otimes H \to H^* \quad ; \quad \alpha(\varphi \otimes h) = \varphi \leftharpoonup h \qquad (3.35)$$

of right-right Hopf modules. In particular, it follows that $\int_{H^*}^l$ is a rank one projective k-module. Similar results hold for the right integral space.

Proof. Remark that the right H-action on H^* is not the usual one. Recall that the usual H-bimodule structure on H^* is given by

$$\langle h \cdot h^* \cdot k, l \rangle = \langle h^*, klh \rangle$$

and we see immediately that

$$h^* \leftharpoonup h = S(h) \cdot h^*$$

Also observe that the right H-coaction on H^* can be rewritten in terms of a dual basis $\{h_i, h_i^* \mid i = 1, \cdots, n\}$ of H:

$$\rho^r(h^*) = \sum_i h_i^* * h^* \otimes h_i$$

The only thing we have to check is the Hopf compatibility relation for H^*, i.e.

$$\rho^r(h^* \leftharpoonup h) = h_{[0]}^* \leftharpoonup h_{(1)} \otimes h_{[1]}^* h_{(2)}$$

for all $h \in H$, $h^* \in H^*$. It suffices to prove that

$$\langle (h^* \leftharpoonup h)_{[0]}, k \rangle (h^* \leftharpoonup h)_{[1]} = \langle h_{[0]}^* \leftharpoonup h_{(1)}, k \rangle h_{[1]}^* h_{(2)} \qquad (3.36)$$

for all $k \in H$. We first compute the left hand side:

$$\rho^r(h^* \leftharpoonup h) = \sum_i (h_i^* * S(h) \cdot h^*) \otimes h_i$$

so

$$\begin{aligned}
\langle (h^* \leftharpoonup h)_{[0]}, k \rangle (h^* \leftharpoonup h)_{[1]} &= \sum_i \langle h_i^* * S(h) \cdot h^*, k \rangle h_i \\
&= \sum_i \langle h_i^*, k_{(1)} \rangle \langle h^*, k_{(2)} S(h) \rangle h_i \\
&= \langle h^*, k_{(2)} S(h) \rangle k_{(1)}
\end{aligned}$$

The right hand side of (3.36) equals

$$\begin{aligned}
\langle h_i^* * h^*, k S(h_{(1)}) \rangle h_i h_{(2)} &= \langle h_i^*, k_{(1)} S(h_{(2)}) \rangle \langle h^*, k_{(2)} S(h_{(1)}) \rangle h_i h_{(3)} \\
&= \langle h^*, k_{(2)} S(h_{(1)}) \rangle k_{(1)} S(h_{(2)}) h_{(3)} = \langle h^*, k_{(2)} S(h_{(1)}) \rangle k_{(1)}
\end{aligned}$$

as needed.

As an application of Proposition 60, we can prove that the antipode of a finitely generated projective Hopf algebra is always bijective.

Proposition 61. *The antipode of a finitely generated projective Hopf algebra is bijective.*

Proof. We know from Proposition 60 that $J = \int_{H^*}^l$ is projective of rank one. This implies that we have an isomorphism

$$J^* \otimes J \to k \quad ; \quad p \otimes \varphi \mapsto p(\varphi)$$

Let $\sum_l p_l \otimes \varphi_l$ be the inverse image of 1:

$$\sum_l p_l(\varphi_l) = 1$$

The isomorphism α of Proposition 60 induces another isomorphism

$$\tilde{\alpha} : H \to J^* \otimes H^* \quad ; \quad \tilde{\alpha}(h) = \sum_l p_l \otimes \alpha(\varphi_l \otimes h) = \sum_l p_l \otimes S(h) \cdot \varphi_l$$

If $S(h) = 0$, then it follows from the above formula that $\tilde{\alpha}(h) = 0$, hence $h = 0$, since $\tilde{\alpha}$ is injective. Hence S is injective.

The fact that S is surjective follows from a local global argument. Let $Q = \mathrm{Coker}\,(S)$. For every prime ideal p of k, $\mathrm{Coker}\,(S_p) = Q_p$, since localization at a prime ideal is an exact functor. Now H_p/pH_p is a finite dimensional Hopf algebra over the field k_p/pk_p, with antipode induced by S_p, the antipode of the localized k_p-Hopf algebra H_p. The antipode of H_p/pH_p is injective, hence bijective, by counting dimensions. Nakayama's Lemma implies that S_p is surjective, for all $p \in \mathrm{Spec}(k)$, and it follows that S is bijective.

Here is another application of Proposition 60:

Proposition 62. *Let H be a finitely generated projective Hopf algebra. Then there exist $\varphi_j \in \int_{H^*}^l$ and $h_j \in H$ such that*

$$\sum_j \langle \varphi_j, h_j \rangle = 1$$

and $t_j \in \int_H^l$ and $h_j^* \in H^*$ such that

$$\sum_j \langle h_j^*, t_j \rangle = 1$$

Proof. Take $\alpha^{-1}(\varepsilon) = \sum_j \varphi_j \otimes \overline{S}(h_j)$ (the antipode is bijective by Proposition 61). Then

$$1_k = \langle \varepsilon_H, 1_H \rangle = \sum_j \langle \varphi_j \leftharpoonup \overline{S}(h_j), 1_H \rangle = \sum_j \langle \varphi_j, h_j \rangle$$

the second statement follows after we apply the first one with H replaced by H^*. $\qquad\square$

The main result is now the following:

Theorem 31. *For a Hopf algebra H, the following assertions are equivalent:*

1. *H/k is Frobenius;*
2. *H^*/k is Frobenius;*
3. *H is finitely generated and projective, and $\int_H^l$ is free of rank one;*
4. *H is finitely generated and projective, and $\int_H^r$ is free of rank one;*

5. H is finitely generated and projective, and $\int_{H^*}^l$ is free of rank one;
6. H is finitely generated and projective, and $\int_{H^*}^r$ is free of rank one;
7. H is finitely generated and projective, and there exist $t \in \int_H^l$ and $\varphi \in \int_{H^*}^l$ such that $\langle \varphi, t \rangle = 1$;
8. H is finitely generated and projective, and there exist $u \in \int_H^r$ and $\psi \in \int_{H^*}^r$ such that $\langle \psi, u \rangle = 1$;

In conditions 7. and 8., the integrals can be chosen in such a way that they are generators of the integral space that they belong to.

Proof. 1. $\Rightarrow$ 3. Theorem 27 implies the existence of $\bar{\nu} \in V_1 = H^*$ and $e \in W_1$ such that $\bar{\nu}(e^1)e^2 = e^1\bar{\nu}(e^2) = 1$. Take $t = p(e) = \varepsilon(e^2)e^1 \in \int_H^l$. We claim that $\int_H^l$ is free with basis $\{t\}$. Take another left integral $u \in \int_H^l$. Then

$$u = ue^1\bar{\nu}(e^2) = e^1\bar{\nu}(e^2u) = e^1\bar{\nu}(\varepsilon(e^2)u)$$
$$= \varepsilon(e^2)e^1\bar{\nu}(u) = \bar{\nu}(u)t$$

and it follows that the map $k \to \int_H^l$ sending $x \in k$ to xt is surjective. This map is also injective: if

$$xt = x\varepsilon(e^2)e^1 = 0$$

then

$$0 = \bar{\nu}(x\varepsilon(e^2)e^1) = \bar{\nu}(e^1)x\varepsilon(e^2) = x\varepsilon(e^2\bar{\nu}(e^1)) = x$$

5. $\Rightarrow$ 2. and 5. $\Rightarrow$ 1. Assume that $\int_{H^*}^l = k\varphi$, with φ a left integral, and consider the Hopf module isomorphism

$$\alpha : k\varphi \otimes H \to H^*$$

from Proposition 60. We first consider the map

$$\Theta : H \to H^* \quad ; \quad \Theta(h) = \alpha(\varphi \otimes h) = S(h) \cdot \varphi$$

α and Θ are right H-colinear, hence left H^*-linear. Θ is therefore an isomorphism of left H^*-modules, and it follows that H^* is Frobenius.
A slightly more subtle argument shows that H is Frobenius: we consider the map

$$\bar{\phi} = \Theta \circ S^{-1} : H \to H^*, \quad \text{i.e. } \bar{\phi}(h) = h \cdot \varphi$$

We know from Proposition 61 that S is bijective, so $\bar{\phi}$ is well-defined, and is a bijection. $\bar{\phi}$ is left H-linear since

$$\bar{\phi}(kh) = (kh) \cdot \varphi = k \cdot (h \cdot \varphi) = k \cdot \bar{\phi}(h)$$

The equivalence of assertions 1.-6. now follows after we apply the above implications 1. $\Rightarrow$ 3. and 5. $\Rightarrow$ 1. with H replaced by H^* (H is finitely generated and projective) or by H^{op} (the Frobenius property is symmetric).

<u>5. ⇒ 7.</u> Let φ be a free generator of $\int_{H*}^l$, and consider the isomorphism $\overline{\phi}: H \to H^*$, $\overline{\phi}(h) = h \cdot \varphi$. Let ϕ be the inverse of $\overline{\phi}$, and put $\phi(\varepsilon) = t$. this means that $\overline{\phi}(t) = t \cdot \varphi = \varepsilon$, or $\varphi(ht) = \varepsilon(h)$ for all $h \in H$, and, in particular, $\langle \varphi, t \rangle = 1$.

t is a left integral, since

$$\langle \overline{\phi}(ht), k \rangle = \langle (ht) \cdot \varphi, k \rangle = \langle t \cdot \varphi, kh \rangle$$
$$= \varepsilon(kh) = \varepsilon(k)\varepsilon(h) = \varepsilon(h)\langle \overline{\phi}(t), k \rangle$$

for all $h, k \in H$, implying that $\overline{\phi}(ht) = \varepsilon(h)\overline{\phi}(t)$, and $ht = \varepsilon(h)t$, as needed. We also have that t is a free generator for $\int_H^l$. If u is another left integral, then

$$\langle \overline{\phi}(u), h \rangle = \langle u \cdot \varphi, h \rangle = \langle \varphi, hu \rangle = \varepsilon(h)\langle \varphi, u \rangle = \langle \varphi, ht \rangle \langle \varphi, u \rangle$$
$$= \langle \varphi, h\varphi(u)t \rangle = \langle (\varphi(u)t) \cdot \varphi, h \rangle = \langle \overline{\phi}(\varphi(u)t), h \rangle$$

implying $\overline{\phi}(u) = \overline{\phi}(\varphi(u)t)$ and $u = \varphi(u)t$.

Assume that $xt = 0$, for some $x \in k$. Then $\overline{\phi}(x\varepsilon) = 0$, hence $x\varepsilon = 0$, and $x = 0$. This proves that t is a free generator for $\int_H^l$.

<u>7. ⇒ 1.</u> The fact that φ is a left integral means that

$$\langle h^*, h_{(1)} \rangle \langle \varphi, h_{(2)} \rangle = \langle h^*, 1 \rangle \langle \varphi, h \rangle$$

for all $h \in H$ and $h^* \in H^*$, and it follows that

$$h_{(1)}\langle \varphi, h_{(2)} \rangle = \langle \varphi, h \rangle 1$$

We now easily compute

$$\langle \varphi, t_{(2)} \rangle \overline{S}(t_{(1)}) = \langle \varphi, t_{(3)} \rangle t_{(2)} \overline{S}(t_{(1)}) = \langle \varphi, t \rangle 1 = 1$$

Now consider the left H-linear maps $\overline{\phi}: H \to H^*$, $\overline{\phi}(h) = h \cdot \varphi$, and $\phi: H^* \to H$, $\phi(h^*) = \langle h^*, \overline{S}(t_{(1)}) \rangle t_{(2)}$. It is straightforward to compute that $\overline{\phi}$ is a left inverse of ϕ. Thus ϕ is injective, and a count of ranks as at the end of the proof of Proposition 61 tells us that ϕ is also surjective, hence H is Frobenius.

Remarks 3. 1. It follows from the preceding Theorem that any finite dimensional Hopf algebra over a field k is Frobenius.

2. In Proposition 58, we have seen that we have a map $p: W_1 \to \int_H^l$, with a right inverse i. The map p is not an isomorphism. To see this, take a Frobenius Hopf algebra H (e.g. any finite dimensional Hopf algebra over a field). As we have seen, $\int_H^l$ is free of rank one over k. Using Proposition 54 and the fact that $H^* \cong H$ as H-modules, we find that

$$W_1 \cong W_2 = \text{Hom}_H(H^*, H) \cong \text{Hom}_H(H, H) \cong H$$

and the rank of W_1 equals the rank of H as a k-module.

3. Let t and φ be as in part 7. of the Theorem. From the proof of 7. $\Rightarrow$ 1., it follows that $(i'(\overline{S}(t)) = t_{(2)} \otimes \overline{S}(t_{(1)}), \varphi)$ is a Frobenius pair for H.

4. As an example, take $H = kG$, with G a finite group. Then G is a finite basis for kG, and let $\{v_\sigma \mid \sigma \in G\}$ be the corresponding dual basis, i.e. $\langle v_\sigma, \tau \rangle = \delta_{\sigma,\tau}$. $t = \sum_{\sigma \in G} \sigma$ and v_1 are generators of the integrals in and on kG. The isomorphism $\overline{\phi}: kG \to kG^*$ is given by $\overline{\phi}(\sigma) = v_{\sigma^{-1}}$. Frobenius pairs for kG, resp. kG^* are respectively

$$\left(\sum_{\sigma \in G} \sigma \otimes \sigma^{-1}, \varphi\right) \quad \text{and} \quad \left(\sum_{\sigma \in G} v_\sigma \otimes v_\sigma, t\right)$$

We generalize the definition of integrals as follows: take $\alpha \in \mathrm{Alg}(H, k)$ and $g \in G(H)$, and define

$$\int_\alpha^l = \{t \in H \mid ht = \alpha(h)t, \text{ for all } h \in H\}$$

$$\int_\alpha^r = \{t \in H \mid th = \alpha(h)t, \text{ for all } h \in H\}$$

$$\int_g^l = \{\varphi \in H^* \mid h^* * \varphi = \langle h^*, g \rangle \varphi, \text{ for all } h^* \in H^*\}$$

$$\int_g^r = \{\varphi \in H^* \mid \varphi * h^* = \langle h^*, g \rangle \varphi, \text{ for all } h^* \in H^*\}$$

Of course we recover the previous definitions if $\alpha = \varepsilon$ and $g = 1$. We have the following generalization of Theorem 31.

Proposition 63. *Let H be a Hopf algebra, and assume that H/k is Frobenius. Then for all $\alpha \in \mathrm{Alg}(H, k)$ and $g \in G(H)$, the integral spaces $\int_\alpha^l, \int_\alpha^l, \int_g^l$ and $\int_g^r$ are free k-modules of rank one.*

Proof. Take $t = \alpha(e^1)e^2$. Arguments almost identical to the ones used in the proof of 1. $\Rightarrow$ 3. in Theorem 31 prove that t is a free generator of $\int_\alpha^l$. The statements for the other integral spaces follow by duality arguments.

Now assume that H is Frobenius, and write $\int_H^l = kt$. It is easy to prove that $th \in \int_H^l$, for all $h \in H$. Indeed,

$$k(th) = (kt)h = \varepsilon(k)th$$

for all $k \in H$. It follows that there exists a unique $\alpha(h) \in k$ such that

$$th = \alpha(h)t$$

$\alpha: H \to k$ is multiplicative, so we can restate our observation by saying that $t \in \int_\alpha^r$. We call α the *distinguished element* of H^*. If $\alpha = \varepsilon$, then we say that H is *unimodular*.

Proposition 64. *Let H be a Frobenius Hopf algebra, and $\alpha \in H^*$ the distinguished element. Then $\int_\alpha^r = \int_H^l$, and H is unimodular if and only if*

$$\int_H^r = \int_H^l$$

Proof. We know from Proposition 63 that $\int_\alpha^r = kt'$ is free of rank one. For all $h \in H$, we have that $ht' \in \int_\alpha^r$, hence we find a unique multiplicative map $\beta : H \to k$ such that

$$ht' = \beta(h)t'$$

for all $h \in H$. Now we have that $t = xt'$ for some $x \in k$, since $t \in \int_\alpha^r$. Thus

$$\varepsilon(h)t = ht = xht' = x\beta(h)t' = \beta(h)t$$

for all $h \in H$. This implies that $\beta = \varepsilon$, since t is a free generator of $\int_H^l$. It follows that $t' \in \int_H^l$, proving the first statement.

If $\alpha = \varepsilon$, then it follows that $\int_H^r = \int_H^l$. Conversely, if $\int_H^r = \int_H^l$, then $t \in \int_H^r$, and this means that the distinguished element is equal to ε.

Frobenius modules and Λ-separability Let R and S be rings. For a left R-linear map $f : M \to N$, we will write $(m)f$ for the image of $m \in M$ under f. For a right linear map, we keep the usual notation $f(m)$, this will make our formalism more transparent. To a bimodule $\Lambda \in {}_S\mathcal{M}_R$, we can associate four new bimodules:

$$
\begin{array}{ll}
\operatorname{Hom}_R(\Lambda, R) \in {}_R\mathcal{M}_S & (r\varphi s)(\lambda) = r\varphi(s\lambda) \\
{}_S\operatorname{Hom}(\Lambda, S) \in {}_R\mathcal{M}_S & (\lambda)(r\varphi s) = ((\lambda r)\varphi)s \\
\operatorname{Hom}_R(\Lambda, \Lambda) \in {}_S\mathcal{M}_S & (s\varphi s')(\lambda) = s\varphi(s'\lambda) \\
{}_S\operatorname{Hom}(\Lambda, \Lambda) \in {}_R\mathcal{M}_R & (\lambda)(r\varphi r') = ((\lambda r)\varphi)r'
\end{array}
$$

Λ is called a *Frobenius bimodule* ([2], [100]) if Λ is finitely generated and projective as a left S-module and a right R-module, and

$$\operatorname{Hom}_R(\Lambda, R) \cong {}_S\operatorname{Hom}(\Lambda, S) \quad \text{in} \quad {}_R\mathcal{M}_S$$

S is called Λ-separable over R, or Λ is called a *separable bimodule* (see [171]) if the map

$$\mu_\Lambda : \Lambda \otimes_R {}_S\operatorname{Hom}(\Lambda, S) \to S, \quad \mu_\Lambda(\lambda \otimes_R \varphi) = (\lambda)\varphi$$

is split as a map of S-bimodules, or, equivalently, if there exists $e = \sum_i \lambda_i \otimes_R \varphi_i \in \Lambda \otimes_R {}^*\Lambda$ such that

$$\sum_i (\lambda_i)\varphi_i = 1 \quad \text{and} \quad se = es$$

for all $s \in S$. We will show that these notions can also be introduced using Frobenius and separable functors, and we will classify all (additive) Frobenius functors between module categories.

The induction functor

$$F = \Lambda \otimes_R \bullet : \ _R\mathcal{M} \to \ _S\mathcal{M}, \quad F(M) = \Lambda \otimes_R M$$

has a right adjoint, the coinduction functor

$$G = \ _S\text{Hom}(\Lambda, \bullet) : \ _S\mathcal{M} \to \ _R\mathcal{M}, \quad G(N) = \ _S\text{Hom}(\Lambda, N)$$

with the left R-action on $G(N) = \ _S\text{Hom}(\Lambda, N)$ given by

$$(\lambda)(rf) = (\lambda r)f$$

for all $\lambda \in \Lambda$, $r \in R$ and $f \in \ _S\text{Hom}(\Lambda, N)$. The unit and counit of the adjunction are

$$\eta_M : \ M \to GF(M) = \ _S\text{Hom}(\Lambda, \Lambda \otimes_R M) \ ; \ (\lambda)(\eta_M(m)) = \lambda \otimes m$$

$$\varepsilon_N : \ FG(N) = \Lambda \otimes_R \ _S\text{Hom}(\Lambda, N) \to N \ ; \ \varepsilon_N(\lambda \otimes f) = (\lambda)f$$

The converse also holds: if (F, G) is an adjoint pair between $_R\mathcal{M}$ and $_S\mathcal{M}$, then F and G are additive (see [11, I.7.2]). F has a right adjoint, and preserves therefore cokernels and arbitrary coproducts (see [11, I.7.1]), and from the Eilenberg-Watts Theorem (see [11, II.2.3]), it follows that $F \cong \Lambda \otimes_R \bullet$ for some $\Lambda \in \ _S\mathcal{M}_R$. From the uniqueness of the adjoint, if follows that $G \cong \ _S\text{Hom}(\Lambda, \bullet)$.

We also consider the functor

$$G_1 : \ _S\mathcal{M} \to \ _R\mathcal{M} \ ; \ G_1(N) = \ ^*\Lambda \otimes_S N$$

We have a natural transformation $\gamma : \ G_1 \to G$ given by

$$\gamma_N : \ ^*\Lambda \otimes_S N \to \ _S\text{Hom}(\Lambda, N) \ ; \ (\lambda)(\gamma_N(f \otimes n)) = (\lambda)fn$$

If Λ is finitely generated and projective as a left S-module, then γ is a natural isomorphism.

Now take an (R, S)-bimodule X, and consider the functor

$$G_2 : \ _S\mathcal{M} \to \ _R\mathcal{M} \ ; \ G_2(N) = X \otimes_S N$$

When is (F, G_2) an adjoint pair? From the uniqueness of the adjoint, we can see that an equivalent question is: when is $G \cong G_2$? Or, in other words, when is the induction functor G_2 representable? The clue to the answer is the following Lemma, where we also consider the functors

$$G_2' = \bullet \otimes_S \Lambda : \ \mathcal{M}_S \to \mathcal{M}_R \ \text{ and } \ F' = \bullet \otimes_R X : \ \mathcal{M}_R \to \mathcal{M}_S$$

Lemma 11. *Let $R, S, \Lambda, X, F, G, G_1, G_2$ be as above. Then we have isomorphisms*

$$\underline{\text{Nat}}(1_{R}\mathcal{M}, G_2 F) \cong \underline{\text{Nat}}(1_{\mathcal{M}_R}, G_2' F')$$
$$\cong {}_R\text{Hom}_R(R, X \otimes_S \Lambda) \cong C_R(X \otimes_S \Lambda)$$
$$= \{\sum_i x_i \otimes_S \lambda_i \in X \otimes_S \Lambda \mid$$
$$\sum_i r x_i \otimes_S \lambda_i = \sum_i x_i \otimes_S \lambda_i r, \text{ for all } r \in R\}$$

$$\underline{\text{Nat}}(F G_2, 1_{S}\mathcal{M}) \cong \underline{\text{Nat}}(F' G_2', 1_{\mathcal{M}_S})$$
$$\cong {}_S\text{Hom}_S(\Lambda \otimes_R X, S) \cong {}_R\text{Hom}_S(X, {}_S\text{Hom}(\Lambda, S))$$
$$\underline{\text{Nat}}(1_{S}\mathcal{M}, F G) \cong \{e \in \Lambda \otimes_R {}_S\text{Hom}(\Lambda, S) \mid se = es, \text{ for all } s \in S\}$$

Proof. For a natural transformation $\alpha : 1_{R}\mathcal{M} \to G_2 F$, consider $\sum_i x_i \otimes_S \lambda_i = \alpha_R(1_R)$. The naturality of α implies that $\sum_i x_i \otimes_S \lambda_i \in C_R(X \otimes_S \Lambda)$. Conversely, given $\sum_i x_i \otimes_S \lambda_i \in C_R(X \otimes_S \Lambda)$, we consider $\alpha \in \underline{\text{Nat}}(1_{\mathcal{M}_R}, G_2 F)$ given by

$$\alpha_M : M \to G_2 F(M) = X \otimes_S \Lambda \otimes_R M, \quad \alpha_M(m) = \sum_i x_i \otimes_S \lambda_i \otimes_R m$$

Given a natural transformation $\beta : F G_2 = \Lambda \otimes_R X \otimes_S \bullet \to 1_{S}\mathcal{M}$, we take

$$\tilde{\beta} = \beta_S : \Lambda \otimes_R X \to S$$

$\tilde{\beta}$ is right S-linear because β is natural. Conversely, given an S-bimodule map $\tilde{\beta} : \Lambda \otimes_R X \to S$, we define a natural transformation β by

$$\beta_N : \Lambda \otimes_R X \otimes_S N \to N; \quad \beta_N = \tilde{\beta} \otimes_S I_N$$

The corresponding map $\Delta : X \to {}_S\text{Hom}(\Lambda, S)$ is given by

$$\Delta(x)(\lambda) = \tilde{\beta}(\lambda \otimes x)$$

Finally consider a natural transformation $\gamma : 1_{S}\mathcal{M} \to F G$, and define $e = \gamma_S(1_S) \in F G(S) = \Lambda \otimes_R {}_S\text{Hom}(\Lambda, S)$. The fact that $es = se$ follows easily from the naturality of γ.

Given $e = \sum_i \lambda_i \otimes_R \varphi_i$, with $es = se$ for all $s \in S$, we define a natural transformation γ as follows:

$$\gamma_N : N \to \Lambda \otimes_R {}_S\text{Hom}(\Lambda, N), \quad \gamma_N(n) = \sum_i \lambda_i \otimes_R \varphi_i \cdot n$$

Given $\varphi \in {}_S\text{Hom}(\Lambda, S)$ and $n \in N$, $\varphi \cdot n \in {}_S\text{Hom}(\Lambda, N)$ is defined by

$$(\lambda)(\varphi \cdot n) = ((\lambda)\varphi)n$$

We leave it to the reader to show that γ_N is left S-linear, and that γ is natural.

The following is an extended version of [141, Theorem 2.1]. We include a very short proof, based on Lemma 11.

Theorem 32. *Let R and S be rings, $\Lambda \in {}_S\mathcal{M}_R$ and $X \in {}_R\mathcal{M}_S$. Then the following are equivalent.*

1. $(F = \Lambda \otimes_R \bullet, G_2 = X \otimes_S \bullet) : {}_R\mathcal{M} \to {}_S\mathcal{M}$ *is an adjoint pair of functors;*
2. $(F' = \bullet \otimes_R X, G'_2 = \bullet \otimes_S \Lambda) : \mathcal{M}_R \to \mathcal{M}_S$ *is an adjoint pair of functors;*
3. $G = {}_S\mathrm{Hom}(\Lambda, \bullet)$ *and* $G_2 = X \otimes_S \bullet$ *are naturally isomorphic;*
4. $G' = \mathrm{Hom}_S(X, \bullet)$ *and* $G'_2 = \bullet \otimes_S \Lambda$ *are naturally isomorphic;*
5. Λ *is finitely generated projective as a left S-module, and*

$$X \cong {}_S\mathrm{Hom}(\Lambda, S) \quad \text{in} \quad {}_R\mathcal{M}_S$$

6. X *is finitely generated projective as a right S-module, and*

$$\Lambda \cong \mathrm{Hom}_S(X, S) \quad \text{in} \quad {}_S\mathcal{M}_R$$

7. *there exists $z = \sum_i x_i \otimes_S \lambda_i \in C_R(X \otimes_S \Lambda)$ and $\omega : \Lambda \otimes_R X \to S$ in ${}_S\mathcal{M}_S$ such that*

$$\lambda = \sum_i \omega(\lambda \otimes x_i)\lambda_i \tag{3.37}$$

$$x = \sum_i x_i\omega(\lambda_i \otimes x) \tag{3.38}$$

for all $x \in X$ and $\lambda \in \Lambda$;

8. *the same condition as 7), but with $z = \sum_i x_i \otimes_S \lambda_i \in X \otimes_S \Lambda$;*

9. *there exist $\Delta : R \to X \otimes_S \Lambda$ in ${}_R\mathcal{M}_R$ and $\varepsilon : X \to {}_S\mathrm{Hom}(\Lambda, S)$ in ${}_R\mathcal{M}_S$ such that, with $\Delta(1_R) = \sum x_i \otimes \lambda_i$,*

$$\lambda = \sum_i (\lambda)(\varepsilon(x_i))\lambda_i \tag{3.39}$$

$$x = \sum_i x_i(\lambda_i)(\varepsilon(x)) \tag{3.40}$$

for all $x \in X$ and $\lambda \in \Lambda$;

10. *there exist $\Delta : R \to X \otimes_S \Lambda$ in ${}_R\mathcal{M}_R$ and $\varepsilon' : \Lambda \to \mathrm{Hom}_S(X, S)$ in ${}_S\mathcal{M}_R$ such that, with $\Delta(1_R) = \sum x_i \otimes \lambda_i$,*

$$\lambda = \sum_i \varepsilon'(\lambda)(x_i)\lambda_i \tag{3.41}$$

$$x = \sum_i x_i\varepsilon'(\lambda_i)(x_i) \tag{3.42}$$

for all $x \in X$ and $\lambda \in \Lambda$;

11. *the same as 9), but we require that ε is an isomorphism;*

12. *the same as 10), but we require that ε' is an isomorphism;*

13. *there exists $\sum_i x_i \otimes_S \lambda_i \in C_R(X \otimes_S \Lambda)$ such that the map*

$$\widetilde{\varepsilon}: {}_S\mathrm{Hom}(\Lambda, S) \to X, \quad \widetilde{\varepsilon}(g) = \sum_i x_i((\lambda_i)g)$$

is an isomorphism in ${}_R\mathcal{M}_S$, and the following condition holds for $\lambda, \lambda' \in \Lambda$:

$$g(\lambda) = g(\lambda'), \text{ for all } g \in {}_S Hom(\Lambda, S) \implies \lambda = \lambda'$$

14. *there exists $\sum_i x_i \otimes_S \lambda_i \in C_R(X \otimes_S \Lambda)$ such that the map*

$$\widetilde{\varepsilon}': \mathrm{Hom}_S(X, S) \to \Lambda, \quad \widetilde{\varepsilon}'(g) = \sum_i g(x_i)\lambda_i$$

is an isomorphism in ${}_R\mathcal{M}_S$, and the following condition holds for $x, x' \in X$:

$$g(x) = g(x'), \text{ for all } g \in \mathrm{Hom}_S(X, S) \implies x = x'$$

Proof. 1. $\Rightarrow$ 7. Let (F, G_2) be an adjoint pair, and let $\eta : 1_{R\mathcal{M}} \to G_2 F$ and $\eta : FG_2 \to 1_{S\mathcal{M}}$ be the unit and counit of the adjunction, and take $\sum_i x_i \otimes_S \lambda_i$ as in Lemma 11. (3.37) and (3.38) follow immediately from the adjointness property of the unit and the counit.

1. $\Leftrightarrow$ 3. follows from the uniqueness of adjoints.

7. $\Rightarrow$ 1. The natural transformations η and ε corresponding to $\sum_i x_i \otimes \lambda_i$ and ω are the unit and counit of the adjunction.

3. $\Rightarrow$ 5. (3.37) tells us that $\{\lambda_i, \omega(\bullet \otimes_R x_i)\}$ is a finite dual basis for Λ as a left S-module.

Let $\gamma_2 : G_2 \to G$ be a natural isomorphism. Obviously $\gamma = \gamma_{2,S} : X \to {}_S\mathrm{Hom}(\Lambda, S)$ is an isomorphism in ${}_R\mathcal{M}$, and we are done if we can show that γ is right S-linear. This follows essentially from the naturality of γ. For any $t \in S$, we consider the map $f_t : S \to S$, $f_t(s) = st$. $f_t \in {}_S\mathcal{M}$, and we have a commutative diagram

$$
\begin{array}{ccc}
X = X \otimes_S S & \xrightarrow{I_X \otimes_S f_t} & X = X \otimes_S S \\
\gamma \downarrow & & \downarrow \gamma \\
{}_S\mathrm{Hom}(\Lambda, S) & \xrightarrow{{}_S\mathrm{Hom}(\Lambda, f_t)} & {}_S\mathrm{Hom}(\Lambda, S)
\end{array}
$$

Now observe that $(I_X \otimes_S f_t)(x) = xt$, and ${}_S\mathrm{Hom}(\Lambda, f_t) = f_t \circ$ -, and the commutativity of the diagram implies that

$$(\lambda)(\gamma(xt)) = (\lambda)(f_t \circ \gamma(x)) = ((\lambda)(\gamma(x)))t = (\lambda)(\gamma(x)t)$$

and γ is right S-linear.

5. $\Rightarrow$ 3. Λ is finitely generated and projective as a left S-module, so we have a natural isomorphism

$$\gamma : \ _S\mathrm{Hom}(\Lambda, S) \otimes \bullet \to \ _S\mathrm{Hom}(\Lambda, \bullet)$$

and from 5. it also follows that

$$_S\mathrm{Hom}(\Lambda, S) \otimes \bullet \cong X \otimes_S \bullet$$

7. $\Rightarrow$ 8. is trivial.

8. $\Rightarrow$ 7. For all $r \in R$, we have

$$rz = \sum_i rx_i \otimes_S \lambda_i$$

$$(3.38) \quad = \sum_{i,j} x_j \omega(\lambda_j \otimes_R rx_i) \otimes_S \lambda_i$$

$$= \sum_{i,j} x_j \otimes_S \omega(\lambda_j r \otimes_R x_i)\lambda_i$$

$$(3.37) \quad = \sum_j x_j \otimes_S \lambda_j r = zr$$

7. $\Rightarrow$ 9. ε is defined by $(\lambda)(\varepsilon(x)) = \omega(\lambda \otimes_S x)$. It is easy to show that ε is left R-linear and right S-linear, and (3.39) and (3.40) follow from (3.37) and (3.38).

9. $\Rightarrow$ 11. The inverse of ε is given by

$$\varepsilon^{-1}(g) = \sum_i x_i((\lambda_i)g)$$

11. $\Rightarrow$ 7. Define ω by $\omega(\lambda \otimes_S x) = (\lambda(\varepsilon(x)))$ and put $\sum_i x_i \otimes_R \lambda_i = \Delta(1_R)$.

9. $\Rightarrow$ 13. It is clear that $\widetilde{\varepsilon}$ is a morphism in $_R\mathcal{M}_S$, and form the proof of 9. $\Rightarrow$ 11., it follows that ε is the inverse of $\widetilde{\varepsilon}$. Assume that $(\lambda)g = (\lambda')g$ for all $g \in \ _S\mathrm{Hom}(\Lambda, S)$. Using (3.39), we find

$$\lambda = \sum_i (\lambda)(\varepsilon(x_i))\lambda_i = \sum_i (\lambda')(\varepsilon(x_i))\lambda_i = \lambda'$$

13. $\Rightarrow$ 9. Assume that $\widetilde{\varepsilon}$ has an inverse, and call it ε. Then for all $x \in X$, we have

$$x = \widetilde{\varepsilon}(\varepsilon(x)) = \sum_i x_i((\lambda_i)\varepsilon(x))$$

and (3.39) follows. For all $g \in \ _S\mathrm{Hom}(\Lambda, S)$, we have

$$(\lambda)g = (\lambda)\Big(\varepsilon(\tilde{\varepsilon}(g))\Big)$$
$$= (\lambda)\Big(\varepsilon(\sum_i x_i((\lambda_i)g))\Big)$$
$$= \sum_i \Big((\lambda)(\varepsilon(x_i))\Big)(\lambda_i)g$$
$$= \Big((\lambda)(\varepsilon(x_i))\lambda_i\Big)g$$

and (3.40) follows also.

The proof of the implications 2. $\Leftrightarrow$ 7. $\Rightarrow$ 10. $\Rightarrow$ 12. $\Rightarrow$ 7., 3. $\Leftrightarrow$ 4. $\Leftrightarrow$ 6. and 10. $\Leftrightarrow$ 14. is similar.

Remarks 4. 1. Theorem 32 together with the Eilenberg-Watts Theorem (cf. [11, Th. II.2.3]) shows that there is a one-to-one correspondence between adjoint pairs between $_R\mathcal{M}$ and $_S\mathcal{M}$, and between $\mathcal{M}_S$ and $\mathcal{M}_R$.

2. The adjunctions in Theorem 32 are equivalences if and only if the unit and counit are natural isomorphisms, and, using Lemma 11, we see that this means that the maps $\omega : \Lambda \otimes_R X \to S$ and $\Delta : R \to X \otimes_S \Lambda$ from part 7. of Theorem 32 have to be isomorphisms. Using (3.39) and (3.40), we see that $(R, S, X, \Lambda, \Delta^{-1}, \omega)$ is a strict Morita context. Thus we recover the classical result due to Morita that (additive) equivalences between module categories correspond to strict Morita contexts. For a detailed discussion of Morita contexts, we refer to [11, Ch. II].

Theorem 32 allows us to determine when the pair (F, G) is Frobenius: this is equivalent to one of the 14 equivalent conditions of the Theorem, combined with one of the 14 conditions, but applied in the situation where R and S, and Λ and X are interchanged. Using the Eilenberg-Watts Theorem once more, we find the following result.

Theorem 33. *[58] Let R and S be rings. There is a one-to-one correspondence between*

- *Frobenius functors between $_R\mathcal{M}$ and $_S\mathcal{M}$;*
- *Frobenius functors between $\mathcal{M}_R$ and $\mathcal{M}_S$;*
- *pairs (X, Λ), with $X \in {}_R\mathcal{M}_S$ and $\Lambda \in {}_S\mathcal{M}_R$ satisfying one of the following equivalent conditions:*

  1. *X is finitely generated and projective on both sides and*

     $$\Lambda \cong \mathrm{Hom}_S(X, S) \cong {}_R\mathrm{Hom}(X, R) \quad \text{in} \quad {}_S\mathcal{M}_R$$

  2. *Λ is finitely generated and projective on both sides and*

     $$X \cong {}_S\mathrm{Hom}(\Lambda, S) \cong \mathrm{Hom}_R(\Lambda, R) \quad \text{in} \quad {}_R\mathcal{M}_S$$

In fact Theorem 33 tells us that a Frobenius pair between $_R\mathcal{M}$ and $_S\mathcal{M}$ is of the type $(\Lambda \otimes_R \bullet, X \otimes_S \bullet)$, where Λ and X are Frobenius bimodules. Let us next look at separable bimodules.

Theorem 34. *Let R and S be rings, and Λ an (S, R)-bimodule. Then the functor $G = {}_S\mathrm{Hom}(\Lambda, \bullet) : {}_S\mathcal{M} \to {}_R\mathcal{M}$ is separable if and only if S is Λ-separable over R.*

Proof. An immediate application of the third part of Lemma 11 and Rafael's Theorem 24.

Remarks 5. 1. It is surprising that there is no algebraic interpretation for the functor $F = \Lambda \otimes_R \bullet$ to be separable - unless Λ is finitely generated and projective as a left R-module, in which case F is also a Hom functor.
2. Let $i : R \to S$ be a ring homomorphism, and let $\Lambda = S$ considered as an (S, R)-bimodule, and $X = S$ considered as an (R, S)-bimodule. Obviously Λ is finitely generated and projective as a left S-module, and $G = G_2$ is the restriction of scalars functors. We then recover some results of the first part of this Section.

3.3 The functor forgetting the C-coaction

Let (A, C, ψ) be a right-right entwining structure. Using the methods developed in the previous Section, we want to study the functor $F : \mathcal{M}(\psi)_A^C \to \mathcal{M}_A$ and its right adjoint. This can be done directly, cf. [30]. But it turns out to be more efficient to start with a more general situation: let A be a ring, and C an A-coring, and look at the forgetful functor

$$F : \mathcal{M}^C \to \mathcal{M}_A$$

Proposition 65. *F has a right adjoint G. For $N \in \mathcal{M}_A$, $G(N) = N \otimes_A C$, with structure*

$$(n \otimes_A c)a = n \otimes_A ca \quad \text{and} \quad \rho^r(n \otimes_A c) = n \otimes_A c_{(1)} \otimes_A c_{(2)}$$

For a morphism $f \in \mathcal{M}_A$, $G(f) = f \otimes_A I_C$.

Proof. One easily verifies that G is a functor. The unit and the counit of the adjunction are defined as follows:

$$\eta : 1_{\mathcal{M}^C} \to GF \quad \eta_M = \rho^r : M \to M \otimes_A C \quad \eta_M(m) = m_{[0]} \otimes_A m_{[1]}$$
$$\varepsilon : FG \to 1_{\mathcal{M}_A} \quad \varepsilon_N = I_N \otimes_A \varepsilon_C : N \otimes_A C \to N \quad \varepsilon_N(n \otimes_A c) = n\varepsilon_C(c)$$

(3.6) and (3.7) are easily verified.

We will now describe $V = \underline{\mathrm{Nat}}(GF, 1_{\mathcal{M}^C})$ and $W = \underline{\mathrm{Nat}}(1_{\mathcal{M}_A}, FG)$. First we need a Lemma.

Lemma 12. *Take $\nu \in V$, and $N \in \mathcal{M}_A$, $G(N) = N \otimes_A C$. Then*

$$\nu_{N \otimes_A C} = I_N \otimes_A \nu_C$$

Proof. For $n \in N$, we consider $f_n : C \to N \otimes_A C$, $f_n(c) = n \otimes_A c$. f_n is a morphism in $\mathcal{M}^C$, and the naturality of ν produces a commutative diagram

$$
\begin{array}{ccc}
C \otimes_A C & \xrightarrow{\ \nu_C\ } & C \\
\downarrow{\scriptstyle f_n \otimes_A I_C} & & \downarrow{\scriptstyle f_n} \\
N \otimes_A C \otimes_A C & \xrightarrow{\ \nu_{N \otimes_A C}\ } & N \otimes_A C
\end{array}
$$

and we find

$$\nu_{N \otimes_A C}(n \otimes_A c) = n \otimes_A \nu_C(c)$$

Proposition 66. *Let C be an A-coring, and put*

$$V_1 = {}^C\mathrm{Hom}^C(C \otimes_A C, C)$$

$$V_2 = \{\theta \in {}_A\mathrm{Hom}_A(C \otimes_A C, A) \mid c_{(1)}\theta(c_{(2)} \otimes_A d) = \theta(c \otimes_A d_{(1)})d_{(2)}, \ \forall c, d \in C\}$$

Then

$$V = \underline{\mathrm{Nat}}(GF, 1_{\mathcal{M}^C}) \cong V_1 \cong V_2$$

Proof. We define $\alpha : V \to V_1$, $\alpha(\nu) = \bar{\nu} = \nu_C$. By definition, $\bar{\nu}$ is right A-linear and a right C-comodule map. The properties on the left-hand side follow from the naturality of ν: for all $a \in A$, we consider the map $f_a : C \to C$, $f_a(c) = ac$, which is a morphism in $\mathcal{M}^C$, so that the naturality of ν gives a commutative diagram

$$
\begin{array}{ccc}
C \otimes_A C & \xrightarrow{\ \bar{\nu}\ } & C \\
\downarrow{\scriptstyle f_a \otimes_A I_C} & & \downarrow{\scriptstyle f_a} \\
C \otimes_A C & \xrightarrow{\ \bar{\nu}\ } & C
\end{array}
$$

Appying the diagram to $c \otimes_A d \in C \otimes_A C$, we find

$$\bar{\nu}(ac \otimes_A d) = a\bar{\nu}(c \otimes_A d)$$

and $\bar{\nu}$ is left A-linear.

The left C-comodule structure map on $C \otimes_A C$ is $\Delta_C \otimes I_C$, and this map is a morphism in $\mathcal{M}^C$, so that we have another commutative diagram

$$
\begin{array}{ccc}
C \otimes_A C & \xrightarrow{\ \bar{\nu}\ } & C \\
\downarrow{\scriptstyle \Delta_C \otimes_A I_C} & & \downarrow{\scriptstyle \Delta_C} \\
C \otimes_A C \otimes_A C & \xrightarrow{\ \nu_{C \otimes_A C}\ } & C \otimes_A C
\end{array}
$$

Taking into account that $\nu_{C \otimes_A C} = I_C \otimes_A \overline{\nu}$ (Lemma 12), we find that $\overline{\nu}$ is a left C-comodule map, and α is well-defined.

Next, we define $\alpha_1 : V_1 \to V_2$:

$$\alpha_1(\overline{\nu}) = \theta = \varepsilon_C \circ \overline{\nu}$$

It is obvious that θ is an (A, A)-bimodule map. From the fact that $\overline{\nu}$ is a C-bicomodule map, we find

$$c_{(1)} \otimes_A \overline{\nu}(c_{(2)} \otimes_A d) = \Delta_C(\overline{\nu}(c \otimes_A d)) = \overline{\nu}(c \otimes_A d_{(1)}) \otimes_A d_{(2)}$$

Applying $I_C \otimes_A \varepsilon_C$ to the first equality and $\varepsilon_C \otimes_A I_C$ to the second one, we find

$$\overline{\nu}(c \otimes_A d) = c_{(1)}\theta(c_{(2)} \otimes_A d) = \theta(c \otimes_A d_{(1)})d_{(2)} \qquad (3.43)$$

and it follows that α_1 is well-defined. We define $\alpha_1^{-1}(\theta) = \overline{\nu}$ defined by (3.43). It is clear from (3.43) and the fact that θ is an (A, A)-bimodule map, that $\overline{\nu}$ is a morphism in $^C\mathcal{M}^C$; (3.43) also tells us that $\alpha_1^{-1}(\alpha_1(\overline{\nu})) = \overline{\nu}$. Conversely,

$$\alpha_1(\alpha_1^{-1}(\theta))(c \otimes_A d) = \varepsilon_C(c_{(1)})\theta(c_{(2)} \otimes_A d) = \theta(c \otimes_A d)$$

We still need to show that α is invertible. For $\overline{\nu} \in V_1$, and $\theta = \alpha_1(\overline{\nu})$, we define

$$\nu = \alpha^{-1}(\overline{\nu}) : GF \to 1_C$$

by

$$\nu_M : M \otimes_A C \to M \ ; \ \nu_M(m \otimes_A c) = m_{[0]}\theta(m_{[1]} \otimes_A c)$$

ν is natural, since for every morphism $f : M \to M'$ in $\mathcal{M}^C$, we have that

$$\nu_{M'}(f \otimes_A I_C)(m \otimes_A c) = \nu_{M'}(f(m) \otimes_A c) = f(m)_{[0]}\theta(f(m)_{[1]} \otimes_A c)$$
$$= f(m_{[0]})\theta(m_{[1]} \otimes_A c) = f(m_{[0]}\theta(m_{[1]} \otimes_A c)) = f(\nu_M(m \otimes_A c))$$

It is clear that $\alpha(\alpha^{-1}(\overline{\nu})) = \overline{\nu}$, since

$$\nu_C(c \otimes_A d) = c_{(1)}\theta(c_{(2)} \otimes_A d) = \overline{\nu}(c \otimes_A d)$$

Finally, let us show that $\alpha^{-1}(\alpha(\nu)) = \nu$. The map $\rho^r : M \to M \otimes_A C$ is in $\mathcal{M}^C$. From Lemma 12, we know that $\nu_{M \otimes_A C} = I_M \otimes_A \overline{\nu}$, so the naturality of ν generates a commutative diagram

$$
\begin{array}{ccc}
M \otimes_A C & \xrightarrow{\ \nu_M\ } & M \\
{\scriptstyle \rho^r \otimes_A I_C}\downarrow & & \downarrow{\scriptstyle \rho^r} \\
M \otimes_A C \otimes_A C & \xrightarrow{\ I_M \otimes_A \overline{\nu}\ } & M \otimes_A C
\end{array}
$$

and we find

$$\rho^r(\nu_M(m \otimes_A c)) = m_{[0]} \otimes_A \overline{\nu}(m_{[1]} \otimes_A c)$$

Apply ε_C to the second factor:

$$\nu_M(m \otimes_A c) = m_{[0]}\theta(m_{[1]} \otimes_A c)$$

This means precisely that $\alpha^{-1}(\alpha(\nu)) = \nu$.

Remark 7. The multiplication in V_2 induced by the multiplication on V can be described in several ways:

$$\begin{aligned}(\theta \cdot \theta')(c \otimes_A d) &= \theta(c_{(1)} \otimes_A c_{(2)})\theta'(c_{(3)} \otimes_A d) \\ &= \theta'(c \otimes_A d_{(1)})\theta(d_{(2)} \otimes_A d_{(3)})\end{aligned} \tag{3.44}$$

Proposition 67. *For an A-coring C, we have*

$$W = \underline{\mathrm{Nat}}(1_{\mathcal{M}_A}, FG) \cong W_1 = {}_A\mathrm{Hom}_A(A, C)$$
$$\cong W_2 = \{z \in C \mid az = za \text{ for all } a \in A\}$$

Proof. We give the definitions of the connecting maps; other details are left to the reader.

$$\beta : W \to W_1 \ ; \ \beta(\zeta) = \zeta_A = \overline{\zeta}$$
$$\beta_1 : W_1 \to W_2 \ ; \ \beta_1(\overline{\zeta}) = \overline{\zeta}(1) = z$$
$$\beta^{-1} : W_1 \to W \ ; \ \beta^{-1}(\overline{\zeta}) = \zeta$$

with

$$\zeta_N : N \to N \otimes_A C \ ; \ \zeta_N(n) = n \otimes_A \overline{\zeta}(1)$$

Theorem 35. *Let A be a ring, and C an A-coring.*

1. *The following assertions are equivalent:*
   a. *$F : \mathcal{M}^C \to \mathcal{M}_A$ is separable;*
   b. *$F' : {}^C\mathcal{M} \to {}_A\mathcal{M}$ is separable;*
   c. *there exists $\theta \in V_2$ such that $\theta \circ \Delta_C = \varepsilon_C$, i.e.*

   $$\theta(c_{(1)} \otimes_A c_{(2)}) = \varepsilon_C(c) \tag{3.45}$$

   for all $c \in C$.
2. *The following assertions are equivalent:*
   a. *$G = \bullet \otimes_A C : \mathcal{M}_A \to \mathcal{M}^C$ is separable;*
   b. *$G' = C \otimes_A \bullet : {}_A\mathcal{M} \to {}^C\mathcal{M}$ is separable;*
   c. *there exists $z \in W_2$ such that $\varepsilon_C(z) = 1_A$.*
3. *The following assertions are equivalent:*
   a. *(F, G) is a Frobenius pair;*
   b. *(F', G') is a Frobenius pair;*

c. *there exists* $\theta \in V_2$ *and* $z \in W_2$ *such that*

$$\theta(z \otimes_A c) = \theta(c \otimes_A z) = \varepsilon_{\mathcal{C}}(c) \qquad (3.46)$$

for all $c \in \mathcal{C}$.

Proof. This is an immediate application of Rafael's Theorem 24, Theorem 23, and Propositions 66 and 67. The equivalence of a. and b. in each case follows from the fact that c. is left-right symmetric. To illustrate the method, we provide the proof of a.$\Rightarrow$c. in part 3. of the Theorem.

If (F, G) is Frobenius, then Theorem 23 gives us $\nu \in V$ and $\zeta \in W$ such that (3.1) and (3.2) hold. Let $\theta = \alpha_1(\alpha(\nu))$ and $z = \beta_1(\beta(\zeta))$ be the corresponding maps in V_2 and W_2. (3.1), with $C = \mathcal{C}$ gives us

$$c = c_{(1)}\theta(c_{(2)} \otimes_A z)$$

Applying $\varepsilon_{\mathcal{C}}$ to this equality, we find $\varepsilon_{\mathcal{C}}(c) = \theta(c \otimes_A z)$. In a similar way, (3.2) implies $\varepsilon_{\mathcal{C}}(c) = \theta(z \otimes_A c)$.

Corollary 15. *We use the same notation as in Theorem 35. If (F, G) is a Frobenius pair, then $\mathcal{C}$ is finitely generated and projective as a (left and right) A-module.*

Proof. For all $c \in \mathcal{C}$, we have

$$c = c_{(1)}\varepsilon_{\mathcal{C}}(c_{(2)}) = c_{(1)}\theta(c_{(2)} \otimes_A z) = \theta(c \otimes_A z_{(1)})z_{(2)} \qquad (3.47)$$

and it follows that

$$\{z_{(2)}, \theta(\bullet \otimes_A z_{(1)})\}$$

is a finite dual basis for $\mathcal{C}$ as a left A-module. In a similar way

$$\{z_{(1)}, \theta(z_{(2)} \otimes_A \bullet)\}$$

is a finite dual basis for $\mathcal{C}$ as a right A-module.

Let us from now on assume that $\mathcal{C}$ is finitely generated and projective as a left A-module. We have seen in Corollary 1 that $R = {}_A\mathrm{Hom}(\mathcal{C}, A)$ is a ring, and that it can be viewed as an object in $\mathcal{M}^{\mathcal{C}} \cong \mathcal{M}_R$. This leads to alternative descriptions of V and W, and to new characterizations for the separability and Frobenius properties of F and G. As in Section 2.7, let

$$\{c_i, f_i \mid i = 1, \cdots, m\}$$

be a finite dual basis for $\mathcal{C}$ as a left A-module.

Proposition 68. *Let $\mathcal{C}$ be an A-coring which is finitely generated and projective as a left A-module. Then*

$$V \cong V_3 = {}_A\mathrm{Hom}^{\mathcal{C}}(\mathcal{C}, R)$$

Proof. $\alpha_2 : V_2 \to V_3$ is defined as follows:

$$\alpha_2(\theta) = \overline{\phi} \text{ with } \overline{\phi}(c) = \sum_i f_i \theta(c_i \otimes_A c)$$

Observe that

$$\overline{\phi}(c)(d) = \sum_i f_i(d)\theta(c_i \otimes_A c) = \theta(d \otimes_A c) \qquad (3.48)$$

Let us show that α_2 is well-defined. $\alpha_2(\theta) = \overline{\phi}$ is an (A, A)-bimodule map, since

$$(a\overline{\phi}(c))(d) = \overline{\phi}(c)(da) = \theta(da \otimes_A c)$$
$$= \theta(d \otimes_A ac) = \overline{\phi}(ac)(d)$$
$$(\overline{\phi}(c)a)(d) = (\overline{\phi}(c)(d))a = \theta(d \otimes_A c)a$$
$$= \theta(d \otimes_A ca) = \overline{\phi}(ca)(d)$$

In order to prove that $\overline{\phi}$ is a right C-comodule map, we have to show that

$$\overline{\phi}(c_{(1)}) \otimes_A c_{(2)} = \overline{\phi}(c)_{[0]} \otimes_A \overline{\phi}(c)_{[1]} \in R \otimes_A C$$

Using (2.96), we find, for all $c, d \in C$:

$$\overline{\phi}(c)_{[0]}(d)\overline{\phi}(c)_{[1]} = d_{(1)}\overline{\phi}(c)(d_{(2)}) = d_{(1)}\theta(d_{(2)} \otimes c)$$
$$= \theta(d \otimes c_{(1)})c_{(2)} = \overline{\phi}(c_{(1)})(d)c_{(2)}$$

as needed. For $\overline{\phi} \in V_3$, we define $\theta = \alpha_2^{-1}(\phi)$ by

$$\theta(d \otimes c) = \overline{\phi}(c)(d)$$

θ is an (A, A)-bimodule map, since

$$\theta(ad \otimes_A c) = \overline{\phi}(c)(ad) = a(\overline{\phi}(c)(d)) = a\theta(d \otimes_A c)$$
$$\theta(d \otimes_A ca) = \overline{\phi}(ca)(d) = (\overline{\phi}(c)a)(d)$$
$$= (\overline{\phi}(c)(d))a = \theta(d \otimes_A c)a$$

Furthermore

$$c_{(1)}\theta(c_{(2)} \otimes_A d) = c_{(1)}\overline{\phi}(d)(c_{(2)})$$
$$(2.96) \qquad = \overline{\phi}(d)_{[0]}(c)\overline{\phi}(d)_{[1]}$$
$$(\overline{\phi} \text{ is a } C \text{ comodule map}) \qquad = \overline{\phi}(d_{(1)})(c)(d_{(2)}) = \theta(c \otimes_A d_{(1)})d_{(2)}$$

and it follows that $\theta \in V_2$. It follows from (3.48) that α_2^{-1} is a left inverse for α_2; it is also a right inverse since

$$\alpha_2(\alpha_2^{-1}(\overline{\phi}))(c)(d) = \alpha_2^{-1}(\overline{\phi})(d \otimes c) = \overline{\phi}(c)(d)$$

Proposition 69. *Let C be an A-coring which is finitely generated and projective as a left A-module. Then*

$$W \cong W_3 = {}_A\mathrm{Hom}^C(R, C)$$

Proof. We define $\beta_2 : W_2 \to W_3$ by $\beta_2(z) = \phi$ with

$$\phi(f) = z_{(1)} f(z_{(2)})$$

ϕ is left and right A-linear since

$$
\begin{aligned}
\phi(af) &= z_{(1)}(af)(z_{(2)}) = z_{(1)} f(z_{(2)}a) \\
&= az_{(1)} f(z_{(2)}) = a\phi(f) \\
\phi(fa) &= z_{(1)}(fa)(z_{(2)}) = z_{(1)}(f(z_{(2)})a) \\
&= (z_{(1)} f(z_{(2)}))a = \phi(f)a
\end{aligned}
$$

We used the fact that, for $z \in V_2$,

$$az_{(1)} \otimes z_{(2)} = a\Delta_C(z) = \Delta_C(az) = \Delta_C(za) = z_{(1)} \otimes z_{(2)}a$$

Let us next show that ϕ is a right C-comodule map. For all $f \in R$, we have

$$
\begin{aligned}
\phi(f_{[0]}) \otimes_A f_{[1]} &= z_{(1)} f_{[0]}(z_{(2)}) \otimes_A f_{[1]} = z_{(1)} \otimes_A f_{[0]}(z_{(2)}) f_{[1]} \\
(2.96) \quad &= z_{(1)} \otimes_A z_{(2)} f(z_{(3)}) \\
(\rho^r \text{ is right } A\text{-linear}) \quad &= \rho^r(z_{(1)} f(z_{(2)})) = \rho^r(\phi(f))
\end{aligned}
$$

$\beta_2^{-1} : W_3 \to W_2$ is defined by

$$\beta_2^{-1}(\phi) = z = \phi(\varepsilon_C)$$

$z \in W_2$ since

$$az = a\phi(\varepsilon_C) = \phi(a\varepsilon_C) = \phi(\varepsilon_C a) = \phi(\varepsilon_C)a = za$$

where we used that

$$(a\varepsilon_C)(c) = \varepsilon_C(ca) = \varepsilon_C(c)a = (\varepsilon_C a)(c)$$

Finally, β_2 and β_2^{-1} are each others inverses, since

$$
\begin{aligned}
\beta_2^{-1}(\beta_2(z)) &= z_{(1)}\varepsilon_C(z_{(2)}) = z \\
\beta_2(\beta_2^{-1}(\phi))(f) &= \phi(\varepsilon_C)_{(1)} f(\phi(\varepsilon_C)_{(2)}) \\
&= \phi(\varepsilon_{C[0]}) f(\varepsilon_{C[1]}) \\
(2.95) \quad &= \phi(f_k) f(c_k) = \phi(f_k f(c_k)) = \phi(f)
\end{aligned}
$$

We can use V_3 and W_3 to give new criteria for F and G to be separable or Frobenius. We begin with the Frobenius property.

Theorem 36. *Let C be an A-coring, and let $F : \mathcal{M}^C \to \mathcal{M}_A$ be the functor forgetting the C-comodule structure, and G its right adjoint. Then the following assertions are equivalent.*

1. *(F, G) is a Frobenius pair;*
2. *C is finitely generated and projective as a left A-module, and there exist $\theta \in V_2$, $z \in W_2$ such that the maps*

$$\overline{\phi} : C \to R, \quad \overline{\phi}(c)(d) = \theta(d \otimes_A c)$$

$$\phi : R \to C, \quad \phi(f) = z_{(1)} f(z_{(2)})$$

   are each others inverses;
3. *C is finitely generated and projective as a right A-module, and there exist $\theta \in V_2$, $z \in W_2$ such that the maps*

$$\overline{\phi}' : C \to R', \quad \overline{\phi}(c)(d) = \theta(c \otimes_A d)$$

$$\phi' : R' \to C, \quad \phi'(f) = f(z_{(1)}) z_{(2)}$$

   are each others inverses;
4. *C is finitely generated and projective as a left A-module, and $C \cong R$ in ${}_A\mathcal{M}^C \cong {}_A\mathcal{M}_R$;*
5. *C is finitely generated and projective as a right A-module, and $C \cong R' = {}_A\mathrm{Hom}(C, A)$ in ${}^C\mathcal{M}_A \cong {}_R\mathcal{M}_A$;*
6. *C is finitely generated and projective as a left A-module, and R/A is Frobenius;*
7. *C is finitely generated and projective as a right A-module, and R'/A is Frobenius.*

Proof. <u>1. $\Rightarrow$ 2.</u> From Theorem 35, we know that there exist $\theta \in V_2$ and $z \in W_2$ satisfying (3.46). Put $\overline{\phi} = \alpha_2(\theta)$, $\phi = \beta_2(z)$. Then for all $f \in R$ and $c \in C$, we have

$$\overline{\phi}(\phi(f))(c) = \overline{\phi}(z_{(1)} f(z_{(2)}))(c) = \theta(c \otimes_A z_{(1)} f(z_{(2)}))$$

$$(\theta \text{ is right } A\text{-linear}) \quad = \theta(c \otimes_A z_{(1)}) f(z_{(2)})$$

$$(f \text{ is left } A\text{-linear}) \quad = f\big(\theta(c \otimes_A z_{(1)}) z_{(2)}\big)$$

$$(3.43) \quad = f\big(c_{(1)} \theta(c_{(2)} \otimes_A z)\big)$$

$$(3.46) \quad = f\big(c_{(1)} \varepsilon_C(c_{(2)})\big) = f(c)$$

and

$$\phi(\overline{\phi}(c)) = z_{(1)} \overline{\phi}(c)(z_{(2)}) = z_{(1)} \theta(z_{(2)} \otimes c)$$

$$(3.43), (3.46) \quad = \theta(z \otimes c_{(1)}) c_{(2)} = \varepsilon_C(c_{(1)}) c_{(2)} = c$$

<u>2. $\Rightarrow$ 4.</u> Obvious; we know from Proposition 38 that ${}_A\mathcal{M}^C \cong {}_A\mathcal{M}_R$.
<u>4. $\Rightarrow$ 1.</u> Let $\phi : R \to C$ in ${}_A\mathcal{M}^C$, with inverse $\overline{\phi}$. Put $\theta = \alpha_2^{-1}(\overline{\phi})$, $z = \beta_2^{-1}(\phi)$. We have to show that (3.46) holds. For all $c \in C$, we have

$$\theta(c \otimes_A z) = \overline{\phi}(\phi(\varepsilon_C))(c) = \varepsilon_C(c)$$

and

$$c = \phi(\overline{\phi}(c)) = z_{(1)}\overline{\phi}(c)(z_{(2)}) = z_{(1)}\theta(z_{(2)} \otimes_A c)$$

Applying ε_C to both sides, we find

$$\varepsilon_C(c) = \theta(z \otimes_A c)$$

4. $\Leftrightarrow$ 6. This follows from 3. in Theorem 28, after we remark that $C \cong \text{Hom}_A(R, A)$. The connecting isomorphism is given by sending $c \in C$ to $F_c : R \to A$, with $F_c(f) = f(c)$. F_c is *right* A-linear, since

$$F_c(fa) = (fa)(c) = f(c)a = F_c(f)a$$

The proof of the equivalences is similar: 3., 5. and 7. are the left-handed versions of 2., 4. and 6.

Now let us do the separability properties; we restrict ourselves to the right-handed versions.

Theorem 37. *Let C be an A-coring, and assume that C is finitely generated and projective as a left A-module.*

1. *The following assertions are equivalent:*
   a. *$F : \mathcal{M}^C \to \mathcal{M}_A$ is separable;*
   b. *there exists $\overline{\phi} \in V_3$ such that $\overline{\phi}(c_{(2)})(c_{(1)}) = \varepsilon_C(c)$;*
   c. *$G : \mathcal{M}_R \to \mathcal{M}_A$ is separable;*
   d. *R/A is separable.*
2. *The following assertions are equivalent:*
   a. *$G : \mathcal{M}_A \to \mathcal{M}^C$ is separable;*
   b. *there exists $\phi \in W_3$ such that $\varepsilon_C(\phi(\varepsilon_C)) = 1_A$;*
   c. *$F : \mathcal{M}_A \to \mathcal{M}_R$ is separable;*
   d. *R/A is a split extension.*

Proof. We give an outline of the proof of Part 1., leaving Part 2. to the reader. *a. $\Leftrightarrow$ b.* follows immediately from Theorem 35 and Proposition 68; *a. $\Leftrightarrow$ c.* follows from Proposition 38, and *c. $\Leftrightarrow$ d.* is Part 1. of Theorem 27.

We will now apply our results to categories of entwined modules. Let (A, C, ψ) be a right-right entwining structure, and put $C = A \otimes C$. First, we give explicit descriptions of the V_i and W_i.

Proposition 70. *Let (A, C, ψ) be a right-right entwining structure.*

1. *If ψ is invertible, then*

$$V_1 \cong V_4 = {}^C_A\text{Hom}^C_A(A \otimes C \otimes C, A \otimes C)$$

2. $V_2 \cong V_5$, *consisting of* $\vartheta \in \mathrm{Hom}(C \otimes C, A)$ *satisfying*

$$\vartheta(c \otimes d)a = a_{\psi\Psi}\vartheta(c^\Psi \otimes d^\psi) \tag{3.49}$$

$$\vartheta(c \otimes d_{(1)}) \otimes d_{(2)} = \vartheta(c_{(2)} \otimes d)_\psi \otimes c^\psi_{(1)} \tag{3.50}$$

for all $c, d \in C$ *and* $a \in A$.

3. *If* C *is finitely generated and projective as a* k-*module, then*

$$V_3 \cong V_6 = {}_A\mathrm{Hom}^C_A(A \otimes C, C^* \otimes A)$$

Proof. 1. As in Section 2.7, we identify $(A \otimes C) \otimes_A (A \otimes C)$ and $A \otimes C \otimes C$. $(a \otimes c) \otimes_A (b \otimes d)$ corresponds to $ab_\psi \otimes c^\psi \otimes d$. The two-sided structure induced on $A \otimes C \otimes C$ is the one presented in (2.80-2.82) (with $A' = A$, $C' = k$). 1. now follows immediately, using Proposition 34.

2. For $\theta : (A \otimes C) \otimes_A (A \otimes C) \cong A \otimes C \otimes C \to A$ in V_2, we put $\vartheta : C \otimes C \to A$, $\vartheta(c \otimes d) = \theta(1 \otimes c \otimes d)$. It is easily checked that $\vartheta \in V_5$.

3. follows after we observe that

$$R = {}_A\mathrm{Hom}(A \otimes C, A) \cong \mathrm{Hom}(C, A) \cong C^* \otimes A$$

The entwined structure on $C^* \otimes A$ induced by the one on R (see (2.91) and (2.95)) is

$$b(c^* \otimes a)b' = \sum_i \langle c^*, c^\psi_i \rangle c^*_i \otimes b_\psi ab' \tag{3.51}$$

$$\rho^r(c^* \otimes a) = \sum_i c^*_i * c^* \otimes a_\psi \otimes c^\psi_i \tag{3.52}$$

$\{1 \otimes c_i, c^*_i \otimes 1 \mid i = 1, \cdots, m\}$ is a dual basis of $A \otimes C$ as a left A-module. The ring structure on R translates into a ring structure (in fact a k-algebra structure) on $C^* \otimes A$. The multiplication we obtain is nothing else then the multiplication on the smash product $C^{*op} \#_R A$.

Proposition 71. *Let* (A, C, ψ) *be a right-right entwining structure.*

1. $W_2 \cong W_5$, *the submodule of* $A \otimes C$ *consisting of* $z = a^1 \otimes c^1 \in A \otimes C$, *satisfying*

$$aa^1 \otimes c^1 = a^1 a_\psi \otimes c^{1\psi} \tag{3.53}$$

2. *If* C *is finitely generated and projective as a* k-*module, then*

$$W_3 \cong W_6 = {}_A\mathrm{Hom}^C_A(C^* \otimes A, A \otimes C)$$

Proof. Similar to the proof of Proposition 70.

We can use Propositions 70 and 71 to reformulate the criteria for separability and Frobenius properties in the case of entwined modules.

Theorem 38. *Let* (A, C, ψ) *be a right-right entwining structure.*

1. *The forgetful functor* $F : \mathcal{M}(\psi)_A^C \to \mathcal{M}_A$ *is separable if and only if there exists* $\vartheta \in V_5$ *such that* $\vartheta \circ \Delta_C = \eta_A \circ \varepsilon_C$. *If* C *is finitely generated and projective as a* k-module, *this is equivalent to the separability of* $C^{*\mathrm{op}} \#_R A$ *over* A.
2. *The induction functor* $G : \mathcal{M}_A \to \mathcal{M}(\psi)_A^C$ *is separable if and only if there exists* $z = a^1 \otimes c^1 \in W_5$ *such that* $\varepsilon_C(c^1)a^1 = 1_A$. *If* C *is finitely generated and projective as a* k-module, *this is equivalent to* $C^{*\mathrm{op}} \#_R A / A$ *being a split extension.*
3. (F, G) *is a Frobenius pair if and only if there exists* $\vartheta \in V_5$ *and* $z = a^1 \otimes c^1 \in W_5$ *such that*

$$a^1 \vartheta(c^1 \otimes c) = (a^1)_\psi \vartheta(c^\psi \otimes c^1) = \varepsilon_C(c)1_A \qquad (3.54)$$

If C *is finitely generated and projective as a* k-module, *this is also equivalent to each of the following assertions:*

a. *there exist* $z = \sum a_l \otimes c_l \in W_5$ *and* $\vartheta \in V_5$ *such that the maps*

$$\phi : C^* \otimes A \to A \otimes C \quad and \quad \overline{\phi} : A \otimes C \to C^* \otimes A$$

given by

$$\phi(c^* \otimes a) = \sum a^1 a_\psi \otimes \langle c^*, c_{(2)}^1 \rangle (c_{(1)}^1)^\psi \qquad (3.55)$$

$$\overline{\phi}(a \otimes c) = \sum_i d_i^* \otimes a_\psi \vartheta(d_i^\psi \otimes c) \qquad (3.56)$$

are each others inverses;
b. $C^* \otimes A$ *and* $A \otimes C$ *are isomorphic as objects in* $_A \mathcal{M}(\psi)_A^C \cong {}_A \mathcal{M}_{C^{*\mathrm{op}} \#_R A}$;
c. $C^{*\mathrm{op}} \#_R A / A$ *is a Frobenius extension.*

If (F, G) is a Frobenius pair, then $A \otimes C$ is finitely generated and projective as a left A-module. In some cases, we can conclude that C is finitely generated and projective as a k-module.

Corollary 16. *Let* (A, C, ψ) *be a right-right entwining structure, and assume that* (F, G) *is a Frobenius pair.*

1. *If* A *is faithfully flat as a* k-module, *then* C *is finitely generated as a* k-module.
2. *If* A *is commutative and faithfully flat as a* k-module, *then* C *is finitely generated projective as a* k-module.
3. *If* k *is a field, then* C *is finite dimensional as a* k-vector space.
4. *If* $A = k$, *then* C *is finitely generated projective as a* k-module.

Proof. 1. Assume that $\vartheta \in V_5$ and $z \in W_5$ satisfy (3.54). The corresponding $\theta \in V_2$ is given by

$$\theta((a \otimes c) \otimes_A (b \otimes d)) = ab_\psi \theta(c^\psi \otimes d)$$

Using (3.47), we find for all $d \in C$ that

$$1 \otimes d = \sum \theta((1 \otimes d) \otimes_A (a^1 \otimes c_{(1)}^1)) \otimes c_{(2)}^1$$
$$= a_\psi^1 \vartheta(d^\psi \otimes c_{(1)}^1) \otimes c_{(2)}^1$$

Let M be the k-module generated by the $c_{(2)}^1$. M is finitely generated, and $1 \otimes d \in A \otimes M$. Since A is faithfully flat, it follows that $d \in M$, hence $M = C$ is finitely generated.

2. From descent theory: if a k-module becomes finitely generated and projective after a faithfully flat commutative base extension, then it is itself finitely generated and projective.

3. Follows immediately from 1): since k is a field, A is faithfully flat as a k-module, and C is projective as a k-module.

4. Follows immediately from 2).

As a special case, we consider the situation where $A = k$ and $\psi = I_C$. Now

$$V_5 = \{\vartheta \in (C \otimes C)^* \mid \vartheta(c \otimes d_{(1)})d_{(2)} = \vartheta(c_{(2)} \otimes d)c_{(1)}\} \quad \text{and} \quad W_5 = C$$

Corollary 17. *Let C be a k-coalgebra.*

1. *The following assertions are equivalent:*
   a. $F : \mathcal{M}^C \to \mathcal{M}$ *is separable;*
   b. $F' : {}^C\mathcal{M} \to \mathcal{M}$ *is separable;*
   c. *there exists $\vartheta \in V_5$ such that $\vartheta \circ \Delta_C = \varepsilon_C$.*
2. *The following assertions are equivalent:*
   a. $G : \mathcal{M} \to \mathcal{M}^C$ *is separable;*
   b. $G' : \mathcal{M} \to {}^C\mathcal{M}$ *is separable;*
   c. *there exists $f \in C$ such that $\varepsilon_C(f) = 1$.*
3. *The following assertions are equivalent:*
   a. (F, G) *is a Frobenius pair;*
   b. (F', G') *is a Frobenius pair;*
   c. *there exist $\vartheta \in V_5$ and $f \in C$ such that*

$$\vartheta(f \otimes c) = \vartheta(c \otimes f) = \varepsilon(c)$$

   for all $c \in C$.
   If C is finitely generated and projective, then these conditions are also equivalent to
   d. *there exist $\vartheta \in V_5$ and $f \in C$ such that*

$$\phi : C^* \to C \quad \text{and} \quad \bar{\phi} : C \to C^*$$

   given by

$$\phi(c^*) = \langle c^*, f_{(2)} \rangle f_{(1)} \tag{3.57}$$

$$\langle \overline{\phi}(c), d \rangle = \vartheta(d \otimes c) \tag{3.58}$$

are each others inverses.

e. $C^* \cong C$ in $_{C^*}\mathcal{M}$;

f. C^* is a Frobenius extension of k.

Condition 1c. means that C is a coseparable coalgebra in the sense of [116]. Now we consider the following problem: let (A, C, ψ) be an entwining structure, and assume that C^*/k is Frobenius. When is the forgetful functor $\mathcal{M}(\psi)_A^C \to \mathcal{M}_A$ Frobenius? The following is a partial answer to this question.

Proposition 72. *Consider* $(A, C, \psi) \in \mathbb{E}_\bullet^\bullet(k)$, *with* ψ *invertible, and* C *finitely generated and projective as a k-module. We also assume that C^* is Frobenius, which implies that there exists $f \in C$ such that the map*

$$\phi_C : \ C^* \to C \ ; \ \ \phi_C(c^*) = \langle c^*, f_{(2)} \rangle f_{(1)}$$

is bijective. If

$$a \otimes f = \psi(f \otimes a) \tag{3.59}$$

for all $a \in A$, then $\mathcal{M}(\psi)_A^C \to \mathcal{M}_A$ is also Frobenius.

Proof. First of all we remark that (3.59) tells us that $f \otimes 1_A \in W_5$. Then we compute the map

$$\phi : \ C^* \otimes A \to A \otimes C$$

from part 3a. of Theorem 38, using (3.55):

$$\phi(c^* \otimes a) = a_\psi \otimes \langle c^*, f_{(2)} \rangle f_{(1)}^\psi$$

ϕ is bijective, since ϕ_C and ψ are bijective. The result is now an immediate application of Theorem 38.

Relative separability Consider the functor $G' : \mathcal{M}(\psi)_A^C \to \mathcal{M}^C$ and the following diagram of forgetful functors:

$$
\begin{array}{ccc}
\mathcal{M}(\psi)_A^C & \xrightarrow{\ F\ } & \mathcal{M}_A \\
{\scriptstyle G'}\Big\downarrow & & \Big\downarrow \\
\mathcal{M}^C & \xrightarrow{\hspace{2cm}} & {}_k\mathcal{M}
\end{array}
\tag{3.60}
$$

The problem is now the following: assume that we have a short exact sequence in $\mathcal{M}(\psi)_A^C$, which is split exact as a sequence of A-modules. When is it split exact as a sequence of C-comodules? This is an application of separability of the second kind as introduced at the end of Section 3.1: we have investigate when F is G'-separable. From Theorem 25, it follows that we have to examine the k-space of natural transformations

$$X = \underline{\mathrm{Nat}}(G'GF, G')$$

Adapting the proof of Proposition 70, we find the following.

Proposition 73. *Let (A, C, ψ) be a right-right entwining structure. We then have isomorphisms*

$$X \cong X_4 = {}_A\mathrm{Hom}_A(A \otimes C \otimes C, A \otimes C)$$
$$\cong X_5 = \{\vartheta : C \otimes C \to A \mid (3.50) \text{ holds}\}$$

Consequently F is G'-separable if and only if there exists a $\vartheta \in X_5$ such that $\vartheta \circ \Delta_C = \eta_A \circ \varepsilon_C$.

3.4 The functor forgetting the A-action

As in the previous Section, let (A, C, ψ) be a right-right entwining structure. We are now interested in the functor $G' : \mathcal{M}(\psi)^C_A \to \mathcal{M}^C$ and its left adjoint F'. It turns out that there exist dual versions of the results obtained in the previous Section. It is possible to obtain these results using *algebroids* over a coalgebra (see [26]), which can be viewed as the formal duals of corings over a ring. However, to avoid technical difficulties, we have chosen for the more direct approach, avoiding algebroids.

Recall that the unit and counit of the adjunction are given by the following formulas

$$\mu : F'G' \to 1_C \quad \text{and} \quad \eta : 1_{\mathcal{M}^C} \to G'F'$$
$$\mu_M : M \otimes A \to M; \quad \mu_M(m \otimes a) = ma$$
$$\eta_N : N \to N \otimes A; \quad \eta_N(n) = n \otimes 1$$

Lemma 13. *Let $M \in {}_A\mathcal{M}(\psi)^C_A$, $N \in {}^C\mathcal{M}(\psi)^C_A$. Then $F'G'(M) \in {}_A\mathcal{M}(\psi)^C_A$ and $G'F'(N) \in {}^C\mathcal{M}(\psi)^C_A$. The left structures are given by*

$$a(m \otimes b) = am \otimes b \quad \text{and} \quad \rho^l(n \otimes b) = \sum n_{[-1]} \otimes n_{[0]} \otimes b$$

for all $a, b \in A$, $m \in M$, $n \in N$. Furthermore μ_M is left A-linear, and ν_N is left C-colinear.

Now write $V' = \underline{\mathrm{Nat}}(G'F', 1_{\mathcal{M}^C})$, $W' = \underline{\mathrm{Nat}}(1_C, F'G')$. Following the philosophy of the previous Sections, we give more explicit descriptions of V' and W'. We will not give full detail of the proofs; the arguments are dual to the ones in the previous Section. We define

$$V'_1 = \{\vartheta \in (C \otimes A)^* \mid \vartheta(c_{(1)} \otimes a_\psi)c^\psi_{(2)} = \vartheta(c_{(2)} \otimes a)c_{(1)}, \text{ for all } c \in C, a \in A\}$$
$$\tag{3.61}$$

Proposition 74. *The map* $\alpha : V' \to V_1'$, $\alpha(\nu') = \varepsilon \circ \nu_C$ *is an isomorphism.*

Proof. We leave verification of the details to the reader; ν' is reconstructed from ϑ as follows: for $N \in \mathcal{M}^C$, $\nu_N' : N \otimes A \to N$ is given by

$$\nu_N'(n \otimes a) = \sum \vartheta(n_{[1]} \otimes a) n_{[0]}$$

Let $e : C \to A \otimes A$ be a k-linear map. We will use the notation (summation understood, as usual):
$$e(c) = e^1(c) \otimes e^2(c)$$

Let W_1' be the k-submodule of $\mathrm{Hom}(C, A \otimes A)$ consisting of maps e satisfying

$$e^1(c_{(1)}) \otimes e^2(c_{(1)}) \otimes c_{(2)} = e^1(c_{(2)})_\Psi \otimes e^2(c_{(2)})_\Psi \otimes c_{(1)}^{\psi\Psi} \qquad (3.62)$$
$$e^1(c) \otimes e^2(c)a = a_\psi e^1(c^\psi) \otimes e^2(c^\psi) \qquad (3.63)$$

Proposition 75. *The map* $\beta : W' \to W_1'$ *given by*

$$\beta(\zeta') = (\varepsilon \otimes I_A \otimes I_A) \circ \zeta_{A \otimes C}' \circ (\eta_A \otimes I_C)$$
$$= (I_A \otimes \varepsilon \otimes I_A) \circ \zeta_{C \otimes A}' \circ (I_C \otimes \eta_A)$$

is an isomorphism. Given $e \in W_1'$, $\zeta' = \beta^{-1}(e)$ *is recovered from* e *as follows: for* $M \in \mathcal{M}(\psi)_A^C$,

$$\zeta_M'(m) = \sum m_{[0]} e^1(m_{[1]}) \otimes e^2(m_{[1]})$$

Proof. Let us show that β is well-defined, leaving the other details to the reader. We have a commutative diagram

$$
\begin{array}{ccccc}
C & \xrightarrow{I_C \otimes \eta_A} & C \otimes A & \xrightarrow{\zeta_{C \otimes A}'} & C \otimes A \otimes A \\
\downarrow{\scriptstyle I_C} & & \downarrow{\scriptstyle \psi} & & \downarrow{\scriptstyle \psi \otimes I_A} \\
C & \xrightarrow{\eta_A \otimes I_C} & A \otimes C & \xrightarrow{\zeta_{A \otimes C}'} & A \otimes C \otimes A
\end{array}
$$

$\overline{\lambda} = \zeta_{C \otimes A}' \circ (I_C \otimes \eta_A)$ is left and right C-colinear. Write

$$\overline{\lambda}(c) = \sum_i c_i \otimes a_i \otimes a_i'$$

Then

$$c_{(1)} \otimes \overline{\lambda}(c_{(2)}) = \sum_i c_{i(1)} \otimes c_{i(2)} \otimes a_i \otimes a_i'$$

Applying ε to the second factor, we find

$$c_{(1)} \otimes e(c_{(2)}) = \overline{\lambda}(c)$$

Using the right C-colinearity of $\overline{\lambda}$, we find

$$\overline{\lambda}(c_{(1)}) \otimes c_{(2)} = \sum_i a_{i\psi} \otimes a'_{i\Psi} \otimes c_i^{\psi\Psi}$$

$$= e^1(c_{(2)})_\psi \otimes e^2(c_{(2)})_\Psi \otimes c_{(1)}^{\psi\Psi}$$

proving (3.62). $\underline{\lambda} = \zeta'_{A \otimes C} \circ (\eta_A \otimes I_C)$ is left and right A-linear, hence

$$e^1(c) \otimes e^2(c)a = (I_A \otimes \varepsilon \otimes I_A)(\zeta'_{A \otimes C}((1 \otimes c)a))$$

$$= (I_A \otimes \varepsilon \otimes I_A)(\zeta'_{A \otimes C}(a_\psi \otimes c^\psi))$$

$$= a_\psi(I_A \otimes \varepsilon \otimes I_A)(\zeta'_{A \otimes C}(1 \otimes c^\psi))$$

$$= a_\psi e^1(c^\psi) \otimes e^2(c^\psi)$$

proving (3.63).

Proposition 76. *Let (A, C, ψ) be a right-right entwining structure.*

1. *$F' = \bullet \otimes A : \mathcal{M}^C \to \mathcal{M}(\psi)_A^C$ is separable if and only if there exists $\vartheta \in V_1'$ such that*

$$\vartheta(c \otimes 1) = \varepsilon(c) \tag{3.64}$$

for all $c \in C$.

2. *$G' : \mathcal{M}(\psi)_A^C \to \mathcal{M}^C$ is separable if and only if there exists $e \in W_1'$ such that*

$$e^1(c)e^2(c) = \varepsilon(c)1 \tag{3.65}$$

for all $c \in C$.

3. *(F', G') is a Frobenius pair if and only if there exists $\vartheta \in V_1'$ and $e \in W_1'$ such that*

$$\varepsilon(c)1 = \vartheta(c_{(1)} \otimes e^1(c_{(2)}))e^2(c_{(2)}) \tag{3.66}$$

$$= \vartheta(c_{(1)}^\psi \otimes e^2(c_{(2)}))e^1(c_{(2)})_\psi \tag{3.67}$$

Proof. We will prove 3. If (F', G') is a Frobenius pair, then there exists $\nu' \in V'$ and $\zeta' \in W'$ such that (3.1) and (3.2) hold. We take $\vartheta \in V_1'$ and $e \in W_1'$ corresponding to ν' and ζ'. We write down (3.1) applied to $n \otimes 1$ with $n \in N \in \mathcal{M}^C$. This gives

$$n \otimes 1 = ((\nu'_N \otimes I_A) \circ \zeta'_{N \otimes A}))(n \otimes 1) = \vartheta(n_{[1]} \otimes e^1(n_{[2]}))n_{[0]} \otimes e^1(n_{[2]})) \tag{3.68}$$

Taking $N = C$, and $n = c$, and applying ε_C to the first factor, we find (3.66). Conversely, if we have $\vartheta \in V_1'$ and $e \in W_1'$ satisfying (3.66), then (3.68) is satisfied for all $N \in \mathcal{M}^C$, and (3.1) follows from the fact that $\nu'_N \otimes I_A$ and $\zeta'_{N \otimes A}$ are right A-linear.

Now we write down (3.1) applied to $m \in M \in \mathcal{M}(\psi)_A^C$. We find

$$m = \left(\nu'_{G'(M\otimes A)} \circ G'(\zeta'_M)\right)(m) = \theta(m^\psi_{[1]} \otimes e^2(m_{[2]}))m_{[0]}e^1(m_{[2]})_\psi \quad (3.69)$$

Take $M = C \otimes A$, $m = c \otimes 1$, and apply ε_C to the first factor. This gives (3.67). Conversely, if we have $\vartheta \in V'_1$ and $e \in W'_1$ satisfying (3.67), then we can show that (3.69) holds for all $M \in \mathcal{M}(\psi)^C_A$: apply (3.67) to the second and third factor in $m_{[0]} \otimes m_{[1]} \otimes 1$, and then apply ε_C to the second factor. Finally remark that (3.69) is equivalent to (3.1).

Inspired by the results in the previous Section, we ask the following question: assuming (F', G') is a Frobenius pair, when is A finitely generated projective as a k-module. We give a partial answer in the next Proposition. We will assume that ψ is bijective; in the Doi-Hopf case, this is true if the underlying Hopf algebra H has a twisted antipode. The inverse of ψ is then given by the formula

$$\psi^{-1}(a \otimes c) = c\bar{S}(a_{(1)}) \otimes a_{(0)}$$

Proposition 77. *Let (A, C, ψ) be a right-right entwining structure. With notation as above, assume that (F', G') is a Frobenius pair. If there exists $c \in C$ such that $\varepsilon(c) = 1$, and if ψ is invertible, with inverse $\varphi = \psi^{-1}$: $A \otimes C \to C \otimes A$, then A is finitely generated and projective as a k-module.*

Proof. Recall (Proposition 15) that (A, C, φ) is a left-left entwining structure. Now fix $c \in C$ such that $\varepsilon(c) = 1$. Then for all $a \in A$

$$
\begin{aligned}
a = \varepsilon(c)a &= \varepsilon(c^\varphi)a_\varphi \\
(3.66) \quad &= \vartheta\big((c^\varphi)_{(1)} \otimes e^1((c^\varphi)_{(2)})\big)e^2((c^\varphi)_{(2)})a_\varphi \\
(2.8) \quad &= \vartheta\big(c^\varphi_{(1)} \otimes e^1(c^\phi_{(2)})\big)e^2(c^\phi_{(2)})a_{\varphi\phi} \\
(3.63) \quad &= \vartheta\big(c^\varphi_{(1)} \otimes a_{\varphi\phi\psi}e^1(c^{\phi\psi}_{(2)})\big)e^2(c^{\phi\psi}_{(2)}) \\
(\varphi = \psi^{-1}) \quad &= \vartheta\big(c^\varphi_{(1)} \otimes a_\varphi e^1(c_{(2)})\big)e^2(c_{(2)})
\end{aligned}
$$

Now write

$$(I \otimes e)\Delta(c) = \sum_{i=1}^m c_i \otimes b_i \otimes a_i \in C \otimes A \otimes A$$

For $i = 1, \cdots, m$, we define $a_i* \in A^*$ by

$$\langle a_i^*, a\rangle = \vartheta\big(c^\varphi_{(1)} \otimes a_\varphi b_i\big)$$

Then $\{a_i, a_i^* \mid i = 1, \cdots, m\}$ is a finite dual basis of A as a k-module.

Assume from now on that A is finitely generated and projective with finite dual basis $\{a_i, a_i^* \mid i = 1, \cdots, m\}$. The proof of the next Lemma is straightforward, and therefore left to the reader.

Lemma 14. *Let (A, C, ψ) be a right-right entwining structure, and assume that A is finitely generated and projective as a k-module. Then $A^* \otimes C \in {}^C\mathcal{M}(\psi)^C_A$. The structure is given by the formulas*

$$(a^* \otimes c)b = \langle a^*, b_\psi a_i \rangle a_i^* \otimes c^\psi \tag{3.70}$$

$$\rho^r(a^* \otimes c) = a^* \otimes c_{(1)} \otimes c_{(2)} \tag{3.71}$$

$$\rho^l(a^* \otimes c) = \langle a^*, a_{i\psi} \rangle c_{(1)}^\psi \otimes a_i^* \otimes c_{(2)} \tag{3.72}$$

We will now give alternative descriptions of V' and W'.

Proposition 78. *Let (A, C, ψ) be a right-right entwining structure, and assume that A is finitely generated and projective as a k-module. Then we have an isomorphism*

$$\beta_1 : W_1' \to W_2' = {}^C\mathrm{Hom}^C_A(A^* \otimes C, C \otimes A)$$

$\beta_1(e) = \Omega$, with

$$\Omega(a^* \otimes c) = \langle a^*, e^1(c_{(2)})_\psi \rangle c_{(1)}^\psi \otimes e^2(c_{(2)})$$

$\beta_1^{-1}(\Omega) = e$ with

$$e(c) = \sum_i a_i \otimes (\varepsilon_C \otimes I_A) \Omega(a^* \otimes c)$$

Proof. We first prove that β_1 is well-defined.

a) $\beta_1(e) = \Omega$ is right A-linear: for all $a^* \in A^*$, $c \in C$ and $b \in A$, we have

$$\Omega((a^* \otimes c)b) = \sum_i \langle a^*, b_\psi a_i \rangle \Omega(a_i^* \otimes c^\psi)$$

$$= \sum_i \langle a^*, b_\psi a_i \rangle \langle a_i^*, e^1((c^\psi)_{(2)})_\Psi \rangle (c^\psi)_{(1)}^\Psi \otimes e^2((c^\psi)_{(2)})$$

$$= \langle a^*, b_\psi e^1((c^\psi)_{(2)})_\Psi \rangle (c^\psi)_{(1)}^\Psi \otimes e^2((c^\psi)_{(2)})$$

$$\overset{(2.3)}{=} \langle a^*, b_{\psi\psi'} e^1(c_{(2)}^\psi)_\Psi \rangle c_{(1)}^{\psi'\Psi} \otimes e^2(c_{(2)}^\psi)$$

$$\overset{(2.1)}{=} \langle a^*, \left(b_\psi e^1(c_{(2)}^\psi)\right)_\Psi \rangle c_{(1)}^\Psi \otimes e^2(c_{(2)}^\psi)$$

$$\overset{(3.63)}{=} \langle a^*, e^1(c_{(2)})_\Psi \rangle c_{(1)}^\Psi \otimes e^2(c_{(2)})b$$

$$= \Omega(a^* \otimes c)b$$

b) $\beta_1(e) = \Omega$ is right C-colinear: for all $a^* \in A^*$ and $c \in C$, we have

$$\rho^r(\Omega(a^* \otimes c)) = \langle a^*, e^1(c_{(2)})_\psi \rangle \rho^r(c_{(1)}^\psi \otimes e^2(c_{(2)}))$$

$$= \langle a^*, e^1(c_{(2)})_\psi \rangle (c_{(1)}^\psi)_{(1)} \otimes e^2(c_{(2)})_\Psi \otimes (c_{(1)}^\psi)_{(2)}^\Psi$$

$$\overset{(2.3)}{=} \langle a^*, e^1(c_{(3)})_{\psi\psi'} \rangle c_{(1)}^{\psi'} \otimes e^2(c_{(3)})_\Psi \otimes c_{(2)}^{\psi\Psi}$$

$$\overset{(3.62)}{=} \langle a^*, e^1(c_{(2)})_{\psi'} \rangle c_{(1)}^{\psi'} \otimes e^2(c_{(2)}) \otimes c_{(3)}$$

$$= \Omega(a^* \otimes c_{(1)}) \otimes c_{(2)}$$

c) $\beta_1(e) = \Omega$ is left C-colinear: for all $a^* \in A^*$ and $c \in C$, we have

$$\rho^l(\Omega(a^* \otimes c)) = \langle a^*, e^1(c_{(2)})_\psi \rangle (c^\psi_{(1)})_{(1)} \otimes (c^\psi_{(1)})_{(2)} \otimes e^2(c_{(2)})$$

$$(2.3) \quad = \langle a^*, e^1(c_{(3)})_{\psi\psi'} \rangle c^{\psi'}_{(1)} \otimes c^\psi_{(2)} \otimes e^2(c_{(3)})$$

$$= \sum_i \langle a^*, a_{i\psi} \rangle \langle a_i^*, e^1(c_{(3)})_{\psi'} \rangle c^{\psi'}_{(1)} \otimes c^\psi_{(2)} \otimes e^2(c_{(3)})$$

$$= \sum_i \langle a^*, a_{i\psi} \rangle c^\psi_{(1)} \otimes \Omega(a_i^* \otimes c_{(2)})$$

We leave it to the reader to show that $\beta_1^{-1}(\Omega) = e$ satisfies (3.62) and (3.63). Let us make clear that β_1 and β_1^{-1} are each others inverses.

$$\beta^{-1}(\beta(e))(c) = \sum_i a_i \otimes (\varepsilon_C \otimes I_A)\langle a_i^*, e^1(c_{(2)})_\psi \rangle c^\psi_{(1)} \otimes e^2(c_{(2)})$$

$$= \sum_i a_i \otimes \langle a_i^*, e^1(c_{(2)})_\psi \rangle \varepsilon_C(c^\psi_{(1)}) \otimes e^2(c_{(2)})$$

$$= \langle a_i^*, e^1(c) \rangle a_i \otimes e^2(c) = e^1(c) \otimes e^2(c)$$

$$\beta(\beta^{-1}(w))(a^* \otimes c) = \langle a^*, (a_i)_\psi \rangle c^\psi_{(1)} \otimes (\varepsilon_C \otimes I_A)\Omega(a_i^* \otimes c_{(2)})$$

$$= (I_C \otimes \varepsilon_C \otimes I_A)(\langle a^*, (a_i)_\psi \rangle c^\psi_{(1)} \otimes \Omega(a_i^* \otimes c_{(2)}))$$

$$= (I_C \otimes \varepsilon_C \otimes I_A)\rho^l(\Omega(a^* \otimes c))$$

$$= \Omega(a^* \otimes c)$$

At the last step, we used that

$$(I_C \otimes \varepsilon_C \otimes I_A)\rho^l(c \otimes a) = c \otimes a$$

for all $c \in C$ and $a \in A$.

Proposition 79. *Let (A, C, ψ) be a right-right entwining structure, and assume that A is finitely generated and projective as a k-module. Then we have an isomorphism*

$$\alpha_1 : V_1' \to V_2' = {}^C\mathrm{Hom}_A^C(C \otimes A, A^* \otimes C)$$

$\alpha_1(\vartheta) = \overline{\Omega}$, *with*

$$\overline{\Omega}(c \otimes a) = \langle \vartheta, c_{(1)} \otimes a_\psi a_i \rangle a_i^* \otimes c^\psi_{(2)}$$

$\alpha_1^{-1}(\overline{\Omega}) = \vartheta$ *with*

$$\vartheta(c \otimes a) = \langle \overline{\Omega}(c \otimes a), 1_A \otimes \varepsilon_C \rangle$$

Proof. We first show that α_1 is well-defined. Take $\vartheta \in V_1'$, and let $\alpha_1(\vartheta) = \overline{\Omega}$.

a) $\overline{\Omega}$ is right A-linear. For all $a, b \in A$ and $c \in C$, we have

$$\overline{\Omega}(c \otimes a)b = \sum_{i,j}\langle\vartheta, c_{(1)} \otimes a_\psi a_i\rangle\langle a_i^*, b_\Psi a_j\rangle a_j^* \otimes c_{(2)}^{\psi\Psi}$$

$$= \sum_{j}\langle\vartheta, c_{(1)} \otimes a_\psi b_\Psi a_j\rangle a_j^* \otimes c_{(2)}^{\psi\Psi}$$

$$(2.1) \quad = \sum_{j}\langle\vartheta, c_{(1)} \otimes (ab)_\psi a_j\rangle a_j^* \otimes c_{(2)}^{\psi}$$

$$= \overline{\Omega}(c \otimes ab)$$

b) $\overline{\Omega}$ is right C-colinear. For all $a \in A$ and $c \in C$, we have

$$\rho^r(\overline{\Omega}(c \otimes a)) = \vartheta(c_{(1)} \otimes a_\psi a_i)a_i^* \otimes (c_{(2)}^{\psi})_{(1)} \otimes (c_{(2)}^{\psi})_{(2)}$$

$$(2.3) \quad = \vartheta(c_{(1)} \otimes a_{\psi\Psi}a_i)a_i^* \otimes c_{(2)}^{\Psi} \otimes c_{(3)}^{\psi}$$

$$= \overline{\Omega}(c_{(1)} \otimes a_\psi) \otimes c_{(2)}^{\psi}$$

c) $\overline{\Omega}$ is left C-colinear. For all $a \in A$ and $c \in C$, we have

$$\rho^l(\overline{\Omega}(c \otimes a)) = \sum_{i,j}\vartheta(c_{(1)} \otimes a_\psi a_i)\langle a_i^*, a_{j\Psi}\rangle(c_{(2)}^{\psi})_{(1)}^{\Psi} \otimes a_j^* \otimes (c_{(2)}^{\psi})_{(2)}$$

$$(2.3) \quad = \sum_{i,j}\vartheta(c_{(1)} \otimes a_{\psi\psi'}a_i)\langle a_i^*, a_{j\Psi}\rangle c_{(2)}^{\psi'\Psi} \otimes a_j^* \otimes c_{(3)}^{\psi}$$

$$= \sum_{j}\vartheta(c_{(1)} \otimes a_{\psi\psi'}a_{j\Psi})c_{(2)}^{\psi'\Psi} \otimes a_j^* \otimes c_{(3)}^{\psi}$$

$$(2.1) \quad = \sum_{j}\vartheta(c_{(1)} \otimes (a_\psi a_j)_{\psi'})c_{(2)}^{\psi'} \otimes a_j^* \otimes c_{(3)}^{\psi}$$

$$(3.61) \quad = \sum_{j}\vartheta(c_{(2)} \otimes a_\psi a_j)c_{(1)} \otimes a_j^* \otimes c_{(3)}^{\psi}$$

$$= c_{(1)} \otimes \overline{\Omega}(c_{(2)} \otimes a)$$

Conversely, given $\overline{\Omega}$, we have to show that $\alpha_1^{-1}(\overline{\Omega}) = \vartheta$ satisfies (3.61). Fix $c \otimes a \in C \otimes A$, and write

$$\overline{\Omega}(c \otimes a) = \sum_{l} b_l^* \otimes d_l \in A^* \otimes C$$

$\overline{\Omega}$ is right C-colinear, so

$$\overline{\Omega}(c_{(1)} \otimes a_\psi) \otimes c_{(2)}^{\psi} = \sum_{l} b_l^* \otimes d_{l(1)} \otimes d_{l(2)}$$

$\overline{\Omega}$ is left C-colinear, so

$$c_{(1)} \otimes \overline{\Omega}(c_{(2)} \otimes a) = \sum_l \langle b_l^*, a_{i\psi} \rangle d_{l(1)}^\psi \otimes a_i^* \otimes d_{l(2)}$$

and we compute that

$$\begin{aligned}
\vartheta(c_{(2)} \otimes a)c_{(1)} &= \langle \overline{\Omega}(c_{(2)} \otimes a), 1_A \otimes \varepsilon \rangle c_{(1)} \\
&= \sum_l \langle b_l^*, a_{i\psi} \rangle \langle a_i^*, 1 \rangle \langle \varepsilon, d_{l(2)} \rangle d_{l(1)}^\psi \\
&= \sum_l \langle b_l^* 1_\psi \rangle d_l^\psi \\
&= \sum_l \langle b_l^* 1 \rangle d_l \\
&= \langle \overline{\Omega}(c_{(1)} \otimes a_\psi), 1_A \otimes \varepsilon_C \rangle c_{(2)}^\psi \\
&= \vartheta(c_{(1)} \otimes a_\psi) c_{(2)}^\psi
\end{aligned}$$

and (3.61) follows. Finally, let us show that α_1 and α_1^{-1} are each others inverses.

$$\begin{aligned}
\alpha_1^{-1}(\alpha_1(\vartheta))(c \otimes a) &= \langle \vartheta(c_{(1)} \otimes a_\psi a_i) a_i^* \otimes c_{(2)}^\psi, 1_A \otimes \varepsilon_C \rangle \\
&= \vartheta(c_{(1)} \otimes a a_i) \langle a_i^*, 1_A \rangle = \vartheta(c \otimes a)
\end{aligned}$$

We know that $\alpha_1(\alpha_1^{-1}(\overline{\Omega}))$ is right A-linear. Hence it suffices to show that

$$\alpha_1(\alpha_1^{-1}(\overline{\Omega}))(c \otimes 1) = c \otimes 1$$

for all $c \in C$. From (3.70), we compute

$$\langle (a^* \otimes c)b, 1_A \otimes \varepsilon_C \rangle = \langle a^*, b \rangle \varepsilon(c) = \langle a^* \otimes c, b \otimes \varepsilon_C \rangle$$

Now write $\overline{\Omega}(c \otimes 1) = \sum_r a_r^* \otimes c_r$. We compute

$$\begin{aligned}
\alpha_1(\alpha_1^{-1}(\overline{\Omega}))(c \otimes 1) &= \langle \overline{\Omega}(c_{(1)} \otimes a_i), 1_A \otimes \varepsilon_C \rangle a_i^* \otimes c_{(2)} \\
&= \langle \overline{\Omega}(c_{(1)} \otimes 1), a_i \otimes \varepsilon_C \rangle a_i^* \otimes c_{(2)} \\
&= \overline{\Omega}(c \otimes 1)_{[0]}, a_i \otimes \varepsilon_C \rangle a_i^* \otimes \overline{\Omega}(c \otimes 1)_{[1]} \\
&= \sum_r \langle a_r^*, a_i \rangle \langle \varepsilon, c_{r(1)} \rangle a_i^* \otimes c_{r(2)} \\
&= \sum_r a_r^* \otimes c_r = \overline{\Omega}(c \otimes 1)
\end{aligned}$$

Theorem 39. Let (A, C, ψ) be a right-right entwining structure, and assume that A is finitely generated and projective as a k-module. With notation as above, we have the following properties:

1. F' is separable if and only if there exists $\overline{\Omega} \in V_2'$ such that

$$\langle \overline{\Omega}(c \otimes 1), 1_A \otimes \varepsilon_C \rangle = \varepsilon_C(c)$$

for all $c \in C$.

2. G' is separable if and only if there exists $\Omega \in W_2'$ such that

$$\sum_i a_i (\varepsilon_C \otimes I_A) \Omega(a_i^* \otimes c) = \varepsilon_C(c) 1$$

for all $c \in C$.

3. The following assertions are equivalent:
   a. (F', G') is a Frobenius pair;
   b. there exist $e \in W_1'$, $\vartheta \in V_1'$ such that $\Omega = \beta_1(e)$ and $\overline{\Omega} = \alpha_1(\vartheta)$ are each others inverses;
   c. $A^* \otimes C$ and $C \otimes A$ are isomorphic objects in ${}^C\mathcal{M}(\psi)_A^C$.

Proof. We will only prove $3a. \Rightarrow 3b$. First we show that Ω is a left inverse of $\overline{\Omega}$. Since $\Omega \circ \overline{\Omega}$ is right A-linear, it suffices to show that

$$\Omega(\overline{\Omega}(c \otimes 1)) = \sum_i \vartheta(c_{(1)} \otimes a_i) \Omega(a_i^* \otimes c_{(2)})$$

$$= \sum_i \vartheta(c_{(1)} \otimes a_i)\langle a_i^*, e^1(c_{(3)})_\psi \rangle c_{(2)}^\psi \otimes e^2(c_{(3)})$$

$$= \vartheta(c_{(1)} \otimes e^1(c_{(3)})_\psi) c_{(2)}^\psi \otimes e^2(c_{(3)})$$

$$(3.61) \quad = \vartheta(c_{(2)} \otimes e^1(c_{(3)})_\psi) c_{(1)}^\psi \otimes e^2(c_{(3)})$$

$$(3.66) \quad = c \otimes 1$$

Now we will show that Ω is a right inverse of $\overline{\Omega}$. Using the fact that $\overline{\Omega} \circ \Omega$ is right C-colinear, we find that it suffices to show that

$$(I_{A^*} \otimes \varepsilon_C)(\overline{\Omega}(\Omega(a^* \otimes c))) = \varepsilon_C(c) a^*$$

for all $c \in C$ and $a^* \in A^*$. Both sides of the equation are in A^*, so we are done if we can show that both sides are equal after we apply them to an arbitrary $a \in A$. Now observe that

$$\overline{\Omega}(\Omega(a^* \otimes c)) = \sum_i \langle a^*, e^1(c_{(2)})_\psi \rangle \vartheta((c_{(1)}^\psi)_{(1)} \otimes e^2(c_{(2)})_\Psi a_i) a_i^* \otimes (c_{(1)}^\psi)_{(2)}^\Psi$$

hence

$$(I_{A^*} \otimes \varepsilon_C)(\overline{\Omega}(\Omega(a^* \otimes c)))(a)$$

$$= \langle a^*, e^1(c_{(2)})_\psi \rangle \vartheta(c_{(1)}^\psi \otimes e^2(c_{(2)})a)$$

$$(3.63) \quad = \langle a^*, (a_\Psi e^1(c_{(2)}^\Psi))_\psi \rangle \vartheta(c_{(1)}^\psi \otimes e^2(c_{(2)}^\Psi))$$

$$(2.1) \quad = \langle a^*, a_{\Psi\psi} e^1(c_{(2)}^\Psi)_{\psi'} \rangle \vartheta(c_{(1)}^{\psi\psi'} \otimes e^2(c_{(2)}^\Psi))$$

$$(2.3) \quad = \langle a^*, a_\psi e^1((c^\psi)_{(2)})_{\psi'} \rangle \vartheta((c^\psi)_{(1)}^{\psi'} \otimes e^2((c^\psi)_{(2)}))$$

$$(3.67) \quad = \langle a^*, a_\psi \rangle \varepsilon(c^\psi) = \langle a^*, a \rangle \varepsilon(c)$$

as needed.

3.5 The general induction functor

Let $(\alpha, \gamma) : (A, C, \psi) \to (A', C', \psi')$ be a morphism of (right-right) entwining structures. The results of the previous two Sections can be extended to the general induction functor

$$F : \mathcal{C} = \mathcal{M}(\psi)_A^C \to \mathcal{C}' = \mathcal{M}(\psi')_{A'}^{C'}$$

and its right adjoint G (see Section 2.5). The idea is the same: we study the natural transformations

$$V = \underline{\mathrm{Nat}}(GF, 1_\mathcal{C}) \quad \text{and} \quad W = \underline{\mathrm{Nat}}(1_{\mathcal{C}'}, FG)$$

Under appropriate assumptions, $\nu \in V$ is completely determined by $\nu_{C \otimes A}$, and $\nu_{C \otimes A}$ is left and right A-linear and C-colinear, more precisely, we get a bijective correspondence

$$V \cong {}_A^C\mathrm{Hom}_A^C(GF(C \otimes A), C \otimes A)$$

and similar results apply to W. The results in this Section have been presented in [43] (in the case of Doi-Koppinen Hopf modules). In order to avoid technical complications, we will assume that the entwining map ψ is bijective, and write $\psi^{-1} = \varphi$. The general case, where ψ is not necessarily bijective, is more technical, and has been discussed by Brzeziński, we refer to [25] for more detail.

We have already seen that $A \otimes C$ is an entwined module, often playing a particular role. However, $A \otimes C$ is not a generator for the category of entwined modules; it has a weaker property, and we can use this to show that $\nu \in V$ is completely determined by $\nu_{A \otimes C}$. this is what we will discuss first.

T-generators Let $T : \mathcal{C} \to \mathcal{D}$ be a functor between two abelian categories. We call $D \in \mathcal{D}$ a T-generator if D generates all $T(C)$, with $C \in \mathcal{C}$. This means that if we have a morphism $f : T(C) \to M$ in $\mathcal{D}$ such that $f \circ g = 0$ for all $g : D \to T(C)$ in $\mathcal{D}$, then $f = 0$. If we put $I = \mathrm{Hom}_\mathcal{D}(D, T(C))$, then $D^{(I)} \to T(C)$ is epic.

Theorem 40. *Consider the morphism* $(I_A, \varepsilon_C) : (A, C, \psi) \to (A, k, I_A)$ *in* $\mathcal{M}(\psi)_A^C$, *let* (F, G) *be the corresponding adjoint pair of functors, and write* $T = GF$. *Then* $A \otimes C$ *is a* T*-generator.*

Proof. Let $f : T(M) = M \otimes C \to N$ in $\mathcal{M}(\psi)_A^C$ be such that $f \circ g = 0$ for all $g : A \otimes C \to M \otimes C$. Take $m \in M$, and consider $g_m : A \otimes C \to M \otimes C$, $g_m(a \otimes c) = am \otimes c$. Then g_m is right A-linear and C-colinear, and for all $c \in C$, we have that

$$f(m \otimes c) = (f \circ g_m)(1_a \otimes c) = 0$$

hence $f = 0$.

Theorem 41. *Let C and $\mathcal{D}$ be abelian categories, $H, H' : C \to \mathcal{D}$ additive functors, $T : C \to C$ a functor, and $\rho : 1_C \to T$ a natural transformation. Assume that the following three conditions hold:*

1. *C is a T-generator for C;*
2. *H preserves epimorphisms;*
3. *$H'(\rho_M)$ is monic, for all $M \in C$.*

Then any natural transformation $\nu : H \to H'$ is completely determined by $\nu_C : H(C) \to H(C')$.

Proof. Let $\nu, \nu' : H \to H'$ be two natural transformations such that $\nu_C = \nu'_C$. We have to show that $\nu_M = \nu'_M$ for all $M \in C$. C is a T-generator, so we have an epimorphism $g : C^{(I)} \to T(M)$. From the naturality of ν and ν', it follows that we have commutative diagrams

$$
\begin{array}{ccc}
H(C^{(I)}) & \xrightarrow{\nu_{C^{(I)}}} & H'(C^{(I)}) \\
{\scriptstyle H(g)}\downarrow & & \downarrow{\scriptstyle H'(g)} \\
H(T(M)) & \xrightarrow{\nu_{T(M)}} & H'(T(M))
\end{array}
\qquad
\begin{array}{ccc}
H(C^{(I)}) & \xrightarrow{\nu'_{C^{(I)}}} & H'(C^{(I)}) \\
{\scriptstyle H(g)}\downarrow & & \downarrow{\scriptstyle H'(g)} \\
H(T(M)) & \xrightarrow{\nu'_{T(M)}} & H'(T(M))
\end{array}
$$

Now

$$\nu_{C^{(I)}} = (\nu_C)^{(I)} = (\nu'_C)^{(I)} = \nu'_{C^{(I)}}$$

hence

$$\nu'_{T(M)} \circ H(g) = \nu_{T(M)} \circ H(g)$$

and

$$\nu'_{T(M)} = \nu_{T(M)}$$

since $H(g)$ is epic. Now consider $\rho_M : M \to T(M)$, and the commutative diagrams

$$
\begin{array}{ccc}
H(M) & \xrightarrow{\nu_M} & H'(M) \\
{\scriptstyle H(\rho_M)}\downarrow & & \downarrow{\scriptstyle H(\rho_M)} \\
H(T(M)) & \xrightarrow{\nu_{T(M)}} & H'(T(M))
\end{array}
\qquad
\begin{array}{ccc}
H(M) & \xrightarrow{\nu'_M} & H'(M) \\
{\scriptstyle H(\rho_M)}\downarrow & & \downarrow{\scriptstyle H(\rho_M)} \\
H(T(M)) & \xrightarrow{\nu'_{T(M)}} & H'(T(M))
\end{array}
$$

We see that

$$H'(\rho_M) \circ \nu_M = H'(\rho_M) \circ \nu'_M$$

and $\nu_M = \nu'_M$ since $H'(\rho_M)$ is monic.

Corollary 18. *Under the assumptions of Theorem 41,* $\underline{\mathrm{Nat}}(F, F')$ *is a set, and the map*

$$\underline{\mathrm{Nat}}(F, F') \to \mathrm{Hom}_{\mathcal{D}}(F(C), F(C')), \quad \nu \mapsto \nu_C$$

is injective.

Separability of F Let $(\alpha, \gamma) : (A, C, \psi) \to (A', C', \psi')$ be a morphism in $\mathbb{E}_\bullet^\bullet(k)$. We will write $\mathcal{C} = \mathcal{M}(\psi)_A^C$ and $\mathcal{C}' = \mathcal{M}(\psi')_{A'}^{C'}$. (F, G) will be the associated pair of adjoint functors between $\mathcal{C}$ and $\mathcal{C}'$. The functor $T : \mathcal{C} \to \mathcal{C}$ will be as above, so that $A \otimes C$ is a T-generator for $\mathcal{C}$. Our aim is now to compute $V = \underline{\mathrm{Nat}}(GF, 1_{\mathcal{C}})$.

Proposition 80. *If G is an exact functor (e.g. C is right C'-coflat), then ν is completely determined by $\nu_{A \otimes C}$.*

Proof. In Theorem 41, take $\mathcal{C} = \mathcal{D} = \mathcal{M}(\psi)_A^C$, $H = GF$, $H' = 1_{\mathcal{C}}$, and $\rho : 1_{\mathcal{C}} \to T$ given by the coaction

$$\rho_M : M \to T(M) = M \otimes C$$

for all $M \in \mathcal{C}$. $A \otimes C$ is a T-generator and $1_{\mathcal{C}}(\rho_M) = \rho_M$ is monic for all M. Indeed, if $\rho_M(m) = \sum m_{[0]} \otimes m_{[1]} = 0$, then $m = \sum \varepsilon(m_{[1]}) m_{[0]} = 0$. Furthermore F preserves epimorphisms since the tensor product is right exact, and G is exact by assumption, so GF preserves epimorphisms. Thus the assumptions of Theorem 41 are satisfied, and the result follows.

If G is exact, then we know from Corollary 18 that V is a set. Actually V is a k-algebra. Addition and scalar multiplication are defined in the obvious way, and the multiplication is given by

$$\nu' \bullet \nu = \nu' \circ \eta \circ \nu \tag{3.73}$$

Lemma 15. *Let $\nu : GF \to 1_{\mathcal{C}}$ be a natural transformation, $M \in \mathcal{C}$, and $N \in {}_k\mathcal{M}$. $N \otimes M$ is an object of $\mathcal{C}$:*

$$\rho_{N \otimes M}^r = I_N \otimes \rho_M^r \; ; \; (n \otimes m)a = n \otimes ma$$

If N is k-flat, then $GF(N \otimes M) = N \otimes GF(M)$ and

$$\nu_{N \otimes M} = I_N \otimes \nu_M$$

Proof. For all $n \in N$, we consider the map $f_n : M \to N \otimes M$, $f_n(m) = n \otimes m$. Then $f_n \in \mathcal{C}$, and the naturality of ν implies that the following diagram is commutative:

$$
\begin{array}{ccc}
GF(M) & \xrightarrow{\;\nu_M\;} & M \\
{\scriptstyle GF(f_m)}\Big\downarrow & & \Big\downarrow{\scriptstyle f_m} \\
GF(N \otimes M) & \xrightarrow{\;\nu_{N \otimes M}\;} & N \otimes M
\end{array}
$$

Using the fact that N is k-flat, we obtain

$$GF(N \otimes M) = (N \otimes M \otimes_A A') \square_{C'} C \cong N \otimes GF(M)$$

For $x \in GF(M)$, we have $GF(f_n)(x) = n \otimes x$, and the commutativity of the diagram tells us that

$$\nu_{N \otimes M}(n \otimes x) = n \otimes \nu_M(x)$$

Theorem 42. *Assume that $(B, D, \phi) \in {}^\bullet_\bullet \mathbb{E}(k)$ is a left-left entwining structure, and that $M \in {}^D_B \mathcal{M}(\phi, \psi)^C_A$ is a two-sided entwined module. Then ν_M is
left D-colinear and left B-linear. In particular, if ψ is invertible, then*

$$\nu_{C \otimes A} \in {}^C_A \mathcal{M}(\psi^{-1}, \psi)^C_A$$

Proof. We automatically have that $\nu_M \in \mathcal{C}$, i.e. ν_M is right A-linear and
right C-colinear. Let us prove first that ν_M is left B-linear. For any $b \in B$,
we consider the map $f_b : M \to M$, $f_b(m) = bm$. From conditions 1) and 4)
in the Definition of two-sided entwining modules, we find that $f_b \in \mathcal{C}$, and
we have a commutative diagram

$$
\begin{array}{ccc}
GF(M) & \xrightarrow{\nu_M} & M \\
\downarrow{\scriptstyle GF(f_b)} & & \downarrow{\scriptstyle f_b} \\
GF(M) & \xrightarrow{\nu_M} & M
\end{array}
$$

Applying the diagram to $x = \sum_i (m_i \otimes a_i') \otimes c_i \in GF(M)$, we find

$$
\begin{aligned}
b\nu_M(x) &= f_b(\nu_M(x)) = \nu_M(GF(f_b)(x)) \\
&= \nu_M\Big(\sum_i (bm_i \otimes a_i') \otimes c_i\Big) = \nu_M(bx)
\end{aligned}
$$

and ν_M is left B-linear.

To show that ν_M is left D-colinear, we consider the left coaction $\rho^l : M \to
D \otimes M$. We will apply Lemma 15 with $N = D$. Conditions 2) and 4) in
the definition of two-sided entwined module entail that $\rho^l \in \mathcal{C}$, so we have a
commutative diagram

$$
\begin{array}{ccc}
GF(M) & \xrightarrow{\nu_M} & M \\
\downarrow{\scriptstyle GF(\rho^l)} & & \downarrow{\scriptstyle \rho^l} \\
GF(D \otimes M) & \xrightarrow{\nu_{D \otimes M}} & D \otimes M
\end{array}
$$

Recall from Lemma 15 that $GF(D \otimes M) = D \otimes GF(M)$, and $\nu_{D \otimes M} = I_D \otimes \nu_M$. For $x = \sum_i (m_i \otimes a'_i) \otimes c_i \in GF(M)$, we obtain

$$(GF)(\rho')(x) = \sum (m_{i[-1]} \otimes m_{i[0]} \otimes a'_i) \otimes c_i$$

and the commutative diagram tells us that

$$\sum x_{[-1]} \otimes \nu_M(x_{[0]}) = \rho'(\nu_M(x))$$

proving that ν_M is left C-colinear.

Corollary 19. *Let* $(\alpha, \gamma) : (A, C, \psi) \to (A', C', \psi')$ *is a morphism of entwining structures, with* ψ *invertible. We then have a well-defined map*

$$f : V = \underline{\mathrm{Nat}}(GF, 1_C) \to V_1 = {}^C_A\mathrm{Hom}^C_A(GF(C \otimes A), C \otimes A)$$

given by $f(\nu) = \nu_{C \otimes A}$. *If* G *is exact, then* f *is injective.*

Next we present an easier description of V_1.

Proposition 81. *As above, let* (α, γ) *be a morphism of entwining structures, and assume that* ψ *is invertible. Let* V_2 *consist of all left and right* A-*linear maps* $\lambda : GF(C \otimes A) \to A$ *satisfying*

$$\lambda(\sum_i (c_i \otimes a'_i) \otimes d_{i(1)}) \otimes d_{i(2)} = \sum_i \lambda((c_{i(2)} \otimes a'_i) \otimes d_i)_\psi \otimes c^\psi_{i(1)} \qquad (3.74)$$

for all $\sum_i (c_i \otimes a'_i) \otimes d_i \in GF(C \otimes A)$. *We have a* k-*linear isomorphism*

$$f_1 : V_1 \to V_2 \; ; \quad f_1(\bar{\nu}) = (\varepsilon \otimes I_A) \circ \bar{\nu}$$

Proof. $\lambda = f_1(\bar{\nu})$ is left and right A-linear since $\bar{\nu}$ and $\varepsilon \otimes I_A$ are left and right A-linear. Take $\sum_i (c_i \otimes a'_i) \otimes d_i \in GF(C \otimes A)$, and write

$$\bar{\nu}(\sum_i (c_i \otimes a'_i) \otimes d_i) = \sum_j c_j \otimes a_j$$

Using the left C-colinearity of $\bar{\nu}$, we obtain

$$\sum_i c_{i(1)} \otimes \bar{\nu}(\sum_i (c_{i(2)} \otimes a'_i) \otimes d_i) = \sum_j c_{j(1)} \otimes c_{j(2)} \otimes a_j$$

and, applying ε_C to the second factor,

$$\sum_i c_{i(1)} \otimes \bar{\lambda}(\sum_i (c_{i(2)} \otimes a'_i) \otimes d_i) = \bar{\nu}(\sum_i (c_i \otimes a'_i) \otimes d_i)$$

$\bar{\nu}$ is also right C-colinear, hence

$$\overline{\nu}(\sum_i (c_i \otimes a_i') \otimes d_{i(1)}) \otimes d_{i(2)} = \sum_j c_{j(1)} \otimes a_{j\psi} \otimes c_{j(2)}^\psi$$

Applying ε_C to the first factor, we obtain

$$\overline{\lambda}(\sum_i (c_i \otimes a_i') \otimes d_{i(1)}) \otimes d_{i(2)} = \sum_j a_{j\psi} \otimes c_j^\psi$$

and we have shown that $\overline{\lambda}$ satisfies (3.74), and f_1 is well-defined. The inverse of f_1 is given by

$$f_1^{-1}(\lambda)(\sum_i (c_i \otimes a_i) \otimes d_i) = \sum_i c_{i(1)} \otimes \lambda(\sum_i (c_{i(2)} \otimes a_i) \otimes d_i)$$

It is obvious that $\overline{\nu} = f_1^{-1}(\lambda)$ is left C-colinear and right A-linear. $\overline{\nu}$ is right C-colinear since

$$\sum_i \overline{\nu}((c_i \otimes a_i') \otimes d_{i(1)}) \otimes d_{i(2)}$$

$$= \sum_i c_{i(1)} \otimes \overline{\lambda}((c_{i(2)} \otimes a_i') \otimes d_{i(1)}) \otimes d_{i(2)}$$

$$(3.74) \quad = \sum_i c_{i(1)} \otimes \overline{\lambda}((c_{i(3)} \otimes a_i') \otimes d_i)_\psi \otimes c_{i(2)}^\psi$$

$$(2.66) \quad = \rho^r(\sum_i c_{i(1)} \otimes \overline{\lambda}((c_{i(2)} \otimes a_i') \otimes d_i))$$

$$= \rho^r(\overline{\nu}((c_i \otimes a_i') \otimes d_i)$$

$\overline{\nu}$ is left A-linear since

$$\overline{\nu}\Big(a(\sum_i (c_i \otimes a_i') \otimes d_i)\Big)$$

$$(2.76) \quad = \overline{\nu}\Big(\sum_i (c_i^\varphi \otimes \alpha(a_\varphi)a_i') \otimes d_i\Big)$$

$$= \sum_i (c_i^\varphi)_{(1)} \otimes \lambda\Big(((c_i^\varphi)_{(2)} \otimes \alpha(a_\varphi)a_i') \otimes d_i\Big)$$

$$(2.8) \quad = \sum_i c_{i(1)}^\varphi \otimes \lambda\Big((c_{i(2)}^\phi \otimes \alpha(a_{\varphi\phi})a_i') \otimes d_i\Big)$$

$$(2.76) \quad = \sum_i c_{i(1)}^\varphi \otimes \lambda\Big(a_\varphi((c_{i(2)}^\phi \otimes a_i') \otimes d_i)\Big)$$

$$= a\Big(\sum_i c_{i(1)} \otimes \lambda((c_{i(1)} \otimes a_i') \otimes d_i)\Big)$$

$$= a\overline{\nu}(\sum_i (c_i \otimes a_i') \otimes d_i)$$

We leave it to the reader to show that $g_1 = f_1^{-1}$.

Theorem 43. *Let* $(\alpha, \gamma) : (A, C, \psi) \to (A', C', \psi')$ *be a morphism of entwining structures. Assume that* ψ *is invertible, and let* V, V_1 *and* V_2 *be defined as above. If* C *is left* C'-*coflat, then* V, V_1 *and* V_2 *are isomorphic as* k-*modules.*

Proof. In view of the previous results, it suffices to show that $f \circ f_1 : V \to V_2$ is surjective. Starting from $\lambda \in V_2$, we have to construct a natural transformation ν, that is, for all $M \in \mathcal{M}(\psi)_A^C$, we have to construct a morphism

$$\nu_M : GF(M) = (M \otimes_A A')\square_{C'}C \to M$$

First we remark that the map

$$\phi : M \otimes_A A' \to M \otimes_A (C \otimes A') \;\; ; \;\; \phi(m \otimes_A a') = m_{[0]} \otimes_A (m_{[1]} \otimes a')$$

is well-defined. Indeed,

$$\phi(ma \otimes_A a') = (ma)_{[0]} \otimes_A ((ma)_{[1]} \otimes a')$$
$$= m_{[0]}a_\psi \otimes_A (m_{[1]}^\psi \otimes a')$$
$$= m_{[0]} \otimes_A (m_{[1]}^{\psi\varphi} \otimes a_{\psi\varphi}a') = \phi(m \otimes_A aa')$$

From the fact that C is left C'-coflat, and using Lemma 3, we find

$$(M \otimes_A (C \otimes A'))\square_{C'}C \cong M \otimes_A ((C \otimes A')\square_{C'}C)$$

and we can consider the map

$$\nu_M = (I_M \otimes_A \lambda) \circ (\phi\square_{C'}I_C) : GF(M) \to M \otimes_A A \cong A$$

given by

$$\nu_M\left(\sum_i (m_i \otimes a'_i) \otimes c_i\right) = \sum_i m_{i[0]}\lambda\big((m_{i[1]} \otimes a'_i) \otimes c_i\big)$$

Let us first show that ν_M is right A-linear

$$\nu_M\left(\left(\sum_i (m_i \otimes a'_i) \otimes c_i\right)a\right) = \nu_M\left(\sum_i (m_i \otimes a'_i\alpha(a_\psi)) \otimes c_i^\psi\right)$$
$$= \sum_i m_{i[0]}\lambda\big((m_{i[1]} \otimes a'_i\alpha(a_\psi)) \otimes c_i^\psi\big)$$
$$(2.74) \quad = \sum_i m_{i[0]}\big(\lambda((m_{i[1]} \otimes a'_i) \otimes c_i)a\big)$$
$$= \nu_M\left(\sum_i (m_i \otimes a'_i) \otimes c_i\right)a$$

ν_M is right C-colinear since

$$\rho^r\left(\nu_M\left(\sum_i (m_i \otimes a_i') \otimes c_i\right)\right) = \sum_i m_{i[0]} \lambda\left((m_{i[2]} \otimes a_i) \otimes c_i\right)_\psi \otimes m_{i[1]}^\psi$$

$$(3.74) \qquad = \sum_i m_{i[0]} \lambda\left((m_{i[1]} \otimes a_i) \otimes c_{i[1]}\right) \otimes c_{i[2]}$$

$$= \nu_M\left((m_i \otimes a_i) \otimes c_{i[1]}\right) \otimes c_{i[2]}$$

as needed.

Let us show that ν is natural. Let $g : M \to N$ be a morphism in $\mathcal{M}(\psi)_A^C$, and take $x = \sum_i (m_i \otimes a_i') \otimes c_i \in (M \otimes_A A') \square_{C'} C$. Then

$$\nu_N(GF(g))(x) = \nu_N\left(\sum_i (g(m_i) \otimes a_i') \otimes c_i\right)$$

$$= \sum_i g(m_{i[0]}) \lambda\left((m_{i[1]} \otimes a_i') \otimes c_i\right)$$

$$= g\left(\sum_i m_{i[0]} \lambda\left((m_{i[1]} \otimes a_i') \otimes c_i\right)\right) = g(\nu_M(x))$$

Finally, we have to show that $f_1(f(\nu)) = \lambda$. Indeed,

$$(\varepsilon_C \otimes I_A)\left(\nu_{C \otimes A}\left(\sum_i (c_i \otimes a_i') \otimes d_i\right)\right)$$

$$= (\varepsilon_C \otimes I_A)\left(\sum_i (c_{i(1)} \otimes 1) \lambda\left((c_{i(2)} \otimes a_i') \otimes d_i\right)\right)$$

$$= \lambda\left(\sum_i (c_i \otimes a_i') \otimes d_i\right)$$

as needed.

Corollary 20. *Let $(\alpha, \gamma) : (A, C, \psi) \to (A', C', \psi')$ be a morphism of entwining structures, and assume that ψ is invertible, and that C is left C'-coflat. The induction functor $F : \mathcal{M}(\psi)_A^C \to \mathcal{M}(\psi')_{A'}^{C'}$ is separable if and only if there exists $\lambda \in V_2$ such that*

$$\lambda\left((c_{(1)} \otimes 1_{A'}) \otimes c_{(2)}\right) = \varepsilon(c) 1_A \tag{3.75}$$

for all $c \in C$ and $a \in A$. F is full and faithful if and only if $\eta_{C \otimes A}$ is an isomorphism.

Proof. If F is separable, then there exists $\nu \in V$ such that $\nu \circ \eta$ is the identity natural transformation, in particular

$$\nu_{C \otimes A} \circ \eta_{C \otimes A} = I_{C \otimes A}$$

Write $\bar{\nu} = f(\nu)$ and $\lambda = f_1(\bar{\nu})$, and apply both sides to $c \otimes 1_A$:

$$\bar{\nu}\left((c_{(1)} \otimes \alpha((1_A)_\psi)) \otimes c_{(2)}^\psi\right) = c \otimes 1_A$$

and (3.75) follows after we apply ε to the first factor.

Conversely, if $\lambda \in V_2$ satisfies (3.75), and ν is the natural transformation corresponding to λ (see Theorem 43), then

$$\nu_M(\eta_M(m)) = \nu_M\Big((m_{[0]} \otimes 1_{A'}) \otimes m_{[1]}\Big)$$
$$= m_{[0]} \otimes \lambda\Big((m_{[1]} \otimes 1_{A'}) \otimes m_{[2]}\Big)$$
$$= m_{[0]}\varepsilon(m_{[1]})1_A = m$$

The second statement is proved in the same way.

Separability of the adjoint of the induction functor As before, let $(\alpha, \gamma) : (A, C, \psi) \to (A', C', \psi')$ be a morphism of entwining structures. We will now assume that ψ' is invertible, and study the k-module of natural transformations

$$\zeta \in W = \underline{\mathrm{Nat}}(1_{C'}, FG)$$

The results below are largely dual to the ones presented above, so we will be somewhat more sketchy. However, there are some differences in the technical assumptions.

Lemma 16. *Let $(\alpha, \gamma) : (A, C, \psi) \to (A', C', \psi')$ be a morphism in $\mathbb{E}_\bullet^\bullet(k)$. Assume that ψ' is bijective, and that $FG(\rho_{M'})$ is a monomorphism, for all $M' \in \mathcal{M}(\psi')_{A'}^{C'}$. Then any $\zeta \in W$ is completely determined by $\zeta_{A' \otimes C'}$.*

Proof. In Theorem 41, we take $\mathcal{C}$ and $\mathcal{D}$ equal to $\mathcal{M}(\psi')_{A'}^{C'}$. For ρ, we take the natural transformation given by the right C'-coaction; for H we take the identity natural transformation, and for H' we take FG. Then all the conditions of Theorem 41 are fulfilled, and our result follows.

Lemma 17. *With assumptions as above, let $\zeta \in W$, and $M' \in \mathcal{M}(\psi')_{A'}^{C'}$. For any flat k-module N, let $N \otimes M' \in \mathcal{M}(\psi')_{A'}^{C'}$ be the entwined module with structure induced by the structure on M'. Then*

$$FG(N \otimes M') \cong N \otimes FG(M')$$

in $\mathcal{M}(\psi')_{A'}^{C'}$, and $\zeta_{N \otimes M'} = I_N \otimes \zeta_{M'}$.

Proof. We omit the proof, as it is similar to the proof of Lemma 15.

Proposition 82. *Assume that $(B', D', \phi') \in {}_\bullet\mathbb{E}(k)$, and that $M' \in {}_{B'}^{D'}\mathcal{M}(\phi', \psi')_{A'}^{C'}$. Then $\zeta_{M'}$ is left D'-colinear and left B'-linear. In particular*

$$\zeta_{A' \otimes C'} \in {}_{A'}^{C'}\mathcal{M}(\psi'^{-1}, \psi')_{A'}^{C'}$$

and, under the assumption that $FG(\rho_{M'})$ is injective for all $M' \in \mathcal{M}(\psi')_{A'}^{C'}$, we have a well-defined monomorphism

$$f : W = \underline{\mathrm{Nat}}(1_{C'}, FG) \to W_1 = {}_{A'}^{C'}\mathrm{Hom}_{A'}^{C'}(A' \otimes C', FG(A' \otimes C'))$$

Proof. We omit the details, since the proof is very similar to the proof of Theorem 42.

W_1 can be described in an easier way. Recall first that

$$FG(A' \otimes C') = ((A' \otimes C')\square_{C'}C) \otimes_A A' \cong (A' \otimes C) \otimes_A A' \qquad (3.76)$$

with structure

$$b'((a' \otimes c) \otimes_A a'')b'' = (b'a' \otimes c) \otimes_A a''b'' \qquad (3.77)$$

$$\rho^r((a' \otimes c) \otimes_A a'') = ((a' \otimes c_{(1)}) \otimes_A a''_{\psi'}) \otimes \gamma(c_{(2)})^{\psi'} \qquad (3.78)$$

$$\rho^l((a' \otimes c) \otimes_A a'') = \gamma(c_{(1)})^{\varphi'} \otimes ((a'_{\varphi'} \otimes c_{(2)}) \otimes_A a'') \qquad (3.79)$$

Given $\overline{\zeta} \in W_1$, we define $e : C' \to (A' \otimes C) \otimes_A A'$ by

$$e(c') = \overline{\zeta}(1_{A'} \otimes c')$$

Obviously e is left and right C'-colinear, and

$$e(c')a' = \overline{\zeta}((1 \otimes c')a') = \overline{\zeta}(a'_{\psi'} \otimes c'^{\psi'}) = a'_{\psi'}e(c'^{\psi'}) \qquad (3.80)$$

Let W_2 be the k-module consisting of C'-bicolinear maps $e : C' \to (A' \otimes C) \otimes_A A'$ satisfying (3.80).
conversely, given $e \in W_2$, we define $\overline{\zeta} \in W_1$ by

$$\overline{\zeta}(a' \otimes c') = a'e(c')$$

We leave it to the reader to show that this defines a bijective correspondence between the maps in W_1 and W_2:

Proposition 83. *The k-modules W_1 and W_2 defined above are isomorphic.*

The following result can be viewed as the dual of Theorem 43.

Theorem 44. *Let $(\alpha, \gamma) : (A, C, \psi) \to (A', C', \psi')$ be a morphism in $\mathbb{E}^\bullet_\bullet(k)$. We assume that ψ' is bijective, A' is flat as a left A-module, and $FG(\rho_{M'})$ is injective for all $M' \in \mathcal{M}(\psi')^{C'}_{A'}$. Then $f : W \to W_1 \cong W_2$ is an isomorphism.*

Remark 8. The condition that $FG(\rho_{M'})$ is injective is automatically fulfilled if k is a field: G is then left exact, and F is exact by assumption.

Proof. In view of Propositions 82 and 83, we have to show that $f_1 \circ f : W \to W_2$ is surjective. For $e \in W_2$, we will construct a natural transformation $\zeta : 1_{C'} \to FG$. We proceed in different steps.
Step 1. For any $M' \in \mathcal{M}(\psi')^{C'}_{A'}$, we have a well-defined map $\phi : M'\square_{C'}(A' \otimes \overline{C}) \to M'\square_{C'}C$ given by

$$\phi(\sum_i m'_i \otimes (a'_i \otimes c_i)) = \sum_i m'_i a'_i \otimes c_i$$

Indeed, if $\sum_i m'_i \otimes (a'_i \otimes c_i) \in M' \Box_{C'}(A' \otimes C)$, then

$$\sum_i m'_{i[0]} \otimes m'_{i[1]} \otimes (a'_i \otimes c_i) = \sum_i m'_i \otimes \gamma(c_{i(1)})^{\varphi'}(a'_{i\varphi'} \otimes c_{i(2)})$$

We apply ψ' to the two middle factors, and then we let the second factor act on the first one; using the fact that φ' is the inverse of ψ', we obtain

$$\sum_i m'_{i[0]} a'_{i\psi'} \otimes m'^{\psi'}_{i[1]} \otimes c_i = \sum_i m'_i a'_i \otimes \gamma(c_{i(1)}) \otimes c_{i(2)}$$

which means exactly that $\sum_i m'_i a'_i \otimes c_i \in M' \Box_{C'} C$.

Step 2. A' is flat as a left A-module, so, by Lemma 3

$$M' \Box_{C'}((A' \otimes C) \otimes_A A') \cong (M' \Box_{C'}(A' \otimes C)) \otimes_A A'$$

Step 3. For $M' \in \mathcal{M}(\psi')^{C'}_{A'}$, we define $\zeta_{M'}$ as the composition

$$\zeta_{M'} : M' \cong M' \Box_{C'} C' \xrightarrow{I_{M'} \otimes e} M' \Box_{C'}((A' \otimes C) \otimes_A A')$$
$$\cong (M' \Box_{C'}(A' \otimes C)) \otimes_A A'$$
$$\xrightarrow{\phi \otimes I_{A'}} (M' \Box_{C'} C) \otimes_A A'$$

More explicitely, this means the following: if we write, according to our traditions,

$$e(c') = (e^1(c') \otimes e^C(c')) \otimes e^2(c') \in (A \otimes C) \otimes_A A'$$

then

$$\zeta_{M'}(m') = (m'_{[0]} e^1(m'_{[1]}) \otimes e^C(m'_{[1]})) \otimes e^2(m'_{[1]}) \in (M' \Box_{C'} C) \otimes_A A'$$

We will use the following more transparent notation:

$$\zeta_{M'}(m') = m'_{[0]} e(m'_{[1]})$$

Using this notation, it is not hard to see that $\zeta_{M'}$ is right A'-linear and right C'-colinear, and that ζ is natural:

Step 4. $\zeta_{M'}$ is right C'-colinear.

$$\rho^r(\zeta_{M'}(m')) = (m'_{[0]} e^1(m'_{[1]}) \otimes e^C(m'_{[1]})_{(1)}) \otimes e^2(m'_{[1]})_{\psi'} \otimes \gamma(e^C(m'_{[1]})_{(1)})^{\psi'}$$
$$(3.78) \qquad = m'_{[0]} \rho^r(e(m'_{[1]}))$$
$$= m'_{[0]} e(m'_{[1]}) \otimes (m'_{[2]} \quad (e \text{ is right } C'\text{-colinear})$$
$$= \zeta_{M'}(m'_{[0]}) \otimes m'_{[1]}$$

Step 5. $\zeta_{M'}$ is right A'-linear.

$$\zeta_{M'}(m'a') = (m'a')_{[0]}e((m'a')_{[1]})$$
$$= m'_{[0]}a'_{\psi'}e(m'^{\psi'}_{[1]})$$
$$(3.80) \qquad = m'_{[0]}e(m'_{[1]})a' = \zeta_{M'}(m')a'$$

Step 6. ζ is natural: for $f : M' \to N'$ in $\mathcal{M}(\psi')^{C'}_{A'}$, we have

$$FG(f)(\zeta_{M'}(m')) = f(m'_{[0]})e(m'_{[1]}) = \zeta_{N'}(f(m))$$

where we use implicitly the fact that f is right C'-colinear and A-linear.

Step 7. $f_1(f(\zeta)) = e$.

$$\zeta_{A'\otimes C'}(1_{A'} \otimes c') = \left((e^1(c'_{(2)})_{\psi'} \otimes c'_{(1)}{}^{\psi'}) \otimes e^C(c'_{(2)})\right) \otimes e^2(c'_{(2)})$$

Using the identification $((A' \otimes C')\square_{C'}C) \otimes_A A' \cong (A' \otimes C) \otimes_A A'$, we find

$$\zeta_{A'\otimes C'}(1_{A'} \otimes c') = (e^1(c'_{(2)})_{\psi'} \otimes \varepsilon(c'_{(1)}{}^{\psi'})e^C(c'_{(2)})) \otimes e^2(c'_{(2)}) = e(c')$$

as needed.

Corollary 21. *Under the assumptions of Theorem 44, the functor G is separable if and only if there exists $e \in W_2$ such that*

$$\varepsilon_C(e^C(c'))e^1(c')e^2(c') = \varepsilon_{C'}(c')1_{A'} \qquad (3.81)$$

for all $c' \in C'$. G is full and faithful if and only if $\varepsilon_{A'\otimes C'}$ is an isomorphism.

Proof. If G is separable, then there exists $\zeta \in W$ such that $\varepsilon \circ \zeta$ is the identity natural transformation, in particular $\varepsilon_{A'\otimes C'} \circ \zeta_{A'\otimes C'} = I_{A'\otimes C'}$. Applying both sides to $1_{A'} \otimes c' \in A' \otimes C'$, and using (2.62), we obtain (3.81). Conversely, if (3.81) holds, then we compute easily that

$$\varepsilon_{M'}(\zeta_{M'}(m')) = \varepsilon_{M'}(m'_{[0]}e(m'_{[1]})$$
$$= \varepsilon_{M'}\left((m'_{[0]}e^1(m'_{[1]}) \otimes e^C(m'_{[1]})) \otimes e^2(m'_{[1]})\right)$$
$$(2.62) \qquad = m'_{[0]}\varepsilon_C(e^C(m'_{[1]}))e^1(m'_{[1]})e^2(m'_{[1]})$$
$$(3.81) \qquad = m'_{[0]}\varepsilon_{C'}(m'_{[1]})1_{A'} = m'$$

for all $m' \in M' \in \mathcal{M}(\psi')^{C'}_{A'}$.

Corollary 22. *Under the assumptions of Theorems 43 and 44, that is, ψ and ψ' are bijective, C is left C'-coflat, and A' is left A-flat, the functors F and G are inverse equivalences if and only if $\eta_{C\otimes A}$ and $\varepsilon_{A\otimes C}$ are bijective.*

Remark 9. The condition that $FG(\rho_{M'})$ is monic for all M' is automatically fulfilled since F and G are exact functors. In the case where k is a field, Corollary 22 has been shown using different methods in [51].

4 Applications

In this Chapter, we will apply the results of the previous two Chapters to some special categories of modules, such as relative Hopf modules, Yetter-Drinfeld modules, Long dimodules and graded modules. We will find new proofs of some existing results, and our techniques also allow us to find new results. We also discuss how Galois theory and descent theory can be introduced using entwined modules and comodules over a coring.

4.1 Relative Hopf modules

Let H be a bialgebra with twisted antipode, and A a right H-comodule algebra. Then $(H, A, H) \in \mathbb{HA}^\bullet_\bullet(k)$, and we have a corresponding entwining structure, (A, H, ψ), with $\psi : H \otimes A \to A \otimes H$ given by the formula

$$\psi(h \otimes a) = a_{[0]} \otimes ha_{[1]}$$

An *integral* $\varphi : H \to A$ is a right H-colinear map. This means that

$$\varphi(h_{(1)}) \otimes h_{(2)} = \varphi(h)_{[0]} \otimes \varphi(h)_{[1]}$$

φ is called a *total integral* [69] if, in addition,

$$\varphi(1_H) = 1_A$$

In the situation where $A = k$, we recover Sweedler's definition of integral in H^*:

$$\varphi(h_{(1)})h_{(2)} = \varphi(h)1_H$$

As we have seen, the separability of the forgetful functor $\mathcal{M}(H)^H_A \to \mathcal{M}_A$ is determined by the k-module V_5 (see Proposition 70). In our particular situation, V_5 consists of k-linear maps $\vartheta : H \otimes H \to A$ satisfying the conditions

$$\vartheta(h \otimes k)a = a_{[0]}\vartheta(ha_{[1]} \otimes ka_{[2]}) \tag{4.1}$$

$$\vartheta(h \otimes k_{(1)}) \otimes k_{(2)} = \vartheta(h_{(2)} \otimes k)_{[0]} \otimes h_{(1)}\vartheta(h_{(2)} \otimes k)_{[1]} \tag{4.2}$$

(cf. (3.49-3.50)). We will now see that there is a close relationship between V_5 and the space of integrals, and consequently how integrals can be used to determine the separability of the forgetful functor.

Theorem 45. *Let H be a Hopf algebra with twisted antipode, and A a right H-comodule algebra. We have a map $p: V_5 \to \mathrm{Hom}^H(H, A)$, $p(\vartheta) = \varphi$ being given by the formula*

$$\varphi(h) = \vartheta(1 \otimes h) \tag{4.3}$$

Given an integral φ, we define $s(\varphi) = \vartheta: H \otimes H \to A$ by

$$\vartheta(h \otimes k) = \varphi(k\overline{S}(h)) \tag{4.4}$$

$\vartheta \in V_5$ if and only if

$$\rho^r(\varphi(H)) \subset Z(A \otimes H) \tag{4.5}$$

Proof. $p(\vartheta) = \varphi$ is right H-colinear since

$$
\varphi(h_{(1)}) \otimes h_{(2)} = \vartheta(1 \otimes h_{(1)}) \otimes h_{(2)}
$$
$$
(4.2) \quad = \vartheta(1 \otimes h)_{[0]} \otimes \vartheta(1 \otimes h)_{[1]} = \rho^r(\varphi(h))
$$

Before we prove the second statement, we observe that (4.5) is equivalent to the following two conditions:

$$\varphi(H) \subset Z(A) \tag{4.6}$$
$$(1 \otimes g)\rho^r(\varphi(h)) = \rho^r(\varphi(h))(1 \otimes g) \tag{4.7}$$

Assuming that (4.6-4.7) hold, and defining ϑ using (4.4), we have

$$
a_{[0]}\vartheta(ha_{[1]} \otimes ka_{[2]}) = a_{[0]}\varphi(ka_{[2]}\overline{S}(ha_{[1]})) = a\varphi(k\overline{S}(h)) = \vartheta(h \otimes k)a
$$

and

$$
\vartheta(h_{(2)} \otimes k)_{[0]} \otimes h_{(1)}\vartheta(h_{(2)} \otimes k)_{[1]}
$$
$$
= \varphi(k\overline{S}(h_{(2)}))_{[0]} \otimes h_{(1)}\varphi(k\overline{S}(h_{(2)}))_{[1]}
$$
$$
(4.7) \quad = \varphi(k\overline{S}(h_{(2)}))_{[0]} \otimes \varphi(k\overline{S}(h_{(2)}))_{[1]}h_{(1)}
$$
$$
= \varphi(k_{(1)}\overline{S}(h_{(3)})) \otimes k_{(2)}\overline{S}(h_{(2)})h_{(1)}
$$
$$
= \varphi(k_{(1)}\overline{S}(h)) \otimes k_{(2)}
$$
$$
= \vartheta(h \otimes k_{(1)}) \otimes k_{(2)}
$$

and it follows that $\vartheta \in V_5$.

Conversely, if $\vartheta = s(\varphi) \in V_5$, then (4.1) implies immediately that $\varphi(H) \subset Z(A)$. Using (4.2), we find that

$$
\varphi(k_{(1)}\overline{S}(h)) \otimes k_{(2)} = \varphi(k_{(1)}\overline{S}(h_{(3)})) \otimes h_{(1)}k_{(2)}\overline{S}(h_{(2)})
$$

and this yields

$$
\varphi(k_{(1)}) \otimes k_{(2)}h = \varphi(k_{(1)}h_{(2)}\overline{S}(h_{(1)})) \otimes k_{(2)}h_{(3)}
$$
$$
= \varphi((kh_{(2)})_{(1)}\overline{S}(h_{(1)})) \otimes (kh_{(2)})_{(2)}
$$
$$
= \varphi((kh_{(4)})_{(1)}\overline{S}(h_{(3)})) \otimes h_{(1)}(kh_{(4)})_{(2)}\overline{S}(h_{(2)})
$$
$$
= \varphi(k_{(1)}h_{(4)}\overline{S}(h_{(3)})) \otimes h_{(1)}k_{(2)}h_{(5)}\overline{S}(h_{(2)})
$$
$$
= \varphi(k_{(1)}) \otimes hk_{(2)}
$$

proving (4.7)

Corollary 23. *Let H be a bialgebra with twisted antipode, and A a right H-comodule algebra. If there exists a total integral $\varphi : H \to A$ such that $\rho^r(\varphi(H)) \subset Z(A \otimes H)$, then the forgetful functor $\mathcal{M}(H)_A^H \to \mathcal{M}_A$ is separable, and, consequently, we have a Maschke type Theorem for this forgetful functor.*

Proof. By Theorem 45, $\vartheta = \varphi(s) \in V_5$. Furthermore

$$\vartheta(h_{(1)} \otimes h_{(2)}) = \varphi(h_{(2)}\overline{S}(h_{(1)})) = \varphi(\varepsilon(h)1_H) = \varepsilon(h)1_A$$

and our result follows from Theorem 38.

The technical condition in Corollary 23 is (4.5). We will now give some alternative sufficient conditions.

Lemma 18. *Let H and A be as in Corollary 23, and assume that $\varphi(H) \subset Z(A)$, and*

$$\varphi(gh) = \varphi(h\overline{S}^2(g)) \tag{4.8}$$

for all $g, h \in H$. Then $\rho^r(\varphi(H)) \subset Z(A \otimes H)$.

Proof. We have seen in the proof of Theorem 45 that it suffices to show that (4.7) holds. We proceed in different steps. Using (4.8), we find, for all $h \in H$:

$$\varphi(h) = \varphi(h1_H) = \varphi(1_H\overline{S}^2(h)) = \varphi(\overline{S}^2(h))$$

Next

$$\varphi(h_{(1)}) \otimes h_{(2)} = \rho^r(\varphi(h)) = \rho^r(\varphi(\overline{S}^2(h)))$$
$$= \varphi(\overline{S}^2(h_{(1)})) \otimes \overline{S}^2(h_{(2)}) = \varphi(h_{(1)}) \otimes \overline{S}^2(h_{(2)}) \tag{4.9}$$

and

$$\varphi(h_{(1)}) \otimes h_{(2)}\overline{S}^2(g) = \varphi(h_{(1)}) \otimes \overline{S}^2(h_{(2)})\overline{S}^2(g)$$
$$= \varphi(h_{(1)}g_{(2)}\overline{S}(g_{(1)})) \otimes \overline{S}^2(h_{(2)})\overline{S}^2(g_{(3)})$$
$$(4.8) \quad = \varphi(\overline{S}(g_{(1)}\overline{S}^2(h_{(1)}g_{(2)}))) \otimes \overline{S}^2(h_{(2)})\overline{S}^2(g_{(3)})$$
$$= \varphi(\overline{S}(g_{(3)}))\overline{S}^2(h_{(1)})\overline{S}^2(g_{(4)})) \otimes \overline{S}^2(g_{(1)})\overline{S}(g_{(2)})\overline{S}^2(h_{(2)})\overline{S}^2(g_{(5)})$$
$$= \varphi\Big((\overline{S}(g_{(2)})\overline{S}^2(h)\overline{S}^2(g_{(3)}))_{(1)}\Big) \otimes \overline{S}^2(g_{(1)})(\overline{S}(g_{(2)})\overline{S}^2(h)\overline{S}^2(g_{(3)}))_{(2)}$$
$$= (1_A \otimes \overline{S}^2(g_{(1)}))\rho^r(\varphi(\overline{S}(g_{(2)})\overline{S}^2(h)\overline{S}^2(g_{(3)})))$$
$$(4.8) \quad = (1_A \otimes \overline{S}^2(g_{(1)}))\rho^r(\varphi(hg_{(3)}\overline{S}(g_{(2)}))))$$
$$= (1_A \otimes \overline{S}^2(g))\rho^r(\varphi(h))$$
$$= \varphi(h_{(1)}) \otimes \overline{S}^2(g)h_{(2)}$$

and we have the following formula, already close to (4.7):

$$\varphi(h_{(1)}) \otimes h_{(2)}\overline{S}^2(g) = \varphi(h_{(1)}) \otimes \overline{S}^2(g)h_{(2)} \tag{4.10}$$

We now compute

$$\varphi(h_{(1)}) \otimes \overline{S}(g)h_{(2)} = \varphi(h_{(1)}) \otimes \overline{S}(g_{(1)})h_{(2)}\overline{S}^2(g_{(2)})\overline{S}(g_{(3)})$$
$$(4.10) \quad = \varphi(h_{(1)}) \otimes \overline{S}(g_{(1)})\overline{S}^2(g_{(2)})h_{(2)}\overline{S}(g_{(3)})$$
$$= \varphi(h_{(1)}) \otimes h_{(2)}\overline{S}(g) \tag{4.11}$$

and, applying the same trick,

$$\varphi(h_{(1)}) \otimes gh_{(2)} = \varphi(h_{(1)}) \otimes g_{(3)}h_{(2)}\overline{S}(g_{(2)})g_{(1)}$$
$$(4.11) \quad = \varphi(h_{(1)}) \otimes g_{(3)}\overline{S}(g_{(2)})h_{(2)}g_{(1)}$$
$$= \varphi(h_{(1)}) \otimes h_{(2)}g$$

and (4.7) follows.

Corollary 24. *Let H be a Hopf algebra, and A a right H-comodule algebra, and $\varphi : H \to A$ an integral. In each of the following situations, we have that $\rho^r(\varphi(H)) \subset Z(A \otimes H)$:*

1. *H is commutative and $\varphi(H) \subset Z(A)$;*
2. *$S^2 = I_H$, $\varphi(H) \subset Z(A)$, and $\varphi(hk) = \varphi(kh)$;*
3. *$\varphi(H) \subset k1_A$.*

Proof. 1. follows immediately from the fact $\rho^r(\varphi(H)) \subset Z(A \otimes H)$ is equivalent to (4.6-4.7). 2. is an immediate consequence of Lemma 18. Finally, if $\varphi(H) \subset k1_A$, then

$$\rho(\varphi(h)) = \varphi(h)1_A \otimes 1_H \in k(1_A \otimes 1_H) \subset Z(A \otimes H)$$

Remark 10. Combining Corollary 24 and Corollary 23, we recover [69, Theorem 1.7] and [70, Theorem 1]: if there exists a total integral φ such that one of the three conditions of Corollary 24 is satisfied, then the forgetful functor $\mathcal{M}(H)_A^H \to \mathcal{M}_A$ is separable, and we have a Maschke type Theorem. We point out that the results in [69], [70] and [41] are stated in the right-left case. Of course, they are easily translated into the right-right case that we discuss here; this also explains why we work with a twisted antipode, while the results in the above references need an antipode.

Classical integrals We now take $A = k$, with trivial H-coaction. Now $\varphi \in H^*$ is an integral if and only if

$$\langle \varphi, h_{(1)} \rangle h_{(2)} = \langle \varphi, h \rangle 1_H$$

for all $h \in H$. This is equivalent to

$$\langle \varphi, h_{(1)} \rangle \langle h^*, h_{(2)} \rangle = \langle \varphi, h \rangle \langle h^*, 1_H \rangle$$

for all $h \in H$ and $h^* \in H^*$, or

$$\varphi * h^* = \langle h^*, 1_H \rangle \varphi$$

which means that φ is a right integral in H^* in the classical sense (cf. [172]). In the sequel, $\int^r_{H^*}$ will denote the k-module consisting of right integrals in H^*. It is now obvious that $\rho^r(\varphi(H)) \subset Z(A \otimes H)$. Recall (see Corollary 17) that V_5 consists of maps $\vartheta \in (H \otimes H)^*$ such that

$$\vartheta(h \otimes k_{(1)})k_{(2)} = \vartheta(h_{(2)} \otimes k)h_{(1)}$$

Using Theorem 45, we now obtain the following result.

Theorem 46. *Let H be a bialgebra with a twisted antipode. We then have maps*

$$p : V_5 \to \int^r_{H^*} \quad \text{and} \quad s : \int^r_{H^*} \to V_5$$

such that s is a section of p (i.e. $s \circ p$ is the identity). Moreover $\mathrm{Im}(s)$ consists of those $\vartheta \in V_5$ that satisfy

$$\vartheta(h \otimes k) = \vartheta(1 \otimes k\overline{S}(h)) \tag{4.12}$$

Proof. We recall that

$$p(\vartheta)(h) = \vartheta(1 \otimes h) \quad \text{and} \quad s(\varphi)(h \otimes k) = \varphi(k\overline{S}(h))$$

It is clear that $(p \circ s)(\varphi) = \varphi$. Obviously, if $\vartheta = s(\varphi)$, then (4.12) holds. If ϑ satisfies (4.12), then $\vartheta = s(p(\vartheta))$.

Corollary 25. *If there exists $\varphi \in \int^r_{H^*}$ such that $\varphi(1_H) = 1$, then $\mathcal{M}^H \to \mathcal{M}$ is separable, this means that H is coseparable as a k-coalgebra.*

Remarks 6. 1. Theorem 46 and Corollary 25 are stated for bialgebras with a twisted antipode. They are also valid for Hopf algebras: it suffices to look at the forgetful functor $^H\mathcal{M}(H)_A \to \mathcal{M}(H)_A$ (using the left-right dictionary), and then take the case $A = k$.
2. The separability of the functor $G : \mathcal{M} \to \mathcal{M}^H$ can also be investigated easily. In fact G is always separable: it suffices to put $f = 1_H$ in condition IIc in Corollary 17.
3. Let H be a finite dimensional Hopf algebra over a field. It follows from the Fundamental Theorem for Hopf modules (see [172]) that the functor $\mathcal{M}^H \to \mathcal{M}$ is Frobenius, and that H^* is a Frobenius extension of k. We sketch the proof, and show how this is related to Part 3. of Corollary 17. First, for every right-right Hopf module M, we have an isomorphism

$$\alpha : M \cong M^{\mathrm{coH}} \otimes H$$

namely

$$\alpha(m) = m_{[0]} S(m_{[1]}) \otimes m_{[2]} \quad ; \quad \alpha^{-1}(m' \otimes h) = m'h$$

This can be restated in the language of Hopf modules:

$$(\eta, \eta, \eta) : \ (k, k, k) \to (H, H, H)$$

is a morphism of Doi-Hopf structures. The corresponding adjoint pair of functors, between $\mathcal{M}$ and $\mathcal{M}(H)_H^H$ is an equivalence of categories.

Secondly, H^* is a right-right Hopf module. The right H-action is given by

$$\langle h^* \leftharpoonup h, k \rangle = \langle h^*, kS(h) \rangle$$

and the right H-coaction is induced by left multiplication in H^*:

$$\rho^r(h^*) = h^*_{[0]} \otimes h^*_{[1]} \quad \Leftrightarrow \quad k^* * h^* = \langle k^*, h^*_{[1]} \rangle h^*_{[0]}$$

for all $k^* \in H^*$. Next,

$$(H^*)^{\text{co}H} = \int_{H^*}^l$$

and we find that

$$H^* \cong \int_{H^*}^l \otimes H$$

as Hopf modules. It also follows that $\int_{H^*}^l$ is one dimensional. Fixing a nonzero left integral φ, we obtain an isomorphism

$$\overline{\phi} : \ H \to H^*, \quad \overline{\phi}(h) = \varphi \leftharpoonup h, \quad \text{or} \quad \langle \overline{\phi}(h), k \rangle = \varphi(kS(h)))$$

of right H-comodules, and therefore also of left H^*-modules, and from Corollary 17 it follows that $\mathcal{M}^H \to \mathcal{M}$ is Frobenius.

Now define $\vartheta : \ H \otimes H \to k$ by

$$\vartheta(k \otimes h) = \varphi(kS(h))$$

An immediate verification shows that $\vartheta \in V_5$:

$$\vartheta(h \otimes k_{(1)})k_{(2)} = \varphi(hS(k_{(1)}))k_{(2)} = \varphi(h_{(2)}S(k_{(1)}))h_{(1)}S(k_{(2)})k_{(3)}$$
$$= \varphi(h_{(2)}S(k))h_{(1)} = \vartheta(h_{(2)} \otimes k)h_{(1)}$$

This is consistent with condition 3d. in Corollary 17.

4. In general, the map $p : \ V_5 \to \text{Hom}^H(H, A)$ is not an isomorphism. To see this, we take $A = k$ a field, and H a finite dimensional Hopf algebra. First observe that

$$V_5 \cong {}_{H^*}\text{Hom}(H, H^*)$$

$\vartheta \in V_5$ corresponds to $\overline{\phi}$ as in Corollary 17:

$$\langle \overline{\phi}(h), k \rangle = \vartheta(k \otimes h)$$

Now we know that $H \cong H^*$ as left H^*-modules; consequently

$$\dim_k(V_5) = \dim_k({}_{H^*}\text{Hom}(H^*, H^*)) = \dim_k(H^*)$$

On the other hand

$$\dim_k(\text{Hom}^H(H, A)) = \dim_k\left(\int_{H^*}^r\right) = 1$$

The Heisenberg double Let H be a bialgebra with twisted antipode, and take $A = H$ in Corollary 23.

Proposition 84. *If H is commutative Hopf algebra, then the forgetful functor $\mathcal{M}(H)_H^H \to \mathcal{M}_H$ is separable.*

Proof. $I_H : H \to H$ is a total integral, and $\Delta(H) \subset Z(H \otimes H)$, since H is commutative.

Now let H be finitely generated and projective; $\mathcal{H}(H) = H \# H^*$ is called the *Heisenberg double* of H.

Corollary 26. *Let H be a commutative finitely generated projective Hopf algebra. Then the Heisenberg double $\mathcal{H}(H)$ is a separable extension of H.*

Proof. From Proposition 21, Theorem 8, and Proposition 84, it follows that

$$_{H^{\text{op}}\# H^*}\mathcal{M} \cong \mathcal{M}(H)_H^H \to \mathcal{M}_H$$

is separable. H is commutative, so $H^{\text{op}} = H$, and it follows from Theorem 27 that $H^{\text{op}}\# H^*$ is a separable extension of H.

Total integrals and relative separability Theorem 45 and Corollary 23 tell us when the existence of a total integral implies the separability of F. What can we conclude if we have a total integral, without any further conditions fulfilled? Let A be an H-comodule algebra, and consider the following commutative diagram of forgetful functors

$$
\begin{array}{ccc}
\mathcal{M}(H)_A^H & \xrightarrow{\;\;F\;\;} & \mathcal{M}_A \\
\Big\downarrow{\scriptstyle G'} & & \Big\downarrow{\scriptstyle G''} \\
\mathcal{M}^H & \longrightarrow & {}_k\mathcal{M}
\end{array}
$$

G will denote the right adjoint of F. Recall (Proposition 73) that

$$X = \underline{\text{Nat}}(G'GF, G') \cong X_5 = \{\theta : H \otimes H \to A \mid (3.50) \text{ holds}\}$$

In our situation, (3.50) takes the form (4.2). We will now show that we have a projection of the k-module X_5 onto the space of integrals $\text{Hom}^H(H, A)$, at least in the situation where H is a Hopf algebra.

Proposition 85. *Let H be a Hopf algebra, and A a right H-comodule algebra. We have a projection p : $X_5 \to \mathrm{Hom}^H(H, A)$, with section s : $\mathrm{Hom}^H(H, A) \to X_5$ given by*

$$p(\vartheta)(h) = \vartheta(1 \otimes h) \quad \text{and} \quad s(\varphi)(h \otimes k) = \varphi(S(h)k)$$

Proof. Recall that X_5 consists of ϑ : $C \otimes C \to A$ satisfying (3.50), or, equivalently, (4.2). As in the proof of Theorem 45, it follows that $p(\vartheta)$ is right H-colinear. Now take φ : $H \to A$ right H-colinear. We have to show that $s(\varphi) = \vartheta$ satisfies (4.2). This is straightforward:

$$\begin{aligned}
\vartheta(h \otimes k_{(1)}) \otimes k_{(2)} &= \varphi(S(h)k_{(1)}) \otimes k_{(2)} \\
&= \varphi(S(h_{(3)})k_{(1)}) \otimes h_{(1)}S(h_{(2)})k_{(2)} \\
&= \varphi(S(h_{(2)})k)_{[0]} \otimes h_{(1)}\varphi(S(h_{(2)})k)_{[1]} \\
&= \vartheta(h_{(2)} \otimes k)_{[0]} \otimes h_{(1)}\vartheta(h_{(2)} \otimes k)_{[1]}
\end{aligned}$$

It is obvious that $p(s(\varphi)) = \varphi$.

We can now state some conditions equivalent to the existence of a total integral. More equivalent conditions can be found in [69, (1.6)] and the forthcoming [47, Theorem 4.20].

Theorem 47. *For a Hopf algebra H and a right H-comodule algebra A, the following assertions are equivalent:*

1. *The forgetful functor F : $\mathcal{M}(H)_A^H \to \mathcal{M}_A$ is G'-separable;*
2. *the forgetful functor $G'' \circ F$: $\mathcal{M}(H)_A^H \to \mathcal{M}$ is G'-separable;*
3. *there exists a map θ : $H \otimes H \to A$ such that $\theta \circ \Delta_H = \eta_A \circ \varepsilon_H$ and (4.2) holds for all $h, k \in H$;*
4. *there exists a total integral φ : $H \to A$;*
5. *every relative Hopf module is relative injective as an H-comodule;*
6. *A is relative injective as an H-comodule.*

Proof. **2. $\Rightarrow$ 1.** follows from Proposition 51.
1. $\Leftrightarrow$ 3. is the final statement of Proposition 73 applied to relative Hopf modules.
3. $\Leftrightarrow$ 4. follows from Proposition 85.
4. $\Rightarrow$ 5. Let M be a relative Hopf module. According to Proposition 50, it suffices to show that the map ρ_M : $M \to M \otimes H$ splits as a map of right H-comodules. The splitting λ_M : $M \otimes H \to M$ is given by $\lambda(m \otimes h) = m_{[0]}\varphi(S(m_{[1]})h)$.
5. $\Rightarrow$ 6. is trivial.
6. $\Rightarrow$ 4. The unit map η_H : $k \to H$ is H-colinear, and split as a k-module map by ε_H. According to the definition of relative injectivity, there exists an H-colinear map φ : $H \to A$ such that $\varphi \circ \eta_H = \eta_A$. φ is then a total integral.
4. $\Rightarrow$ 5. Let φ be then a total integral. The functor F has right adjoint G,

and G'' has a right adjoint K'' given by $K''(N) = \text{Hom}(A, N)$. Thus $G \circ K''$ is a right adjoint of $G'' \circ F$. We will apply Rafael's Theorem 25: we define a natural transformation $\nu : G'GK''G''F \to G'$ as follows:

$$\nu_M : \text{Hom}(A, M) \otimes H \to M \ ; \ \nu_M(f \otimes h) = f(1_A)_{[0]}\varphi(S(f(1_A)_{[1]})h)$$

It is straightforward to verify that f is natural. Finally

$$\begin{aligned}
\nu_M(G'(\eta_M)(m)) &= \nu_M(m_{[0]} \bullet \otimes m_{[1]}) \\
&= m_{[0]}\varphi(S(m_{[1]})m_{[2]}) \\
&= m\varphi(1_H) = m
\end{aligned}$$

Here $m\bullet : A \to M$, $m \bullet (a) = ma$.

Assume that we have a total integral φ. Then F will be separable if $\vartheta = s(\varphi)$ not only satisfies (4.2) but also (4.1).

Proposition 86. *Take $\vartheta = s(\varphi) \in X_5$. $\vartheta \in V_5$ if and only if φ satisfies the condition*

$$\varphi(h)a = a_{[0]}\varphi(S(a_{[1]})ha_{[2]}) \tag{4.13}$$

for all $h \in H$ and $a \in A$. Consequently p restricts to a projection

$$p_{|V_5} : V_5 \to \{\varphi \in \text{Hom}^H(H, A) \mid (4.13) \text{ holds}\}$$

and $F : \mathcal{M}(H)_A^H \to \mathcal{M}_A$ is separable if and only if there exists a total integral satisfying condition (4.13).

Proof. Assume that φ satisfies (4.13). Then

$$\begin{aligned}
a_{[0]}\vartheta(ha_{[1]} \otimes ka_{[2]}) &= a_{[0]}\varphi(S(ha_{[1]})ka_{[2]}) \\
&= a_{[0]}\varphi(S(a_{[1]})S(h)ka_{[2]}) = \varphi(S(h)k)a \\
&= \vartheta(h \otimes k)a
\end{aligned}$$

and (4.1) follows. The rest is left to the reader.

The Frobenius property

Proposition 87. *Let H be a Frobenius Hopf algebra, and A a right H-comodule algebra. The functor $F : \mathcal{M}(H)_A^H \to \mathcal{M}_H$ and its right adjoint G form a Frobenius pair of functors.*

Proof. This is a consequence of Proposition 72: take $f = t$, with t a free generator of $\int_H^r$. Then $\phi_H : H^* \to H$, $\phi_H(h^*) = \langle h^*, t_{(2)} \rangle t_{(1)}$ is bijective, and

$$\psi(t \otimes a) = a_{[0]} \otimes ta_{[1]} = \varepsilon(a_{[1]})a_{[0]} \otimes t = a \otimes t$$

as needed.

Corollary 27. *Let k be a field, H a Hopf algebra, and A a right H-comodule algebra. Then (F, G) is a Frobenius pair if and only if H is finite dimensional.*

4.2 Hopf-Galois extensions

Schneider's affineness Theorems Throughout this Section, H will be a Hopf algebra with bijective antipode over a commutative ring k. At some places, we will assume that k is a field, we will mention this explicitly. A will be a right H-comodule algebra; we will always assume that A and H are flat as k-modules. B will be the algebra of coinvariants of A, $B = A^{coH}$. As H is at the same time a left and right H-module coalgebra (via left and right comultiplication), we have that $(H, A, H) \in {}_\bullet\mathbb{DK}^\bullet(k)$ and $(H, A, H) \in \mathbb{DK}^\bullet_\bullet(k)$, and we can consider the categories ${}_A\mathcal{M}(H)^H$ and $\mathcal{M}(H)^H_A$ of left-right and right-right relative Hopf modules. Let us first look at the right-right case. We have the following morphism in $\mathbb{DK}^\bullet_\bullet(k)$:

$$(\eta_H, i, \eta_H) : \ (k, B, k) \to (H, A, H)$$

and a pair of adjoint functors (cf. Theorem 15):

$$F = \bullet \otimes_B A : \ \mathcal{M}_B \to \mathcal{M}(H)^H_A \ \text{ and } \ G = (\bullet)^{coH} : \ \mathcal{M}(H)^H_A \to \mathcal{M}_B$$

For a right B-module M, the right H-coaction on $M \otimes_B A$ is $I_M \otimes \rho_A$, and the right A-action is given by $(m \otimes a)a' = m \otimes aa'$, for all $m \in M$ and $a, a' \in A$. We describe the unit and the counit of the adjunction: for $M \in \mathcal{M}_B$ and $N \in \mathcal{M}(H)^H_A$,

$$\mu_N : \ N^{coH} \otimes_B A \to N \ ; \ \ \mu_N(n \otimes a) = na$$

$$\eta_M : \ M \to (M \otimes_B A)^{coH} \ ; \ \ \eta_M(m) = m \otimes 1$$

Recall that $A \otimes H \in \mathcal{M}(H)^H_A$, the right H-coaction is just $I_A \otimes \Delta_H$, while the right A-action is

$$(a \otimes h)b = ab_{[0]} \otimes hb_{[1]}$$

Using the fact A is k-flat, we can easily show that

$$(A \otimes H)^{coH} \cong A \otimes H^{coH} = A \otimes k = A$$

so that the counit map $\mu_{A \otimes H} : \ (A \otimes H)^{coH} \otimes_B A \to A \otimes H$ translates into a map

$$\text{can} : \ A \otimes_B A \to A \otimes H$$

We find easily that

$$\text{can}(a \otimes b) = (a \otimes 1)b = ab_{[0]} \otimes b_{[1]} \tag{4.14}$$

Similar constructions can be carried out in the left-right case. We have a pair of adjoint functors

$$F' = A \otimes_B \bullet : \ {}_B\mathcal{M} \to {}_A\mathcal{M}(H)^H \ \text{ and } \ G' = (\bullet)^{coH} : \ {}_A\mathcal{M}(H)^H \to {}_B\mathcal{M}$$

$A \otimes H \in {}_A\mathcal{M}(H)^H$, and the adjunction map $\mu'_{A \otimes H}$ now defines a map

$$\text{can}' : A \otimes_B A \to A \otimes H$$

given by

$$\text{can}'(a \otimes b) = a(b \otimes 1) = a_{[0]}b \otimes a_{[1]} \qquad (4.15)$$

Lemma 19. *The map* $\Phi : A \otimes H \to A \otimes H$ *given by*

$$\Phi(a \otimes h) = a_{[0]} \otimes a_{[1]}S(h)$$

is an isomorphism. Furthermore $\text{can}' = \Phi \circ \text{can}$, *so* can *is an isomorphism if and only if* can' *is an isomorphism.*

Proof. We easily compute that

$$\Phi^{-1}(a \otimes h) = a_{[0]} \otimes \overline{S}(h)a_{[1]}$$

and all the rest follows.

A is called a *Hopf-Galois extension* of B if can (or can') is an isomorphism. If (F, G) or (F', G') is a pair of inverse equivalences of categories, then clearly can and can' are isomorphisms. In this Sections, we give some affineness Theorems, giving additional sufficient conditions for (F, G) or (F', G') to be pairs of inverse equivalences. For general pairs of adjoint functors, such sufficient conditions are given in Corollary 22. In the particular situation under discussion here, we can relax these conditions. Our first result tells that we have a pair of inverse equivalences if and only if A is a *faithfully flat* Galois extension. It is originally due to Doi and Takeuchi [73]; a more general version where H is replaced by a quotient coalgebra has been presented by Schneider in [165, Theorem 3.7].

Theorem 48. *Let* H *be a Hopf algebra with bijective antipode,* A *a right* H-*comodule algebra, and assume that* A *and* H *are* k-*flat. Then the following are equivalent:*

1. A *is faithfully flat as a left* B-*module, and* A *is a Hopf-Galois extension of* B;
2. (F, G) *is a pair of inverse equivalences between the categories* $\mathcal{M}_B$ *and* $\mathcal{M}(H)^H_A$.

Proof. $\underline{2. \Rightarrow 1.}$ We have already seen that A is a Hopf-Galois extension of B. Let $M \to M'$ be an injective map between right B-modules. Since $\mathcal{M}_B$ and $\mathcal{M}(H)^H_A$ are equivalent, $F(M) = M \otimes_B A \to F(M') = M' \otimes_B A$ is monic in $\mathcal{M}(H)^H_A$, and a fortiori in $\mathcal{M}_A$ since monics in $\mathcal{M}(H)^H_A$ are injective maps. Thus A is left B-flat. Faithful flatness is treated in a similar way: assume that we have a sequence

$$0 \to M' \to M \to M'' \to 0$$

such that

$$0 \to M' \otimes_B A \to M \otimes_B A \to M'' \otimes_B A \to 0$$

is exact in $\mathcal{M}_A$. The three A-modules have the structure of relative Hopf module, and the sequence is exact in $\mathcal{M}(H)_A^H$. It stays exact after we apply G (since we have equivalence of categories), hence the original sequence in $\mathcal{M}_B$ is exact.

1. $\Rightarrow$ 2. First we will prove that, for any $N \in \mathcal{M}(H)_A^H$, the adjunction map μ_N is an isomorphism. If X is a right A-module, then the map

$$\text{can}_X = I_X \otimes_A \text{can} : \ X \otimes_B A \cong X \otimes_A A \otimes_B A \to X \otimes H = X \otimes_A A \otimes H$$

given by

$$\text{can}_X (x \otimes a) = x a_{[0]} \otimes a_{[1]}$$

is bijective because can is bijective. Now we have a commutative diagram

$$
\begin{array}{ccccc}
1 \longrightarrow & N^{\text{coH}} \otimes_B A & \longrightarrow & N \otimes_B A & \xrightarrow[I_N \otimes \eta_H \otimes I_A]{\rho_N \otimes I_A} & N \otimes H \otimes_B A \\
 & \Big\downarrow{\mu_N} & & \Big\downarrow{\text{can}_N} & & \Big\downarrow{\text{can}^{N \otimes H}} \\
1 \longrightarrow & N & \xrightarrow{\Delta_N} & N \otimes H & \xrightarrow[I_N \otimes \Delta_H]{\rho_N \otimes I_N} & N \otimes H \otimes H
\end{array}
$$

The bottom row is exact, since the coequalizer of $\rho_N \otimes I_N$ and $I_N \otimes \Delta_H$ is $N \square_H H \cong N$. The top row is also exact, since N^{coH} is the coequalizer of ρ_N and $I_N \otimes \eta_H$, and since A is flat as a left B-module. can_N and $\text{can}_{N \otimes H}$ are isomorphisms, so μ_N is an isomorphism by the Five Lemma.

Now take $M \in \mathcal{M}_B$. We want to show that $\varepsilon_M : \ M \to (M \otimes B_A)^{\text{coH}}$ is an isomorphism. Define

$$i_1, i_2 : \ M \otimes_B A \to M \otimes_B A \otimes_B A$$

as follows:

$$i_1(m \otimes a) = m \otimes 1 \otimes a \ \text{ and } \ i_2(m \otimes a) = m \otimes a \otimes 1$$

for all $m \in M$ and $a \in A$. We have a commutative diagram

$$
\begin{array}{ccccc}
1 \longrightarrow & M & \xrightarrow{I_M \otimes \eta_A} & M \otimes_B A & \xrightarrow[i_2]{i_1} & M \otimes_B A \otimes_B A \\
 & \Big\downarrow{\varepsilon_M} & & \Big\| & & \Big\downarrow{I_M \otimes \text{can}} \\
1 \longrightarrow & (M \otimes_B A)^{\text{coH}} & \xrightarrow{\subset} & M \otimes_B A & \xrightarrow[I_M \otimes \rho_A]{I_M \otimes \rho_A} & M \otimes_B A \otimes H
\end{array}
$$

Indeed, we compute easily that

$$(I_M \otimes \mathrm{can})(i_2(m \otimes a)) = m \otimes \mathrm{can}(a \otimes 1) = m \otimes a \otimes 1$$
$$= (I_M \otimes I_A \otimes \eta_H)(m \otimes a)$$
$$(I_M \otimes \mathrm{can})(i_1(m \otimes a)) = m \otimes \mathrm{can}(1 \otimes a) = m \otimes a_{[0]} \otimes a_{[1]}$$
$$= (I_M \otimes \rho_A)(m \otimes a)$$

The top row is exact because A is faithfully flat as a left B-module. The bottom row is exact, by definition of coinvariants. can is an isomorphism, so ε_N is an isomorphism, again by the Five Lemma.

If there exists a total integral $\varphi : H \to A$ and if H is projective over k, then it suffices to verify that can is surjective, and we do not have to verify that A is faithfully flat over B. This is the context of the following imprimitivity Theorem.

Theorem 49. *[165, Theorem 3.5] As before, we assume that H is a Hopf algebra with bijective antipode. A is a right H-comodule algebra, and H and A are k-flat. If can is surjective, and if there exists a total integral $\varphi : H \to A$, then the adjoint pair (F, G) between $\mathcal{M}_B$ and $\mathcal{M}(H)_A^H$ is a pair of inverse equivalences.*

Before we give the proof of the Theorem, we need some Lemmas. We follow [165]. Let M be a left H-comodule. Then M is also a right H-comodule, via $m \to m_{[0]} \otimes S(m_{[-1]})$. Applying the induction functor $A \otimes \bullet : \mathcal{M}^H \to {}_A\mathcal{M}(H)^H$, we find that $A \otimes M \in {}_A\mathcal{M}(H)^H$, with structure

$$a(b \otimes m) = ab \otimes m \quad \text{and} \quad \rho^r(b \otimes m) = b_{[0]} \otimes m_{[0]} \otimes b_{[1]}S(m_{[-1]})$$

Lemma 20. *With notation as above, we have for any left H-comodule M that*

$$(A \otimes M)^{\mathrm{co}H} = A \square_H M$$

Proof. Take $x = \sum_i a_i \otimes m_i \in (A \otimes M)^{\mathrm{co}H}$. Then

$$\rho^r(x) = \sum_i a_{i[0]} \otimes m_{i[0]} \otimes a_{i[1]}S(m_{i[-1]}) = \sum_i a_i \otimes m_i \otimes 1$$

Apply ρ_M to the second factor, and then multiply the fourth factor on the right by the second factor. We then find

$$\sum_i a_{i[0]} \otimes m_{i[0]} \otimes a_{i[1]}S(m_{i[-2]})m_{i[-1]} = \sum_i a_i \otimes m_{i[0]} \otimes m_{i[-1]}$$

Switching the second and the third factor, we find

$$\sum_i a_{i[0]} \otimes a_{i[1]} \otimes m = \sum_i a_i \otimes m_{i[-1]} \otimes m_{i[0]}$$

and $x \in A \square_H M$. The converse inclusion can be done in a similar way.

Lemma 21. *Take $N \in \mathcal{M}(H)_A^H$. N is a left H-comodule via $n \to S(n_{[1]}) \otimes n_{[0]}$, and we have well-defined maps*

$$i: \ N^{coH} \to A\square_H N \ ; \quad i(n) = 1 \otimes n$$

$$p: \ A\square_H N \to N^{coH} \ ; \quad p(\sum_i a_i \otimes n_i) = \sum_i a_i n_i$$

such that $p \circ i = I_{N^{coH}}$.

Proof. It is obvious that i is well-defined; let us show that p is well-defined. Take $\sum_i a_i \otimes n_i \in A\square_H N$. Then

$$\sum_i a_{i[0]} \otimes a_{i[1]} \otimes n_i = \sum_i a_i \otimes S(n_{i[1]}) \otimes n_{i[0]}$$

Applying ρ_N to the last factor, we find

$$\sum_i a_{i[0]} \otimes a_{i[1]} \otimes n_{i[0]} \otimes n_{i[1]} = \sum_i a_i \otimes S(n_{i[2]}) \otimes n_{i[0]} \otimes n_{i[1]}$$

and consequently

$$\rho_N(\sum_i a_i n_i) = \sum_i n_{i[0]} a_{i[0]} \otimes n_{i[1]} a_{i[1]}$$

$$= \sum_i n_{i[0]} a_{i[0]} \otimes n_{i[1]} S(n_{i[2]}) = \sum_i a_i n_i \otimes 1$$

as needed. It is clear that $p \circ i = I_{N^{coH}}$.

Lemma 22. *Assume that A and H are k-flat. The following are equivalent.*

1. *A is right H-coflat;*
2. *$G = (\bullet)^{coH} : \mathcal{M}(H)_A^H \to \mathcal{M}_B$ is an exact functor;*
3. *$G' = (\bullet)^{coH} : {}_A\mathcal{M}(H)^H \to {}_B\mathcal{M}$ is an exact functor.*

Proof. 1. $\Rightarrow$ 2. It is clear that G is left exact. Assume that $f : \ N \to N'$ is surjective in $\mathcal{M}(H)_A^H$. $I_A\square_H f$ is surjective because A is right H-coflat. Now looking at the commutative diagram

$$
\begin{array}{ccc}
A\square_H N & \xrightarrow{\ I_A\square_H f\ } & A\square_H N' \\[4pt]
p \big\downarrow \big\uparrow i & & p \big\downarrow \big\uparrow i \\[4pt]
N^{coH} & \xrightarrow{\quad f \quad} & N'^{coH}
\end{array}
$$

we find that $f : \ N^{coH} \to N'^{coH}$ is surjective.

3. $\Rightarrow$ 1. Using Lemma 20, we see that $A\square_H \bullet$ is the composition of the functors

$$_H\mathcal{M} \xrightarrow{A\otimes\bullet} {}_A\mathcal{M}(H)^H \xrightarrow{G'} {}_B\mathcal{M} \longrightarrow {}_k\mathcal{M}$$

G' is exact, by assumption. $A \otimes \bullet$ is exact since A is k-flat. It follows that $A\square_H\bullet$ is exact, and A is right H-coflat.

1. $\Rightarrow$ 3. An application of the left-right dictionary. We have $(H, A, H) \in {}_\bullet\mathbb{DK}^\bullet(k)$. According to Proposition 16, $(H^{\mathrm{op}}, A^{\mathrm{op}}, H) \in \mathbb{DK}^\bullet_\bullet(k)$ and the categories $_A\mathcal{M}(H)^H$ and $\mathcal{M}(H^{\mathrm{op}})^H_{A^{\mathrm{op}}}$ are isomorphic. The right H^{op}-action on the coalgebra H is given by right multiplication in H^{op}, as needed. A^{op} is still right H-coflat (we did not change the coaction and comultiplication), so we can 1) $\Rightarrow$ 2) to $(H^{\mathrm{op}}, A^{\mathrm{op}}, H)$, and find that G' is exact.

2. $\Rightarrow$ 1. This is done in a similar way: we use the left-right dictionary, and apply 3. $\Rightarrow$ 1. to $(H^{\mathrm{op}}, A^{\mathrm{op}}, H) \in {}_\bullet\mathbb{DK}^\bullet(k)$.

Lemma 23. *Assume that A is a right H-comodule algebra, and that $\varphi : H \to A$ is a total integral. For any $M \in \mathcal{M}_B$, the adjunction map $\eta_M : M \to (M \otimes_B A)^{\mathrm{co}H}$ is an isomorphism.*

Proof. We have a well-defined map

$$t: A \to B \;\; ; \;\; t(a) = a_{[0]}\varphi(S(a_{[1]}))$$

Indeed,

$$\begin{aligned}
\rho^r(t(a)) &= a_{[0]}\varphi(S(a_{[2]}))_{[0]} \otimes a_{[1]}\varphi(S(a_{[2]}))_{[1]} \\
&= a_{[0]}\varphi(S(a_{[3]})) \otimes a_{[1]}S(a_{[2]}) \\
&= a_{[0]}\varphi(S(a_{[1]})) \otimes 1_H = t(a) \otimes 1_H
\end{aligned}$$

and $t(a) \in A^{\mathrm{co}H} = B$. Now define

$$\psi_M : (M \otimes_B A)^{\mathrm{co}H} \to M \;\; : \;\; \psi_M(\sum_i m_i \otimes a_i) = \sum_i m_i t(a_i)$$

A direct computation shows that ψ_M is the inverse of η_M:

$$\psi_M(\eta_M(m)) = \psi(m \otimes 1_A) = mt(1_A) = m1_A\varphi(S(1_H)) = m$$

and

$$\begin{aligned}
\eta_M(\psi_M(\sum_i m_i \otimes a_i)) &= \sum_i m_i t(a_i) \otimes 1_A \\
(t(a_i) \in B) \quad &= \sum_i m_i \otimes t(a_i) = \sum_i m_i \otimes a_{i[0]}\varphi(S(a_{i[1]})) \\
&= \sum_i m_i \otimes a_i\varphi(S(1_H)) = \sum_i m_i \otimes a_i
\end{aligned}$$

At the last step, we used that $\sum_i m_i \otimes a_i \in (M \otimes_B A)^{\mathrm{co}H}$.

Proof. (of Theorem 49) We have shown in Lemma 23 that η_M is bijective. We still need to show that μ_N is an isomorphism, for all $N \in \mathcal{M}(H)_A^H$. We prove this first in the case where $N = V \otimes A$, where V is an arbitrary k-module, and the structure on N is induced by the structure on A, i.e.

$$(v \otimes a)b = v \otimes b \quad \text{and} \quad \rho_{V \otimes A} = I_V \otimes \rho_A$$

By Lemma 23, we have

$$(V \otimes A)^{\mathrm{co}H} \cong (V \otimes B \otimes_B A)^{\mathrm{co}H} \cong V \otimes B$$

and we have a commutative diagram

$$
\begin{array}{ccc}
(V \otimes A)^{\mathrm{co}H} \otimes_B A & \xrightarrow{\mu_{V \otimes A}} & V \otimes A \\
\cong \Big\uparrow & & \Big\uparrow I_{V \otimes A} \\
V \otimes B \otimes_B A & \xrightarrow[\cong]{} & V \otimes A
\end{array}
$$

and we see that $\mu_{V \otimes A}$ is an isomorphism.

Let $\varphi : H \to A$ be a total integral. We have seen that $\rho_A : A \to A \otimes H$ has a section ν_A, namely

$$\nu_A(a \otimes h) = a_{[0]}\varphi(S(a_{[1]})h)$$

(cf. Theorem 47 and Proposition 50). $N \otimes A \otimes H \in \mathcal{M}(H)_A^H$, with structure induced by the structure on $A \otimes H$, i.e.

$$(n \otimes a \otimes h)b = n \otimes ab_{0]} \otimes hb_{[1]} \quad \text{and} \quad \rho^r = I_N \otimes I_A \otimes \Delta_H$$

The map

$$f : N \otimes A \otimes H \to N \; ; \; f(n \otimes a \otimes h) = n_{[0]}a_{[0]}\varphi(a_{[1]}S(n_{[1]})h)$$

is H-colinear. It is a k-split epimorphism, since

$$f(n_{[0]} \otimes 1 \otimes n_{[1]}) = n_{[0]}\varphi(S(n_{[1]})n_{[2]}) = n$$

H is projective as a k-module, so $A \otimes H$ is projective as a left A-module. The map can : $A \otimes A \to A \otimes H$ is a left A-linear epimorphism, so it has an A-linear splitting, and a fortiori a k-linear splitting. $N \otimes A \otimes A \in \mathcal{M}(H)_A^H$, with structure defined on the third factor:

$$(n \otimes a \otimes a')b = n \otimes a \otimes a'b \quad \text{and} \quad \rho^r = I_N \otimes I_A \otimes \rho_A$$

It is easy to check that

$$I_N \otimes \text{can} : \ N \otimes A \otimes A \to N \otimes A \otimes H$$

is a morphism in $\mathcal{M}(H)_A^H$, which is surjective, and k-split. Therefore

$$g = f \circ (I_N \otimes \text{can}) : \ N \otimes A \otimes A \to N$$

is surjective and k-split. Writing $\text{Ker}\,(g) = N'$, we obtain an exact sequence

$$0 \longrightarrow N' \longrightarrow N \otimes A \otimes A \overset{g}{\longrightarrow} N \longrightarrow 0 \qquad (4.16)$$

in $\mathcal{M}(H)_A^H$ which is split as a sequence of k-modules. Because $\mathcal{M}(H)_A^H \to \mathcal{M}_A$ is relative separable (see Proposition 50), (4.16) is also a split exact sequence of H-comodules. If we repeat the argument with N replaced by N', we find another exact sequence in $\mathcal{M}(H)_A^H$ and split in $\mathcal{M}^H$, namely

$$0 \longrightarrow N'' \longrightarrow N' \otimes A \otimes A \overset{g'}{\longrightarrow} N' \longrightarrow 0 \qquad (4.17)$$

Now write $N_1 = N \otimes A \otimes A$, $N_2 = N' \otimes A \otimes A$ Combining the two sequences, we find an exact sequence

$$N_2 \overset{g'}{\longrightarrow} N_1 \overset{g}{\longrightarrow} N \longrightarrow 0 \qquad (4.18)$$

in $\mathcal{M}(H)_A^H$. Since (4.16-4.17) are split exact in $\mathcal{M}^H$, they stay exact after we take H-coinvariants, and, combining them, we find an exact sequence in $\mathcal{M}_B$,

$$N_2^{coH} \longrightarrow N_1^{coH} \longrightarrow N^{coH} \longrightarrow 0$$

The tensor product is always right exact, so we have an exact sequence

$$N_2^{coH} \otimes_B A \longrightarrow N_1^{coH} \otimes_B A \longrightarrow N^{coH} \otimes_B A \longrightarrow 0$$

in $\mathcal{M}(H)_A^H$. Now we have the following commutative diagram with exact sequences in $\mathcal{M}(H)_A^H$:

$$
\begin{array}{ccccccc}
N_2^{coH} \otimes_B A & \longrightarrow & N_1^{coH} \otimes_B A & \longrightarrow & N^{coH} \otimes_B A & \longrightarrow & 0 \\
\Big\downarrow{\mu_{N_2}} & & \Big\downarrow{\mu_{N_1}} & & \Big\downarrow{\mu_N} & & \\
N_2 & \overset{g'}{\longrightarrow} & N_1 & \overset{g}{\longrightarrow} & N & \longrightarrow & 0
\end{array}
$$

Since the structure on $N_1 = N \otimes A \otimes A$ and $N_2 = N' \otimes A \otimes A$ is induced by the structure on the third factor A, we know that μ_{N_1} and μ_{N_2} are isomorphisms, and μ_N is an isomorphism.

Theorem 50. *[165, Theorem 1] Let H be a Hopf algebra with bijective antipode over a field k, and A a right H-comodule algebra. The following assertions are equivalent.*

1. *There exists a total integral* $\varphi : H \to A$, *and the map* can $: A \otimes_B A \to A \otimes H$ *is surjective;*
2. *The functors* F *and* G *are a pair of inverse equivalences between the categories* $\mathcal{M}(H)_A^H$ *and* $\mathcal{M}_B$;
3. *The functors* F' *and* G' *are a pair of inverse equivalences between the categories* $_A\mathcal{M}(H)^H$ *and* $_B\mathcal{M}$;
4. A *is a Hopf-Galois extension of* B, *and is faithfully flat as a left* B-*module;*
5. A *is a Hopf-Galois extension of* B, *and is faithfully flat as a right* B-*module.*

Proof. $\underline{1. \Rightarrow 2.}$ follows from Theorem 49, and $\underline{2. \Leftrightarrow 4.}$ follows from Theorem 48 (over a field k, every module is projective).

$\underline{4. \Rightarrow 1.}$ We have to show that A is relatively injective, or, equivalently, that A is right H-coflat, by Theorem 1. As can is bijective, can$'$: $A \otimes_B A \to A \otimes H$ is also bijective. can$'$ is a morphism in $_A\mathcal{M}(H)^H$, and it is therefore a morphism in $_B\mathcal{L}^H$. Recall that the right H-comodule structure on $A \otimes_B A$ is determined by the first factor, i.e. $\rho^r(a \otimes a') = (a_{[0]} \otimes a') \otimes a_{[1]}$. Keeping this in mind, we have a left B-linear map

$$(A\square_H V) \otimes_B A \to (A \otimes_B A)\square_H V \;\; ; \;\; \left(\sum_i a_i \otimes v_i\right) \otimes b \mapsto \sum_i (a_i \otimes b) \otimes v_i$$

for every left H-comodule V. This map is an isomorphism, because A is flat as a left B-module. Using the fact that also can$'$ is an isomorphism, we have a sequence of left B-module isomorphisms

$$(A\square_H V) \otimes_B A \cong (A \otimes_B A)\square_H V \cong (A \otimes H)\square_H V \cong A \otimes V$$

The functor

$$^C\mathcal{M} \to {}_k\mathcal{M} \;\; ; \;\; V \mapsto A \otimes V \cong (A\square_H V) \otimes_B A$$

is exact because k is a field; since A is faithfully flat as a left B-module, it follows that

$$^C\mathcal{M} \to {}_k\mathcal{M} \;\; ; \;\; V \mapsto A\square_H V$$

is also exact, and this proves that A is right H-coflat.

Hopf-Galois extensions and separability properties We use the same notation as before: H is a (k-flat) Hopf algebra, A is a right H-comodule algebra, and $B = A^{\mathrm{co}H}$.

Proposition 88. *If* A/B *is separable, then the functor* $\bullet \otimes H :$ $\mathcal{M}_A \to \mathcal{M}(H)_A^H$ *is separable.*

Proof. Let $e = e^1 \otimes e^2 \in A \otimes_B A$ be the separability idempotent. We apply part 2. of Theorem 38. Take

$$z = \mathrm{can}(e) = e^1 e^2_{[0]} \otimes e^2_{[1]}$$

$z \in W_5$, since, for all $a \in A$,

$$\begin{aligned} e^1 e^2_{[0]} a_\psi \otimes e^{2\psi}_{[1]} &= e^1 e^2_{[0]} a_{[0]} \otimes e^2_{[1]} a_{[1]} \\ &= e^1 (e^2 a)_{[0]} \otimes (e^2 a)_{[1]} \\ &= a e^1 e^2_{[0]} \otimes e^2_{[1]} \end{aligned}$$

Furthermore z is normalized, since

$$e^1 e^2_{[0]} \varepsilon(e^2_{[1]}) = e^1 e^2 = 1$$

The converse property holds if A is a Hopf-Galois extension of B:

Proposition 89. *If A is a Hopf-Galois extension of B, and the functor $\bullet \otimes H : \mathcal{M}_A \to \mathcal{M}(H)_A^H$ is separable, then A/B is separable.*

Proof. Let $z \in W_5$ be normalized. Then $\mathrm{can}^{-1}(z) = e$ is a separability idempotent.

Example 16. If A is a strongly G-graded ring, then A is separable over A_e if and only if the induction functor $\bullet \otimes kG : \mathcal{M}_A \to \mathcal{M}(kG)_A^{kG} = \mathrm{gr}\text{-}A$ is separable.

We will next study the separability of the functor $F : \mathcal{M}_B \to \mathcal{M}(H)_A^H$ and its adjoint G.

Proposition 90. *If k is left H-coflat, then the functor $F : \mathcal{M}_B \to \mathcal{M}(H)_A^H$ is separable.*

Proof. We apply Theorem 43 and its Corollary 20. We have that $k \otimes B = B$, $GF(k \otimes B) = A^{\mathrm{co}H} = B$, so

$$V_1 = {}_B\mathrm{Hom}_B(B, B)$$

and $I_B \in V_1$ satisfies the required normalizing property.

If k is a field, then k is left H-coflat if and only if k is injective as a left H-module, by Theorem 1. This condition is equivalent to the existence of a total integral $H \to k$, by Theorem 47, and this is equivalent to H being cosemisimple. So we have the following Corollary.

Corollary 28. *If H is a cosemisimple algebra over a field k, then the functor $F : \mathcal{M}_B \to \mathcal{M}(H)_A^H$ is separable.*

Proposition 91. *Assume that A is flat as a left B-module, and that A is a Hopf-Galois extension of B. Then the functor $G = (\bullet)^{coH} : \mathcal{M}(H)_A^H \to \mathcal{M}_B$ is separable.*

Proof. The functor F is exact, since A is left B-flat. G is left exact, so FG is left exact, and $FG(\rho_N)$ is injective for every relative Hopf module $N \in \mathcal{M}(H)_A^H$. Thus the assumptions in Theorem 44 and Corollary 21 are fulfilled, and we can conclude that G is separable if $\mu_{A\otimes H}$ has a right inverse which is left and right A-linear and left and right H-colinear. Now can has a two-sided inverse, so $\mu_{A\otimes H}$ has a two-sided inverse, and this inverse is left and right H-colinear and A-linear, since $\mu_{A\otimes H}$ itself is left and right H-colinear and A-linear.

Under some additional flatness assumptions, Proposition 91 has a converse:

Theorem 51. *Assume that k is right H-coflat and that A is left B-flat. The following assertions are equivalent:*

1. *A is a Hopf-Galois extension of B;*
2. *$G = (\bullet)^{coH} : \mathcal{M}(H)_A^H \to \mathcal{M}_B$ is separable;*
3. *G is full and faithful.*

Proof. $\underline{1. \Rightarrow 2.}$ follows from Proposition 91, and $\underline{3. \Rightarrow 1.}$ is obvious. $\underline{2. \Rightarrow 3.}$ If G is separable, then there exists a natural transformation $\zeta : 1_{\mathcal{M}(H)_A^H} \to FG$ such that $\mu \circ \zeta$ is the identity natural transformation; in particular $\mu_{A\otimes H} \circ \zeta_{A\otimes H} = I_{A\otimes H}$, and $\mu_{A\otimes H}$ is surjective. $A \in \mathcal{M}(H)_A^H$, the A-action is given by right multiplication. We claim that A is a T-generator of $\mathcal{M}(H)_A^H$ (see Section 3.5). By Theorem 40, we know that $A \otimes H$ is a T-generator, so it suffices to show that A generates $A \otimes H$. For every $a \in A$, we consider

$$\psi_a : A \to A \otimes H \; ; \; \psi_a(b) = \text{can}(a \otimes b) = ab_{[0]} \otimes b_{[1]}$$

It is easily verified that $\psi_a \in \mathcal{M}(H)_A^H$:

$$\psi_a(bb') = ab_{[0]}b'_{[0]} \otimes b_{[1]}b'_{[1]} = \psi_a(b)b'$$

and

$$(\psi_a \otimes I)\rho^r(b) = ab_{[0]} \otimes b_{[1]} \otimes b_{[2]} = \rho^r(\psi_a(b))$$

can is surjective, so for any $x \in A \otimes H$, we can find $y = \sum_i a_i \otimes b_i \in A \otimes_B A$ such that

$$\text{can}(y) = \sum_i \psi_{a_i}(b_i) = x$$

If $f : A \otimes H \to N$ in $\mathcal{M}(H)_A^H$ is such that $f \circ g = 0$ for all $g : A \to A \otimes H$ in $\mathcal{M}(H)_A^H$, then

$$f(x) = \sum_i f(\psi_{a_i}(b_i)) = 0$$

for all $x \in A \otimes H$, and $f = 0$. This proves that A generates $A \otimes H$.
Now $\mu_A : FG(A) = A^{\text{co}H} \otimes_B A \cong A \to A$ is the identity map on A, and

$$\zeta_A \circ \mu_A = I_A = I_{FG(A)}$$

We conclude that $\zeta \circ \mu$ and 1_{FG} coincide on the T-generator A, hence they coincide, by Theorem 41. This proves that G is fully faithful.

4.3 Doi's $[H, C]$-modules

Let H be a bialgebra with twisted antipode. H is a right H-comodule algebra: take $\rho^r = \Delta_H$. Consider a right H-comodule algebra C. Then (H, H, C) is a right-right Doi-Hopf structure, and the corresponding Doi-Hopf modules are $[C, H]$-modules in the sense of [67] (up to left-right conventions). In this Section, we examine the functor $\mathcal{M}(H)_H^C \to \mathcal{M}^C$ and its adjoint. As one might expect, the results are in duality with the results of Section 4.1.

A right H-linear map $\phi : C \to H$ is called an *integral*. An integral is called *total* if $\varepsilon_H \circ \phi = \varepsilon_C$. There is a close relationship between integrals and the space W_1' introduced in Section 3.4. In our situation, $\psi : C \otimes H \to H \otimes C$ is given by

$$\psi(c \otimes h) = h_{(1)} \otimes ch_{(2)}$$

Using the notation of Section 3.4, we have that V_1' consists of those $\vartheta \in (C \otimes H)^*$ that satisfy

$$\vartheta(c_{(1)} \otimes h_{(1)})c_{(2)}h_{(2)} = \vartheta(c_{(2)} \otimes h)c_{(1)} \tag{4.19}$$

for all $c \in C$ and $h \in H$. W_1' consists of maps $e \in \text{Hom}(C, H \otimes H)$ satisfying

$$e^1(c_{(1)}) \otimes e^2(c_{(1)}) \otimes c_{(2)}$$
$$= e^1(c_{(2)})_{(1)} \otimes e^2(c_{(2)})_{(1)} \otimes c_{(1)}e^1(c_{(2)})_{(2)}e^2(c_{(2)})_{(2)} \tag{4.20}$$
$$e^1(c) \otimes e^2(c)h = h_{(1)}e^1(ch_{(2)}) \otimes e^2(ch_{(2)}) \tag{4.21}$$

Proposition 92. *We have a map* $p : W_1' \to \text{Hom}^C(C, H)$ *given by*

$$p(e) = \phi = (\varepsilon_H \otimes I_H) \circ e$$

Proof. We see immediately that

$$\phi(c) = \varepsilon_H(e^1(c))e^2(c)$$

Applying ε_H to the first factor in (4.21), we find that $\phi(c)h = \phi(ch)$, and ϕ is right H-linear, as needed.

Proposition 93. *Let* $\phi: C \to H$ *be right* H-*linear. We define* $s(\phi) = e:$ $C \to H \otimes H$ *by*

$$e(c) = \overline{S}(\phi(c)_{(2)}) \otimes \phi(c)_{(1)}$$

Then $e \in W'_1$ *if and only if the following two conditions hold for all* $c \in C$ *and* $h \in H$:

$$\phi(c_{(2)}) \otimes c_{(1)} = \phi(c_{(1)}) \otimes c_{(2)} \qquad (4.22)$$

$$h_{(2)} \otimes \phi(ch_{(1)}) = h_{(1)} \otimes \phi(ch_{(2)}) \qquad (4.23)$$

Proof. Let A_1 and A_2 be respectively the left and right hand side of (4.20). Then

$$A_1 = \overline{S}(\phi(c_{(1)})_{(2)}) \otimes \phi(c_{(1)})_{(1)} \otimes c_{(2)}$$

$$A_2 = \overline{S}(\phi(c_{(2)})_{(2)})_{(1)} \otimes \phi(c_{(1)})_{(1)(1)} \otimes c_{(1)}\overline{S}(\phi(c_{(2)})_{(2)})_{(2)}\phi(c_{(1)})_{(1)(2)}$$

$$= \overline{S}(\phi(c_{(2)})_{(4)}) \otimes \phi(c_{(2)})_{(1)} \otimes c_{(1)}\overline{S}(\phi(c_{(2)})_{(3)})\phi(c_{(2)})_{(2)}$$

$$= \overline{S}(\phi(c_{(2)})_{(2)}) \otimes \phi(c_{(2)})_{(1)} \otimes c_{(1)}$$

Assuming that $A_1 = A_2$, apply $\varepsilon_H \otimes I_H \otimes I_C$ to both sides. Then we find (4.22). Conversely, if (4.22) holds, then (4.20) follows easily after we apply $(\overline{S} \otimes I_H) \circ \Delta_H^{\mathrm{cop}}$ to the first factor of both sides of (4.22).

Now let B_1 and B_2 be the left and right hand side of (4.21):

$$B_1 = \overline{S}(\phi(c)_{(2)}) \otimes \phi(c)_{(1)}h$$

$$B_2 = h_{(1)}\overline{S}(\phi(ch_{(2)})_{(2)}) \otimes \phi(ch_{(2)})_{(1)}$$

Assume that $B_1 = B_2$, apply Δ_H^{cop} to the second factor, and then multiply the first two factors. We obtain

$$\overline{S}(\phi(c)_{(3)})\phi(c)_{(2)}h_{(2)} \otimes \phi(c)_{(1)}h_{(1)} = h_{(2)} \otimes \phi(c)h_{(1)} = h_{(2)} \otimes \phi(ch_{(1)})$$

$$= h_{(1)}\overline{S}(\phi(ch_{(2)})_{(3)})\phi(ch_{(2)})_{(2)} \otimes \phi(ch_{(2)})_{(1)} = h_{(1)} \otimes \phi(ch_{(2)})$$

and (4.23) follows. Conversely, (4.23) implies that

$$B_2 = h_{(2)}\overline{S}(\phi(ch_{(1)})_{(2)}) \otimes \phi(ch_{(1)})_{(1)}$$

$$= h_{(3)}\overline{S}(\phi(c)_{(2)}h_{(2)}) \otimes \phi(c)_{(1)}h_{(1)}$$

$$= \overline{S}(\phi(c)_{(2)}) \otimes \phi(c)_{(1)}h = B_1$$

finishing our proof.

Corollary 29. *Let* H *be a bialgebra with twisted antipode, and* C *a right* H-*module coalgebra. If there exists a total integral* $\phi: C \to H$ *satisfying (4.22) and (4.23), then the forgetful functor* $\mathcal{M}(H)_H^C \to \mathcal{M}^C$ *is separable.*

Proof. According to Proposition 93, $s(\phi) = e \in W'_1$. We easily compute that

$$e^1(c)e^2(c) = \overline{S}(\phi(c)_{(2)})\phi(c)_{(1)} = \varepsilon_H(\phi(c)) = \varepsilon_C(c)$$

and our result follows from part 2) of Proposition 76

4.4 Yetter-Drinfeld modules

Let K and H be bialgebras, A a (K, H)-bicomodule algebra, and C an (H, K)-bimodule coalgebra. A left-right *Yetter-Drinfeld module* is a k-module M together with a left A-action and a right C-coaction satisfying the compatibility relation

$$(a_{[0]}m)_{[0]} \otimes (a_{[0]}m)_{[1]}a_{[-1]} = a_{[0]}m_{[0]} \otimes a_{[1]}m_{[1]} \tag{4.24}$$

In a similar way, we define right-left Yetter-Drinfeld modules: now we need a right A-action and a left C-coaction such that

$$a_{[1]}(ma_{[0]})_{[-1]} \otimes (ma_{[0]})_{[0]} = m_{[-1]}a_{[-1]} \otimes m_{[0]}a_{[0]} \tag{4.25}$$

Our notation for the category of left-right Yetter-Drinfeld modules and A-linear C-colinear maps will be $_A\mathcal{YD}(K, H)^C$. A similar notation will be used in the right-left case.

If K has a twisted antipode $\overline{S}_K$, then (4.24) is clearly equivalent to

$$\rho^r(am) = a_{[0]}m_{[0]} \otimes a_{[1]}m_{[1]}\overline{S}_K(a_{[-1]}) \tag{4.26}$$

We will then call (K, H, A, C) a left-right *Yetter-Drinfeld structure*. A morphism

$$(K, H, A, C) \to (K', H', A', C')$$

between two Yetter-Drinfeld structures consists of a fourtuple $(\kappa, \hbar, \alpha, \gamma)$, with $\kappa : K \to K'$, $\hbar : H \to H'$ bialgebra maps, $\alpha : A \to A'$ an algebra map, and $\gamma : C \to C'$ a coalgebra map such that

$$\rho^{lr}_{A'} \circ \alpha = (\kappa \otimes \alpha \otimes \hbar) \circ \rho^{lr}_A$$

and

$$\gamma(hck) = \hbar(h)\gamma(c)\kappa(k)$$

for all $h \in H$, $c \in C$, and $k \in K$. $_\bullet \mathbb{YD}^\bullet(k)$ is the category of left-right Yetter-Drinfeld structures.

In a similar way, if H has a twisted antipode $\overline{S}_H$, then (4.25) is equivalent to

$$\rho^l(ma) = \overline{S}_H(a_{[1]})m_{[-1]}a_{[-1]} \otimes m_{[0]}a_{[0]} \tag{4.27}$$

and we call (K, H, A, C) a right-left Yetter-Drinfeld structure. The category of right-left Yetter-Drinfeld structures is denoted by $^\bullet\mathbb{YD}_\bullet(k)$.

Proposition 94. *We have a functor*

$$F : \,_\bullet\mathbb{YD}^\bullet(k) \to \,_\bullet\mathbb{DK}^\bullet(k)$$

defined as follows:

$$F(K, H, A, C) = (K^{\mathrm{op}} \otimes H, A, C)$$

where the right $K^{op} \otimes H$-coaction on A and the left $K^{op} \otimes H$-action on C are given by the formulas

$$\rho^r_{K^{op} \otimes H}(a) = a_{[0]} \otimes \overline{S}_K(a_{[-1]}) \otimes a_{[1]} \qquad (4.28)$$

$$(k \otimes h) \rhd c = hck \qquad (4.29)$$

for all $a \in A$, $c \in C$, $h \in H$ and $k \in K$. Moreover, we have an isomorphism of categories

$$_A\mathcal{YD}(K, H)^C \cong {}_A\mathcal{M}(K^{op} \otimes H)^C \qquad (4.30)$$

Let (A, C, ψ) be the corresponding entwining structure (cf. Proposition 17). The map $\psi: A \otimes C \to A \otimes C$ is then given by the formula

$$\psi(a \otimes c) = a_{[0]} \otimes a_{[1]} c \overline{S}_K(a_{[-1]}) \qquad (4.31)$$

Proof. A routine verification. We will use the notation

$$\rho^r_{K^{op} \otimes H}(a) = a_{\{0\}} \otimes a_{\{1\}}$$

A is a right $K^{op} \otimes H$-comodule algebra since

$$a_{\{0\}} \otimes \Delta_{K^{op} \otimes H}(a_{\{1\}}) = a_{[0]} \otimes \left(\overline{S}_K(a_{[-1]}) \otimes a_{[1]}\right) \otimes \left(\overline{S}_K(a_{[-2]}) \otimes a_{[2]}\right)$$
$$= \rho^r_{K^{op} \otimes H}(a_{\{0\}}) \otimes a_{\{1\}}$$

and

$$\rho^r_{K^{op} \otimes H}(ab) = a_{[0]}b_{[0]} \otimes \overline{S}_K(a_{[-1]}b_{[-1]}) \otimes a_{[1]}b_{[1]}$$
$$= a_{[0]}b_{[0]} \otimes \overline{S}_K(b_{[-1]})\overline{S}_K(a_{[-1]}) \otimes a_{[1]}b_{[1]}$$
$$= \rho^r_{K^{op} \otimes H}(a)\rho^r_{K^{op} \otimes H}(b)$$

Clearly C is a left $K^{op} \otimes H$-module. C is a $K^{op} \otimes H$-module coalgebra, since

$$\Delta_C((k \otimes h) \rhd c) = \Delta_C(hck)$$
$$= h_{(1)}c_{(1)}k_{(1)} \otimes h_{(2)}c_{(2)}k_{(2)}$$
$$= (k_{(1)} \otimes h_{(1)}) \rhd c_{(1)} \otimes (k_{(2)} \otimes h_{(2)}) \rhd c_{(2)}$$

$M \in {}_A\mathcal{YD}(K, H)^C$ if and only if ${}_A\mathcal{M}(K^{op} \otimes H)^C$. Indeed, (2.19) amounts to

$$\rho^r(am) = a_{[0]}m_{[0]} \otimes (\overline{S}_K(a_{[-1]}) \otimes a_{[1]}) \rhd m_{[1]}$$
$$= a_{[0]}m_{[0]} \otimes a_{[1]}m_{[1]}(\overline{S}_K(a_{[-1]}))$$

and this is exactly (4.26)

Remarks 7. 1. If H has an antipode, then $K^{op} \otimes H$ also has an antipode, and the map ψ is bijective.
2. If $A = C = H = K$, then we obtain the classical Yetter-Drinfeld modules, also named *crossed modules* or *quantum Yang-Baxter modules*. In this situation, ψ is bijective if and only if H has a bijective antipode.

Using our left-right dictionary, we can find the right-left version of Proposition 94. First observe that we have an isomorphism of categories

$$_\bullet \mathrm{YD}^\bullet(k) \cong {}^\bullet \mathrm{YD}_\bullet(k)$$

We send (K, H, A, C) to $(H^{\mathrm{opcop}}, K^{\mathrm{opcop}}, A^{\mathrm{op}}, C^{\mathrm{cop}})$. Moreover,

$$_A\mathcal{YD}(K, H)^C \cong {}^{C^{\mathrm{cop}}}\mathcal{YD}(H^{\mathrm{opcop}}, K^{\mathrm{opcop}})_{A^{\mathrm{op}}}$$

Using this isomorphism, and Proposition 94, we obtain a functor

$$F' : {}^\bullet \mathrm{YD}_\bullet(k) \to {}^\bullet \mathrm{DK}_\bullet(k)$$

such that the diagram

$$
\begin{array}{ccc}
_\bullet \mathrm{YD}^\bullet(k) & \xrightarrow{\ F\ } & _\bullet \mathrm{DK}^\bullet(k) \\
\Big\downarrow{\cong} & & \Big\downarrow{\cong} \\
{}^\bullet \mathrm{YD}_\bullet(k) & \xrightarrow{\ F'\ } & {}^\bullet \mathrm{DK}_\bullet(k)
\end{array}
$$

commutes, namely

$$F'(K, H, A, C) = (K \otimes H^{\mathrm{op}}, A, C)$$

with

$$\rho^l(a) = a_{[-1]} \otimes \overline{S}_H(a_{[1]}) \otimes a_{[0]} \quad \text{and} \quad c \triangleleft (k \otimes h) = hck$$

We can also introduce one-sided Yetter-Drinfeld modules. Let A be a (K, H)-bimodule algebra, and C a (K, H)-bimodule coalgebra. The compatibility conditions for left-left and right-right Yetter-Drinfeld modules are respectively

$$(a_{[0]}m)_{[-1]}a_{[1]} \otimes (a_{[0]}m)_{[0]} = a_{[-1]}m_{[-1]} \otimes a_{[0]}m_{[0]} \tag{4.32}$$

$$(ma_{[0]})_{[0]} \otimes a_{[-1]}(ma_{[0]})_{[1]} = m_{[0]}a_{[0]} \otimes m_{[1]}a_{[1]} \tag{4.33}$$

If H (resp. K) is a Hopf algebra, these relations are equivalent to

$$\rho^l(am) = a_{[-1]}m_{[-1]}S_H(a_{[1]}) \otimes a_{[0]}m_{[0]} \tag{4.34}$$

$$\rho^r(ma) = m_{[0]}a_{[0]} \otimes S_K(a_{[-1]})m_{[1]}a_{[1]} \tag{4.35}$$

In this situation, (K, H, A, C) is called a left-left (resp. a right-right) Yetter-Drinfeld structure. $\mathrm{YD}_\bullet^\bullet(k)$ and $_\bullet^\bullet\mathrm{YD}(k)$ are the categories of right-right and left-left Yetter-Drinfeld module structures. The categories of left-left (resp. right-right) Yetter-Drinfeld modules are denoted by $_A^C\mathcal{YD}(K, H)$ (resp. $\mathcal{YD}(K, H)_A^C$).

The above results can be extended easily to the one-sided case. For example, we have a functor $F : \mathrm{YD}_\bullet^\bullet(k) \to \mathrm{DK}_\bullet^\bullet(k)$, mapping (K, H, A, C) to $(K^{\mathrm{op}} \otimes H, A, C)$, with

$$\rho^r_{K^{\mathrm{op}} \otimes H}(a) = a_{[0]} \otimes S_K(a_{[-1]}) \otimes a_{[1]} \quad \text{and} \quad c \triangleleft (k \otimes h) = kch$$

and we have an isomorphism of categories

$$\mathcal{YD}(K, H)^C_A \cong \mathcal{M}(K^{\mathrm{op}} \otimes H)^C_A$$

Proposition 95. *The categories* $\mathbb{YD}^\bullet_\bullet(k)$, $^\bullet_\bullet\mathbb{YD}(k)$, $_\bullet\mathbb{YD}^\bullet(k)$, *and* $^\bullet\mathbb{YD}_\bullet(k)$
*are isomorphic. The corresponding categories of Yetter-Drinfeld modules are
also isomorphic.*

Proof. We have already seen that $_\bullet\mathbb{YD}^\bullet(k) \cong {}^\bullet\mathbb{YD}_\bullet(k)$. Let us define the
isomorphism

$$P : \ \mathbb{YD}^\bullet_\bullet(k) \to {}_\bullet\mathbb{YD}^\bullet(k)$$

Take $(K, H, A, C) \in \mathbb{YD}^\bullet_\bullet(k)$. This means that K has an antipode, A is a
(K, H)-bicomodule algebra and C is a (K, H)-bimodule coalgebra. We define

$$P(K, H, A, C) = (K^{\mathrm{op}}, H^{\mathrm{op}}, A^{\mathrm{op}}, C)$$

K^{op} has a twisted antipode, as needed. A^{op} is a $(K^{\mathrm{op}}, H^{\mathrm{op}})$-bicomodule (the
"op" does not matter), and a $(K^{\mathrm{op}}, H^{\mathrm{op}})$-comodule algebra (we did put the
"op" everywhere). C is a (K, H)-bimodule, and therefore an $(H^{\mathrm{op}}, K^{\mathrm{op}})$-
bimodule. C is an $(H^{\mathrm{op}}, K^{\mathrm{op}})$-bimodule coalgebra since

$$\Delta(h \cdot c \cdot k) = \Delta(kch) = k_{(1)} c_{(1)} h_{(1)} \otimes k_{(2)} c_{(2)} h_{(2)} = h_{(1)} \cdot c_{(1)} \cdot k_{(1)} \otimes h_{(2)} \cdot c_{(2)} \cdot k_{(2)}$$

Remark 11. We have a commutative diagram of functors (cf. Proposition 15)

$$
\begin{array}{ccc}
\mathbb{YD}^\bullet_\bullet(k) & \xrightarrow{\ P\ } & {}_\bullet\mathbb{YD}^\bullet(k) \\
{\scriptstyle F}\downarrow & & \downarrow{\scriptstyle F} \\
\mathbb{DK}^\bullet_\bullet(k) & \longrightarrow & {}_\bullet\mathbb{DK}^\bullet(k)
\end{array}
$$

Applying Theorem 19, we obtain

Corollary 30. *If* C *is flat as a* k-module, then $_A\mathcal{YD}(K, H)^C$ *is a Grothendieck
category.*

Now we take $A = C = H = K$, so that we have classical Yetter-Drinfeld
modules. We view A and C as bialgebras, respectively $A = H$ and $C = H^{\mathrm{op}}$.

Proposition 96. *Let* H *be a bialgebra with twisted antipode. Then the category of Yetter-Drinfeld modules* $_H\mathcal{YD}(H, H)^H$ *is a monoidal category.*

Proof. It suffices to verify (2.111) and (2.113). From (4.31), we know that, for all $a, c \in H$:

$$a_\psi \otimes c^\psi = a_{(2)} \otimes a_{(3)} c \overline{S}(a_{(1)})$$

We easily compute that

$$a_{(1)\psi} \otimes a_{(2)\Psi} \otimes d^\Psi c^\psi = a_{(2)} \otimes a_{(5)} \otimes a_{(6)} d \overline{S}(a_{(4)}) a_{(3)} c \overline{S}(a_{(1)})$$

$$= a_{(2)} \otimes a_{(3)} \otimes a_{(4)} dc \overline{S}(a_{(1)}) = \Delta(a_\psi) \otimes (dc)^\psi$$

as needed (the multiplication on C is opposite to the one on H). Finally

$$\varepsilon_A(a_\psi) 1_C^\psi = \varepsilon_A(a_{(2)}) a_{(3)} 1_C \overline{S}(a_{(1)}) = \varepsilon_A(a) 1_C$$

From Corollary 4, we find

Corollary 31. *The category of Yetter-Drinfeld modules over a field k is a Grothendieck category with enough injective objects.*

The Drinfeld double Now let $(K, H, A, C) \in {}_\bullet \mathbb{YD}^\bullet(k)$ (K has a twisted antipode), and assume that C is finitely generated and projective as a k-module, with finite dual basis $\{d_i, d_i^* \mid i = 1, \cdots, n\}$. We have $(A, C, \psi) \in {}_\bullet \mathbb{E}^\bullet(k)$ (see (4.31)), and Theorem 8 delivers a smash product structure $(A, C^*, R) \in \mathbb{S}(k)$. We write

$$D_1 = A \#_R C^* = A \bowtie C^*$$

(2.30) tells us that

$$R(c^* \otimes a) = \sum_i \langle c^*, d_i^\psi \rangle a_\psi \otimes d_i^*$$

$$= \sum_i \langle c^*, a_{[1]} d_i \overline{S}_K(a_{[-1]}) \rangle a_{[0]} \otimes d_i^*$$

$$= \sum_i \langle \overline{S}(a_{[-1]}) c^* a_{[1]}, d_i \rangle a_{[0]} \otimes d_i^*$$

$$= a_{[0]} \otimes \overline{S}(a_{[-1]}) c^* a_{[1]}$$

Thus the multiplication on D_1 is given by the formula

$$(a \bowtie c^*)(b \bowtie d^*) = a b_{[0]} \bowtie (\overline{S}_K(b_{[-1]} c^* b_{[1]}) * d^* \qquad (4.36)$$

Now assume that $(K, H, A, C) \in {}^\bullet \mathbb{YD}_\bullet(k)$ (H has a twisted antipode). Now we find $(A, C, \varphi) \in {}^\bullet \mathbb{E}_\bullet(k)$, and, if C is finitely generated and projective, a smash product structure (C^*, A, S) (see Theorem 9). We now find $S : A \otimes C^* \to C^* \otimes A$ given by

$$S(a \otimes c^*) = a_{[-1]} c^* \overline{S}_H(a_{[1]}) \otimes a_{[0]}$$

The multiplication on

$$D_2 = C^* \#_S A = C^* \bowtie A$$

is the following

$$(c^* \bowtie a)(d^* \bowtie b) = c^* * a_{[-1]}d^* \overline{S}_H(a_{[1]}) \bowtie a_{[0]}b$$

Now assume that both H and K have a twisted antipode. It is easy to check that $S = R^{-1}$, and we conclude from Proposition 23 that $A \bowtie C^* \cong C^* \bowtie A$. We call $A \bowtie C^*$ the *Drinfeld double* associated to (K, H, A, C).

The classical situation is $K = H = A = C$, where H is a bialgebra with twisted antipode. The classical Drinfeld double is

$$D(H) = H^* \bowtie H$$

with multiplication rule

$$(h^* \bowtie h)(k^* \bowtie k) = h^* * (h_{(1)}k^* \overline{S}(h_{(3)})) \bowtie h_{(2)}k \qquad (4.37)$$

(Compare to [108, IX.4.2]). Let us make clear also that the Drinfeld double that we introduced also coincides with the one of Majid's book [128]. First observe that

$$\langle hk^*k, l \rangle = \langle k^*, klh \rangle = \langle k^*_{(1)}, k \rangle \langle k^*_{(2)}, l \rangle \langle k^*_{(3)}, h \rangle$$

hence

$$hk^*k = \langle k^*_{(1)}, k \rangle \langle k^*_{(3)}, h \rangle k^*_{(2)}$$

and (4.37) can be rewritten as

$$(h^* \bowtie h)(k^* \bowtie k) = \langle k^*_{(1)}, \overline{S}(h_{(3)}) \rangle \langle k^*_{(3)}, h_{(1)} \rangle h^* * k^*_{(2)} \bowtie h_{(2)}k \qquad (4.38)$$

which is the multiplication rule found in [128, Exercise 7.1.2]. If we regard H^* and H as subalgebras of $D(H) = H^* \bowtie H$ via $h \mapsto \varepsilon \bowtie h$ and $h^* \mapsto h^* \otimes 1$, then (4.38) can be written as a commutation rule:

$$hk^* = \langle k^*_{(1)}, \overline{S}(h_{(3)}) \rangle \langle k^*_{(3)}, h_{(1)} \rangle k^*_{(2)} h_{(2)}$$

and this can be written in a more symmetric form:

$$\langle k^*_{(1)}, h_{(2)} \rangle h_{(1)} k^*_{(2)} = \langle k^*_{(2)}, h_{(1)} \rangle k^*_{(1)} h_{(2)} \qquad (4.39)$$

Proposition 97. *Let H be a finitely generated projective Hopf algebra with invertible antipode. Then $D(H) = H^{*\mathrm{cop}} \bowtie H$ is a Hopf algebra. The antipode is given by*

$$S_{D(H)}(h^* \bowtie h) = (\varepsilon \bowtie S(h))(\overline{S}^*(h^*) \bowtie 1) \qquad (4.40)$$

Proof. This is an immediate application of Proposition 39. Using (4.38), we find

$$R(h \otimes h^*) = \langle h^*_{(1)}, \overline{S}(h_{(3)}) \rangle \langle h^*_{(3)}, h_{(1)} \rangle h^*_{(2)} \otimes h_{(2)} \qquad (4.41)$$

and

$$h^*_{(2)R} \otimes h_{(1)R} \otimes h^*_{(1)r} \otimes h_{(2)r}$$
$$= \langle h^*_{(4)}, \overline{S}(h_{(3)}) \rangle \langle h^*_{(6)}, h_{(1)} \rangle \langle h^*_{(1)}, \overline{S}(h_{(6)}) \rangle \langle h^*_{(3)}, h_{(4)} \rangle$$
$$h^*_{(5)} \otimes h_{(2)} \otimes h^*_{(2)} \otimes h_{(5)}$$
$$= \langle h^*_{(1)}, \overline{S}(h_{(4)}) \rangle \langle h^*_{(4)}, h_{(1)} \rangle h^*_{(3)} \otimes h_{(2)} \otimes h^*_{(2)} \otimes h_{(3)}$$
$$= h^*_{R(2)} \otimes h_{R(1)} \otimes h^*_{R(1)} \otimes h_{R(2)}$$

proving (2.100). (2.101) is easy.

Frobenius properties Since Yetter-Drinfeld modules are Doi-Koppinen Hopf modules, and the Drinfeld double is a smash product, we can use the results of Sections 3.2 and 3.3 to obtain criteria for the forgetful functor $_H\mathcal{YD}(H,H)^H \to {}_H\mathcal{M}$, or the extension $D(H)/H$ to be separable of Frobenius. We will apply these results in this Section; in particular, we will show that there is a connection between Frobenius properties for the Drinfeld double and the unimodularity of the underlying Hopf algebra. Our first result is due to Radford [154] in the case where k is a field; in the case where k is a commutative ring, it appeared in [49].

Theorem 52. *If H is a Frobenius Hopf algebra, then $D(H)$ is Frobenius and unimodular.*

Proof. Using Theorem 31, we find free generators t and φ for $\int_H^l$ and $\int_{H^*}^l$ such that $\langle \varphi, t \rangle = 1$. Let α be the distinguished element in H^*, i.e.

$$th = \alpha(h)t$$

for all $h \in H$ (cf. Proposition 63). The multiplication in $D(H)$ can be written in the following way:

$$(h^* \bowtie h)(k^* \bowtie k) = h^* * h_{(1)} \cdot k^* \cdot \overline{S}(h_{(3)}) \bowtie h_{(2)}k \qquad (4.42)$$

$\{k_i, k_i^* \mid i = 1, \cdots, n\}$ will be a dual basis for H. We will first compute $\int_{D(H)}^l$. A left integral y in $D(H)$ can be written under the form

$$y = \sum_{i=1}^n h_i^* \bowtie k_i$$

for some $h_i^* \in H^*$. For all $h^* \in H^*$, we have

$$(h^* \bowtie 1)y = \langle h^*, 1 \rangle y$$

or

$$\sum_{i=1}^n h^* * h_i^* \bowtie k_i = \langle h^*, 1 \rangle \sum_{i=1}^n h_i^* \bowtie k_i$$

For a fixed index j, we apply k_j^* to the second factor. This gives us

$$h^* * \sum_{i=1}^n \langle k_j^*, k_i \rangle h_i^* = \sum_{i=1}^n \langle h^*, 1 \rangle \langle k_j^*, k_i \rangle h_i^*$$

and

$$\sum_{i=1}^n \langle k_j^*, k_i \rangle h_i^* \in \int_{H^\bullet}^l$$

so that we can write

$$\sum_{i=1}^n \langle k_j^*, k_i \rangle h_i^* = x_j \varphi$$

for some $x_j \in k$. Then

$$y = \sum_{i,j=1}^n \langle k_j^*, k_i \rangle h_i^* \bowtie k_j = \sum_{j=1}^n x_j \varphi \otimes k_j = \varphi \otimes k$$

where we wrote $k = \sum_{j=1}^n x_j k_j$. Next

$$(\varepsilon \bowtie h)y = \varepsilon(h)y$$

for all $h \in H$, so

$$\varepsilon(h)\varphi \bowtie k = (\varepsilon \bowtie h)(\varphi \bowtie k) = h_{(1)} \cdot \varphi \cdot \overline{S}(h_{(3)}) \bowtie h_{(2)} k$$

We apply the first factor to t, and find, using $\langle \varphi, t \rangle = 1$,

$$\varepsilon(h)k = \langle \varphi, \overline{S}(h_{(3)})th_{(1)} \rangle h_{(2)} k$$
$$= \langle \varphi, \varepsilon(h_{(3)})\alpha(h_{(1)})t \rangle h_{(2)} k = \langle \alpha, h_{(1)} \rangle h_{(2)} k$$

and

$$\langle \alpha, S(h) \rangle k = hk$$

Apply S to both sides:

$$\langle \alpha, S(h) \rangle S(k) = S(k)S(h)$$

This holds for all $h \in H$, and since S is bijective, we have

$$\langle \alpha, h \rangle S(k) = S(k)h$$

This implies that $S(k) \in \int_\alpha^r = \int_H^l$ and $k \in \int_H^r$, and we have shown that

$$\int_{D(H)}^l \subset \int_{H^\bullet}^l \bowtie \int_H^r = k(\varphi \bowtie u)$$

where $u = \overline{S}(t)$. This inclusion is an equality. To see this, we put

$$I = \{x \in k \mid x(\varphi \bowtie u) \in \int_{D(H)}^l \}$$

Clearly I is an ideal of k. Using Proposition 62, we find $y_j \in \int_{D(H)}^l$ and $y_j^* \in D(H)^*$ such that

$$\sum_{j=1}^m \langle y_j^*, y_j \rangle = 1$$

We can write $y_j = x_j(\varphi \bowtie u)$, with $x_j \in I$; we obtain

$$1 = \sum_{j=1}^m \langle y_j^*, (\varphi \bowtie u) \rangle x_j \in I$$

and $I = k$, or

$$\int_{D(H)}^l = k(\varphi \bowtie u)$$

This shows that $D(H)$ is Frobenius; to complete the proof, it suffices to show that $\varphi \bowtie u$ is also a right integral in $D(H)$. To see this, we proceed as follows. $\varphi \bowtie u$ is a left integral, so

$$\varepsilon(h)(\varphi \bowtie u) = (\varepsilon \bowtie h)(\varphi \bowtie u) = h_{(1)} \cdot \varphi \cdot \overline{S}(h_{(3)}) \bowtie \alpha(S(h_{(2)}))u$$

for all $h \in H$, and this implies

$$\langle \alpha, S(h_{(2)}) \rangle h_{(1)} \cdot \varphi \cdot \overline{S}(h_{(3)}) = \varepsilon(h)\varphi \tag{4.43}$$

(4.43) holds in any Frobenius Hopf algebra. We write down (4.43) with H replaced by H^*. Let g be the distinguished element in H, then (4.43) takes the form

$$\begin{aligned}
\langle h^*, 1 \rangle t &= \langle S^*(h_{(2)}^*), g \rangle h_{(1)}^* \cdot t \cdot \overline{S}^*(h_{(3)}^*) \\
&= \langle h_{(2)}^*, S(g) \rangle \langle h_{(1)}^*, t_{(3)} \rangle \langle h_{(3)}^*, \overline{S}(t_{(1)}) \rangle t_{(2)} \\
&= \langle h^*, t_{(3)} S(g) \overline{S}(t_{(1)}) \rangle t_{(2)} \tag{4.44}
\end{aligned}$$

Now $t = S(u)$, so

$$\begin{aligned}
\langle S^*(h^*), 1 \rangle S(u) &= \langle h^*, 1 \rangle t = \langle h^*, S(u_{(1)})S(g)u_{(3)} \rangle S(u_{(2)}) \\
&= \langle S^*(h^*), \overline{S}(u_{(3)})gu_{(1)} \rangle S(u_{(2)})
\end{aligned}$$

S and S^* are bijective, so

$$\langle h^*, 1 \rangle u = \langle h^*, \overline{S}(u_{(3)})gu_{(1)} \rangle u_{(2)} \tag{4.45}$$

Now we can prove that $\varphi \bowtie u$ is a right integral:

$$(\varphi \bowtie u)(\varepsilon \bowtie h) = \varphi \bowtie uh = \varepsilon(h)(\varphi \bowtie u)$$

and

$$
\begin{aligned}
(\varphi \bowtie u)(h^* \bowtie a) &= \varphi * (u_{(1)} \cdot h^* \cdot \overline{S}(u_{(3)})) \bowtie u_{(2)} \\
&= \varphi \langle u_{(1)} \cdot h^* \cdot \overline{S}(u_{(3)}), g \rangle \bowtie u_{(2)} \\
&= \varphi \bowtie \langle h^*, \overline{S}(u_{(3)}) g u_{(1)} \rangle a n u_{(2)} \\
(4.45) \qquad &= \langle h^*, 1 \rangle \varphi \bowtie u
\end{aligned}
$$

We are now able to give necessary and sufficient conditions for the Drinfeld double $D(H)$ to be Frobenius over H.

Theorem 53. *For a Hopf algebra H with bijective antipode over a commutative ring k, the following conditions are equivalent:*

1. *H is finitely generated and projective, and $D(H)/H$ is Frobenius;*
2. *H is Frobenius over k and unimodular;*
3. *H is finitely generated and projective, and there exists $t \in H$ such that the map*

$$\phi_H : H^* \to H \; ; \quad \phi_H(h^*) = \langle h^*, t_{(2)} \rangle t_{(1)} \tag{4.46}$$

is bijective, and

$$h_{(2)} \otimes th_{(1)} = h_{(1)} \otimes h_{(2)} t \tag{4.47}$$

for all $h \in H$.

Proof. $\underline{1. \Rightarrow 2.}$ It follows from Corollary 12 that H/k is Frobenius (take into account that the Drinfeld double is a particular case of the smash product). From Theorem 52, it follows that $D(H)$ is unimodular. We will use the same notation as in the proof of Theorem 52, namely

$$\int_H^l = kt \; ; \quad \int_H^r = ku \; ; \quad \int_{H^*}^l = k\varphi \; ; \quad \int_{H^*}^r = k\psi$$

with

$$u = S(t) \; ; \quad \psi = \overline{S}^*(\varphi) \; ; \quad \langle \varphi, t \rangle = 1$$

Recall from Proposition 55 that we have an isomorphism of functors $\Phi : F' \to F$, where $F, F' : \mathcal{M}_H \to \mathcal{M}_{D(H)}$ with

$$F(M) = M \otimes H^* \quad \text{and} \quad F'(M) = H \otimes M$$

for every right H-module M. According to (3.31), the right $D(H)$-action on $F(M) = M \otimes H^*$ is given by

$$(m \otimes k^*) \triangleleft (h^* \bowtie h) = mh_{\overline{R}} \otimes (k^* * h^*)_{\overline{R}} \tag{4.48}$$

and the right $D(H)$-action on $F'(M) = H \otimes M$ is (cf. (3.32))

$$(k \otimes m) \triangleleft (h^* \bowtie h) = \sum_i \langle h^* * k_{iR}^*, k \rangle k_i \otimes m h_R \qquad (4.49)$$

$M = kt$ is a right H-module, and we have an isomorphism

$$\Phi_M : H \otimes kt \to kt \otimes H^*$$

We write

$$\Phi_M^{-1}(t \otimes \psi) = h_0 \otimes t$$

with $h_0 \in H$. In Theorem 52, we have seen that $\varphi \bowtie u$ is a free generator of $\int_{D(H)}^l = \int_{D(H)}^r$. This implies that

$$S_{D(H)}(\varphi \bowtie u) = (\varepsilon \bowtie S(u))(\overline{S}^*(\varphi) \bowtie 1) = (\varepsilon \bowtie t)(\psi \bowtie 1)$$

is also a free generator of $\int_{D(H)}^l = \int_{D(H)}^r$. In particular, we find for all $h \in H$:

$$\begin{aligned}
\varepsilon(h)(\varepsilon \bowtie t)(\psi \bowtie 1) &= (\varepsilon \bowtie t)(\psi \bowtie 1)(\varepsilon \bowtie h) \\
&= (\varepsilon \bowtie t)(\psi \bowtie h) \\
&= (\varepsilon \bowtie t)(\varepsilon \bowtie h_{\overline{R}})(\psi_{\overline{R}} \bowtie 1) \\
&= (\varepsilon \bowtie t h_{\overline{R}})(\psi_{\overline{R}} \bowtie 1)
\end{aligned}$$

We can rewrite this as

$$\varepsilon(h) R(t \otimes \psi) = R(t h_{\overline{R}} \otimes \psi_{\overline{R}})$$

or, since R is invertible,

$$\varepsilon(h) t \otimes \psi = t h_{\overline{R}} \otimes \psi_{\overline{R}}$$

Using this formula, we compute

$$\begin{aligned}
(t \otimes \psi) \triangleleft (h^* \otimes h) &= t h_{\overline{R}} \otimes (\psi * h^*)_{\overline{R}} \\
&= \langle h^*, 1 \rangle t h_{\overline{R}} \otimes \psi_{\overline{R}} = \langle h^*, 1 \rangle \langle \varepsilon, h \rangle t \otimes \psi
\end{aligned}$$

We also used (4.48) and the fact that ψ is a right integral. The clue point is now that Φ_M and $\Phi_{M^{-1}}$ are right $D(H)$-linear. This implies that

$$\Phi_{M^{-1}}((t \otimes \psi) \triangleleft (h^* \bowtie h)) = (h_0 \otimes \psi) \triangleleft (h^* \bowtie h)$$

or

$$\langle h^*, 1 \rangle \langle \varepsilon, h \rangle h_0 \otimes t = (h_0 \otimes \psi) \triangleleft (h^* \bowtie h) \qquad (4.50)$$

Take $h = 1$. Then

$$\begin{aligned}
\langle h^*, 1 \rangle h_0 \otimes t &= \sum_i \langle h^*, k_i^* , h_0 \rangle k_i \otimes t \\
&= \sum_i \langle h^*, h_{0(1)} \rangle h_{0(2)} \otimes t = \langle h^*, 1 \rangle h_0 \otimes t
\end{aligned}$$

Applying ε to the first factor, we find

$$\langle h^*, 1 \rangle \langle \varepsilon, h_0 \rangle t = \langle h^*, h_0 \rangle t$$

and, since kt is free of rank one,

$$\langle h^*, 1 \rangle \langle \varepsilon, h_0 \rangle = \langle h^*, h_0 \rangle$$

for all $h^* \in H^*$. This implies that

$$h_0 = \varepsilon(h_0) 1_H \in k 1_H$$

and we can write $h_0 = x_0 1_H$ with $x_0 \in k$. Now we apply (4.50) with $h^* = \varepsilon$. For all $h \in H$, we have that

$$\varepsilon(h) x_0 1_H \otimes t = \sum_i x_0 \langle k^*_{iR}, 1_H \rangle k_i \otimes t h_R$$

$$= \sum_i x_0 \langle \overline{S}(h_{(1)}) \cdot k^*_i \cdot h_{(3)}, 1_H \rangle k_i \otimes t h_{(2)}$$

$$= x_0 h_{(3)} \overline{S}(h_{(1)}) \otimes t h_{(2)}$$

We apply ε to the first factor. This yields

$$x_0 \varepsilon(h) t = x_0 \alpha(h) t \quad \text{and} \quad x_0 (\varepsilon(h) - \alpha(h)) = 0$$

for all $h \in H$, since t is a free generator of $\int^l_H$. Now

$$0 = \Phi_M((\varepsilon(h) - \alpha(h)) x_0 1_H \otimes t) = (\varepsilon(h) - \alpha(h)) t \otimes \psi$$

so $\varepsilon(h) = \alpha(h)$ for all $h \in H$, since $t \otimes \psi$ is a free generator $\int^l_{D(H)}$. This shows that H is unimodular.

$2. \Rightarrow 3.$ H/k is finitely generated and projective, since H/k is Frobenius. Let t be a free generator of $\int^l_H = \int^r_H$. We know from the proof of Theorem 31 that ϕ_H is bijective. Furthermore

$$h_{(2)} \otimes t h_{(1)} = h_{(1)} \otimes h_{(2)} t = h \otimes t$$

since t is at the same time a left and right integral.

$3. \Rightarrow 1.$ We compute easily that

$$\psi(h \otimes t) = h_{(2)} \otimes h_{(3)} t \overline{S}(h_{(1)}) = h_{(3)} \otimes t h_{(2)} \overline{S}(h_{(1)}) = h \otimes t$$

and it follows from Proposition 72 (applied to $\mathcal{M}(\psi \circ \tau)^H_{H^{op}} \cong {}_H\mathcal{M}(\psi)^H \cong {}_H\mathcal{YD}(H, H)^H \cong {}_{D(H)}\mathcal{M}$ that $D(H)/H$ is Frobenius.

Remark 12. An imprimitivity Theorem for Yetter-Drinfeld modules has been proved recently by the second author and Menini, we refer to [130].

4.5 Long dimodules

Let A be an algebra and C a coalgebra, and consider the identity

$$I_{A \otimes C} : A \otimes C \to A \otimes C$$

It is obvious that $(A, C, I_{A \otimes C}) \in {}_{\bullet}\mathbb{E}^{\bullet}$. The corresponding entwined modules satisfy the compatibility relation

$$\rho^r(am) = am_{[0]} \otimes m_{[1]}$$

i.e. the left A-action is right C-colinear, or, equivalently, the right C-coaction is left A-linear. The objects of

$$_A\mathcal{M}(I_{A \otimes C})^C = {}_A\mathcal{L}^C$$

are called (generalized) *Long dimodules*. If C is finitely generated and projective, then

$$_A\mathcal{L}^C \cong {}_{A \otimes C^*}\mathcal{M}$$

where $A \otimes C^*$ is the "trivial" smash product: the "braiding" map $R : C^* \otimes A \to A \otimes C^*$ is nothing else then the switch map. Long dimodules can be viewed as a special case of Doi-Koppinen Hopf modules: let H be any bialgebra (e.g. $H = k$), and let H coact trivially on A and act trivially on C. In Chapter 7, we will study the case where $A = C = H$ is a bialgebra, in the context of nonlinear equations. In the situation where H is commutative and cocommutative, Long dimodules are then the same thing as Yetter-Drinfeld modules. Long dimodules were first introduced in [118] to study *H-Azumaya algebras* and the *Brauer-Long group* of a finitely generated projective, commutative, cocommutative Hopf algebra. We refer to [35] for actual computation of the Brauer-Long group. One can generalize the Brauer-Long group to the situation where one works over a general Hopf algebra (not necessarily commutative or cocommutative), see [52]. Then one has to consider Yetter-Drinfeld modules instead of dimodules. This is related to the fact that Yetter-Drinfeld modules form a braided monoidal category, and that one can define the Brauer group of a braided monoidal category (see [182]).

Obviously, we can also consider right-right, left-left and right-left versions of Long dimodules. We will now discuss the results of Section 3.3 for categories of Long dimodules.

Proposition 98. *For an algebra A and a coalgebra C, the following conditions are equivalent:*

1. $_A\mathcal{L}^C \to {}_A\mathcal{L}$ *is separable;*
2. $\mathcal{L}_A^C \to \mathcal{L}_A$ *is separable;*
3. *there exists $\vartheta : C \otimes C \to A$ such that*
   a. $\mathrm{Im}\,\vartheta \subset Z(A);$

    b. $\vartheta(c \otimes d_{(1)}) \otimes d_{(2)} = \vartheta(c_{(2)} \otimes d) \otimes c_{(1)}$;

    c. $\vartheta(\Delta(c)) = \varepsilon(c)1_A$

    for all $c \in C$ and $a \in A$.

4. $_A^C\mathcal{L} \to {}_A\mathcal{L}$ is separable;

5. $^C\mathcal{L}_A \to \mathcal{L}_A$ is separable;

6. there exists $\vartheta' : C \otimes C \to A$ such that

    a. $\operatorname{Im}\vartheta' \subset Z(A)$;

    b. $\vartheta'(c \otimes d_{(2)}) \otimes d_{(1)} = \vartheta(c_{(1)} \otimes d) \otimes c_{(2)}$;

    c. $\vartheta(\Delta^{\mathrm{cop}}(c)) = \varepsilon(c)1_A$

    for all $c \in C$ and $a \in A$.

In particular, if C is coseparable as a coalgebra, then $_A\mathcal{L}^C \to {}_A\mathcal{L}$ is a separable functor.

Proof. $\underline{2. \Leftrightarrow 3.}$ follows immediately from Theorem 38. $\underline{1. \Leftrightarrow 2.}$ follows from the fact that 3) is the same for A as for A^{op}.

$\underline{3. \Leftrightarrow 6.}$ follows after we put $\vartheta' = \vartheta \circ \tau$. The equivalence of 4., 5. and 6. is the same as the equivalence of 1., 2. and 3., but with C replaced by C^{cop}. The final statement follows after we look at part 1. of Corollary 17.

Proposition 99. *Let H be a bialgebra. Then $_H\mathcal{L}^H \to {}_H\mathcal{L}$ is a separable functor if and only if H is coseparable as a coalgebra.*

Proof. One direction has already been shown in Proposition 98. Conversely, assume that $_H\mathcal{L}^H \to {}_H\mathcal{L}$ is separable. Then there exists $\vartheta : H \otimes H \to H$ satisfying the requirements of Proposition 98. It is easy to prove that $\theta = \varepsilon_H \circ \vartheta$ is a Larson coseparability idempotent, and the result follows using part I of Corollary 17.

Similar results apply to the other forgetful functor.

Proposition 100. *Let A be an algebra and C a coalgebra. The following assertions are equivalent.*

1. $\mathcal{L}_A^C \to \mathcal{L}^C$ is separable;

2. $^C\mathcal{L}_A \to {}^C\mathcal{L}$ is separable;

3. there exists $e : C \to A \otimes A$, $e(c) = e^1(c) \otimes e^2(c)$, such that

    a. $e(c_{(1)}) \otimes c_{(2)} = e(c_{(2)}) \otimes c_{(1)}$;

    b. $e^1(c) \otimes e^2(c)a = ae^1(c) \otimes e^2(c)$;

    c. $e^1(c)e^2(c) = \varepsilon(c)1_A$,

    for all $c \in C$ and $a \in A$.

4. $_A^C\mathcal{L} \to {}^C\mathcal{L}$ is separable;

5. $_A\mathcal{L}^C \to \mathcal{L}^C$ is separable;

6. there exists $e' : C \to A \otimes A$, $e(c) = e^1(c) \otimes e^2(c)$, such that

    a. $e(c_{(1)}) \otimes c_{(2)} = e(c_{(2)}) \otimes c_{(1)}$;

b. $e^1(c)a \otimes e^2(c) = e^1(c) \otimes ae^2(c)$;

c. $e^2(c)e^1(c) = \varepsilon(c)1_A$,

for all $c \in C$ and $a \in A$.

If A is a separable k-algebra, then $\mathcal{L}_A^C \to \mathcal{L}^C$ is separable. The converse holds if $A = C = H$ is a bialgebra.

Proof. Similar to the proof of Propositions 98 and 99, but now using part 1. of Proposition 76.

4.6 Modules graded by G-sets

Let G be a group, and X a right G-set. Let A be a G-graded k-algebra:

$$A = \oplus_{\sigma \in G} A_\sigma, \text{ with } A_\sigma A_\tau \subset A_{\sigma\tau}$$

for all $\sigma, \tau \in G$. Then kG is a Hopf algebra, A is a right kG-comodule algebra, and kX is a right kG-module coalgebra, and we have $(kG, A, kX) \in \mathbb{DK}_\bullet^\circ(k)$. The entwining map $\psi : kX \otimes A \to A \otimes kX$ is given by

$$\psi(x \otimes a_\sigma) = a_\sigma \otimes x\sigma$$

for $x \in X$ and $a_\sigma \in A_\sigma$ homogeneous of degree σ. As we have seen, the corresponding category of entwined modules $\mathcal{M}(kG)_A^{kX}$ is isomorphic to the category gr-(G, A, X) of right A-modules graded by kX. This means that $M \in \mathcal{M}(kG)_A^{kX} \cong$ gr-(G, A, X) can be written as

$$M = \oplus_{x \in X} M_x, \text{ with } M_x A_\sigma \subset M_{x\sigma}$$

for all $x \in X$ and $\sigma \in G$.

Proposition 101.   *With notation as above, the forgetful functor $F : \mathcal{M}(kG)_A^{kX} \to \mathcal{M}_A$ is separable.*

Proof. This is a direct application of Theorem 38. Consider $\vartheta : kX \otimes kX \to A$ defined by

$$\vartheta(x \otimes y) = \delta_{xy}1_A$$

ϑ satisfies (3.49) and (3.50). For $a_\sigma \in A_\sigma$ and $x, y \in X$, the right hand side of (3.49) is

$$a_\sigma \vartheta(x\sigma \otimes y\sigma) = \delta_{x\sigma,y\sigma} a_\sigma 1_A = \delta_{x,y} a_\sigma 1_A = \vartheta(x \otimes y)a$$

and this is the left hand side of (3.49). (3.50) takes the form

$$\vartheta(x \otimes y) \otimes y = \psi(x \otimes \vartheta(x \otimes y))$$

and this is OK since $\deg(\vartheta(x \otimes y)) = e$. Finally $\vartheta(\Delta(x)) = \vartheta(x \otimes x) = \varepsilon(x)1_A$.

Let us now investigate the separability of the adjoint functor

$$G: \mathcal{M}_A \rightarrow \mathcal{M}(kG)_A^{kX}$$

Recall from that the separability of G is determined by the elements of the k-module

$$W \cong W_5 = \{z = a^1 \otimes c^1 \in A \otimes kX \mid aa^1 \otimes c = a^1 a_\psi \otimes c^{1\psi}\}$$

Write z under the form

$$z = \sum_{x \in X} z(x) \otimes x$$

with only a finite number of $z(x)$ different from 0. $z \in W_5$ if and only if for all $\sigma \in G$ and $a_\sigma \in A_\sigma$, we have

$$\sum_{x \in X} a_\sigma z(x) \otimes x = \sum_{x \in X} z(x) a_\sigma \otimes x\sigma$$

The right hand side equals $\sum_{x \in X} z(x\sigma^{-1}) a_\sigma \otimes x$. Using the fact that X is a free basis for kX, we conclude that W_5 consists of families $(z(x))_{x \in X} \subset A$ such that $z(x) \neq 0$ for a finite number of $x \in X$ and

$$a_\sigma z(x) = z(x\sigma^{-1}) a_\sigma$$

for all $x \in X$, $\sigma \in G$ and $a_\sigma \in A_\sigma$. From the second part of Theorem 38, we can now conclude the following result, which was originally proved in [145] (for $X = G$) and [159] (in general); the present proof appeared in [41].

Proposition 102. *Let G be a group, X a right G-set, and A a G-graded k-algebra. The functor $G = \bullet \otimes kX : \mathcal{M}_A \rightarrow \mathcal{M}(kG)_A^{kX}$ is separable if and only if there exists a family $(z(x))_{x \in X} \subset A$ with all but a finite number of $z(x)$ equal zero such that*

$$a_\sigma z(x) = z(x\sigma^{-1}) a_\sigma \quad \text{and} \quad \sum_{x \in X} z(x) = 1_A$$

for all $\sigma \in G$ and $a_\sigma \in A_\sigma$.

Corollary 32. *With notation as in Proposition 102, we have*

1. *If there exists a finite G-subset $X' \subset X$ such that $|X'|$ is invertible in k, then the functor $G = \bullet \otimes kX$ is separable.*
2. *Conversely, if A is G-strongly graded, and G is separable, then there exists a finite G-subset $X' \subset X$.*

Proof. 1. Put

$$z(x) = \begin{cases} |X'|^{-1} 1_A & \text{if } x \in X' \\ 0 & \text{if } x \notin X' \end{cases}$$

2. Assume that $(z(x))_{x \in X} \subset A$ satisfies the conditions of Proposition 102. We claim that
$$X' = \{x \in X \mid z(x) \neq 0\}$$
is a finite G-subset of X. As A is strongly graded, so for every $\sigma \in G$, $A_\sigma A_{\sigma^{-1}} = A_e$, and there exist $b_1, \cdots, b_m \in A_\sigma$ and $b'_1, \cdots, b'_m \in A_{\sigma^{-1}}$ such that
$$\sum_{j=1}^{m} b_j b'_j = 1$$

Now for every $x \in X'$, we have
$$z(x) = \sum_{j=1}^{m} b_j b'_j z(x) = \sum_{j=1}^{m} b_j z(x\sigma) b'_j$$
and $z(x) \neq 0$ implies $z(x\sigma) \neq 0$, and $x\sigma \in X'$.

Corollary 33. *Let k be a field of characteristic zero, G a group, X a right G-set, and A a strongly G-graded k-algebra. The following statements are equivalent:*

1. *The functor $G = \bullet \otimes kX : \mathcal{M}_A \to$ gr-(G, X, A) is separable.*
2. *there exists a finite G-subset X' of X;*
3. *There exists $x \in X$ with finite G-orbit $\mathcal{O}(x)$.*

Remarks 8. 1. If A is not strongly graded, then the separability of the functor G does not imply that X has a finite G-subset. For example, take X any G-set without finite orbit, and A an arbitrary k-algebra with trivial G-grading. For any fixed $y \in X$, the family $(z(x) = \delta_{x,y} 1_A$ meets the requirements of Proposition 102, so G is separable.
2. Let $X = G = \mathbb{Z}$, and A a strongly G-graded k-algebra. Then the functor $G : \mathcal{M}_A \to$ gr-A is not separable.

Now take $X = G$, where the action is the usual multiplication of G. Then the category gr-(G, G, A), also denoted by gr-A is the category of G-graded A-modules, and we obtain the following properties:

1. if $F : \mathcal{M}_A \to$ gr-A is separable then G is finite;
2. if G is finite and $|G|$ is invertible in k, then F is a separable functor.

Let us now examine when (F, G) is a Frobenius pair.

Theorem 54. *[49] Let G be a group, X a right G-set, and A a G-graded ring. The functor $F : M(kG)_A^{kX} \to \mathcal{M}_A$ and its right adjoint $G = \bullet \otimes kX$ form a Frobenius pair if and only if X is finite.*

Proof. First assume that (F, G) is a Frobenius pair. From Theorem 36, we know that $\mathcal{C} = A \otimes kX$ is finitely generated and projective as a left A-module, and this is only possible if X is finite.

To prove the converse, we apply part III of Theorem 38. We have already seen that the map $\vartheta : kX \otimes kX \to A$

$$\vartheta(x \otimes y) = \delta_{x,y} 1_A$$

belongs to V_5, and that $z = 1_A \otimes \sum_{x \in X} x \in W_5$. It is immediate to verify that ϑ and z satisfy (3.54).

4.7 Two-sided entwined modules revisited

Let H be a Hopf algebra. In [161], Schauenburg has shown that the category of Yetter-Drinfeld modules is equivalent to a category of two-sided two-cosided Hopf modules. Surprisingly, this category is not a special cases of the two-sided categories of entwined modules that we introduced in Section 2.6, and that played an important role in the development of the theory of Chapter 3. The left-right compatibility conditions are different: in fact Schauenburg needs the Hopf module compatibility conditions, while in Section 2.6, the left-right compatibility conditions were just Long's compatibility conditions. A common feature of the two types of two-sided modules is that they can both be viewed as one-sided Doi-Hopf modules (and a fortiori of entwined modules, see [14] and [162]).

Compatible entwining structures Let (A, C, ψ) and (B, C, λ) be two right-right entwining structures, and consider the map

$$\theta = (I_A \otimes \varphi) \circ (\psi \otimes I_B) : \ C \otimes A \otimes B \to A \otimes B \otimes C$$

that is,

$$\theta(c \otimes a \otimes b) = a_\psi \otimes b_\lambda \otimes c^{\psi \lambda}$$

We say that (A, C, ψ) and (B, C, λ) are *compatible* if $(A \otimes B, C, \theta)$ is a right-right entwining structure.

Proposition 103. (A, C, ψ) *and* (B, C, λ) *are compatible if and only if*

$$a_\psi \otimes b_\lambda \otimes c^{\psi \lambda} = a_\psi \otimes b_\lambda \otimes c^{\lambda \psi} \tag{4.51}$$

for all $a \in A$, $b \in B$ *and* $c \in C$.

Proof. Assume first that (4.51) holds. Then

$$((a \otimes b)(a' \otimes b'))_\theta \otimes c^\theta = (aa' \otimes bb')_\theta \otimes c^\theta$$
$$= (aa')_\psi \otimes (bb')_\lambda \otimes c^{\psi\lambda}$$
$$(2.1) \qquad = a_\psi a'_{\psi'} \otimes b_\lambda b'_\Lambda \otimes c^{\psi\Psi\lambda\Lambda}$$
$$(4.51) \qquad = a_\psi a'_{\psi'} \otimes b_\lambda b'_\Lambda \otimes c^{\psi\lambda\Psi\Lambda}$$
$$= (a \otimes b)_\theta (a' \otimes b')_\Theta \otimes c^{\theta\Theta}$$

This means that θ satisfies (2.1). The other requirements (2.2-2.4) are obvious.

Conversely, if $(A \otimes B, C, \theta)$ is a right-right entwining structure, then θ satisfies (2.1), and the above computation shows that this means that

$$a_\psi a'_{\psi'} \otimes b_\lambda b'_\Lambda \otimes c^{\psi\Psi\lambda\Lambda} = a_\psi a'_{\psi'} \otimes b_\lambda b'_\Lambda \otimes c^{\psi\lambda\Psi\Lambda}$$

After we take $a = 1_A$ and $b' = 1_B$, we find (4.51).

Example 17. Let (H, A, C) and (K, B, C) be right-right Doi-Koppinen structures. If C is a right $H \otimes K$-module coalgebra, then the corresponding entwining structures (A, C, ψ) and (B, C, λ) are compatible.

Proof. Recall that

$$\psi(c \otimes a) = a_{[0]} \otimes ca_{[1]} \quad \text{and} \quad \lambda(c \otimes b) = b_{[0]} \otimes cb_{[1]}$$

We easily find that

$$a_\psi \otimes b_\lambda \otimes c^{\psi\lambda} = a_{[0]} \otimes b_{[0]} \otimes (ca_{[1]})b_{[1]}$$
$$= a_{[0]} \otimes b_{[0]} \otimes (cb_{[1]})a_{[1]}$$
$$= a_\psi \otimes b_\lambda \otimes c^{\lambda\psi}$$

If (A, C, ψ) and (B, C, λ) are compatible, then the category $\mathcal{M}(\theta)^C_{A\otimes B}$ consists of k-modules together with a right C-coaction and compatible right A and B-actions such that $M \in \mathcal{M}(H)^C_A$ and $M \in \mathcal{M}(H)^C_B$.

Now take $(A, C, \psi) \in {}_\bullet\mathbb{E}^\bullet(k)$, and $(B, C, \lambda) \in \mathbb{E}^\bullet_\bullet(k)$, respectively a left-right and right-right entwining structure. From Proposition 14, we know that $(A^{\text{op}}, C, \psi \circ \tau) \in \mathbb{E}^\bullet_\bullet(k)$, and we call (A, C, ψ) and (B, C, λ) compatible if and only if $(A^{\text{op}}, C, \psi \circ \tau)$ and (B, C, λ) are compatible. We then consider the category

$$_A\mathcal{M}({}_\psi \lambda)^C_B$$

consisting of k-modules M with a right C-coaction and an (A, B)-bimodule structure such that

$$M \in {}_A\mathcal{M}(\psi)^C \quad \text{and} \quad M \in \mathcal{M}(\lambda)^C_B$$

Example 18. $(A, C, I_{A \otimes C}) \in {}_{\bullet}\mathbb{E}^{\bullet}(k)$ is compatible with every right-right entwining structure $(B, C, \lambda) \in \mathbb{E}^{\bullet}_{\bullet}(k)$. The corresponding modules are (A, B)-bimodules, right-right (B, C, λ) entwined modules, and left-right (A, C)-Long dimodules.

The same game can be played with the coalgebras: (A, C, ψ), $(A, D, \kappa) \in \mathbb{E}^{\bullet}_{\bullet}(k)$ are called compatible if

$$\theta = (\psi \otimes I_D) \circ (I_C \otimes \kappa) : \ C \otimes D \otimes A \to A \otimes C \otimes D$$

makes $(A, C \otimes D, \theta)$ into an object of $\mathbb{E}^{\bullet}_{\bullet}(k)$. The proof of the next result is left to the reader.

Proposition 104. (A, C, ψ), $(A, D, \kappa) \in \mathbb{E}^{\bullet}_{\bullet}(k)$ *are compatible if and only if*

$$a_{\psi\kappa} \otimes c^{\psi} \otimes d^{\kappa} = a_{\kappa\psi} \otimes c^{\psi} \otimes d^{\kappa} \tag{4.52}$$

for all $a \in A$, $c \in C$ *and* $d \in D$.

$(A, C, \psi) \in {}^{\bullet}\mathbb{E}_{\bullet}$ and $(A, D, \kappa) \in \mathbb{E}^{\bullet}_{\bullet}(k)$ are called compatible if $(A, C^{\mathrm{cop}}, \tau \circ \psi)$ and (A, D, κ) are compatible. We can then consider the category

$$^{C}\mathcal{M}({}^{\psi}\kappa)^{D}_{B}$$

Example 19. Take $(H, A, C) \in {}^{\bullet}\mathbb{DK}_{\bullet}(k)$ and $(K, A, D) \in \mathbb{DK}^{\bullet}_{\bullet}(k)$. If A is an (H, C)-bicomodule algebra, then the associated entwining structures are compatible.

Our next step is to combine the two constructions. Consider

$$(A, C, \psi), (B, D, \varphi), (B, C, \lambda), (A, D, \kappa) \in \mathbb{E}^{\bullet}_{\bullet}(k)$$

These 4 structures are pairwise compatible if and only if

$$\theta = (I_A \otimes \lambda \otimes I_D) \circ (\psi \otimes \varphi) \circ (I_C \otimes \kappa \otimes I_B) : \ C \otimes D \otimes A \otimes B \to A \otimes B \otimes C \otimes D$$

makes $(A \otimes B, C \otimes D, \theta)$ into an object of $\mathbb{E}^{\bullet}_{\bullet}(k)$.
Using left-right arguments as before, let

$$(A, C, \psi) \in {}^{\bullet}_{\bullet}\mathbb{E}(k) \ ; \ (B, D, \varphi) \in \mathbb{E}^{\bullet}_{\bullet}(k)$$

$$(B, C, \lambda) \in {}^{\bullet}\mathbb{E}_{\bullet}(k) \ ; \ (A, D, \kappa) \in {}_{\bullet}\mathbb{E}^{\bullet}(k)$$

We can then consider the category

$$^{C}_{A}\mathcal{M}\left({}^{\lambda}_{\psi}\varphi_{\kappa}\right)^{D}_{B}$$

Examples 7. 1. Let $\kappa = I_{A\otimes C}$, $\lambda = I_{B\otimes C}$. Then we recover the two-sided entwined modules from Section 2.6.

2. Let H, K, L, M be bialgebras, and

– A an (H, K)-comodule algebra;
– B an (L, M)-comodule algebra;
– C an (H, L)-module coalgebra;
– D a (K, M)-module coalgebra.

Then we have Doi-Koppinen structures

$$(H, A, C) \in {}^\bullet_\bullet\mathbb{DK}(k) \ ; \ (M, B, D) \in \mathbb{DK}^\bullet_\bullet(k)$$

$$(L, B, C) \in {}^\bullet\mathbb{DK}_\bullet(k) \ ; \ (K, A, D) \in {}_\bullet\mathbb{DK}^\bullet(k)$$

and the corresponding entwining structures are compatible. We obtain the following category of two-sided Doi-Hopf modules

$$\begin{smallmatrix}C\\A\end{smallmatrix}\mathcal{M}\left(H \begin{smallmatrix}L\\K\end{smallmatrix} M\right)^D_B$$

3. Let us consider some particular case of Example 2); these will be of some use in the Proposition 105 and 105. Let H be a bialgebra. Via left and right multiplication an comultiplication, H is an (H, H)-bicomodule algebra and an (H, H)-bimodule coalgebra, and we can consider the category

$$\begin{smallmatrix}H\\H\end{smallmatrix}\mathcal{M}\left(H \begin{smallmatrix}H\\H\end{smallmatrix} H\right)^H_H$$

Its objects are k-modules with a left and right H-action and H-coaction, such that they are at the same time left-left, left-right, right-left and right-right Hopf modules.

Now k viewed as a bialgebra acts and coacts trivially on H, and we find that H is an (H, k)-bicomodule algebra, a (k, H)-bicomodule algebra, an (H, k)-bimodule coalgebra, and a (k, H)-bimodule coalgebra, and we can consider the category

$$\begin{smallmatrix}H\\H\end{smallmatrix}\mathcal{M}\left(H \begin{smallmatrix}k\\k\end{smallmatrix} H\right)^H_H$$

Its objects are left-left and right-right Hopf modules, and left-right, right-left Long dimodules.

Finally, consider the category

$$\begin{smallmatrix}H\\H\end{smallmatrix}\mathcal{M}\left(H \begin{smallmatrix}k\\k\end{smallmatrix} k\right)^H_H$$

Its objects are left-left Hopf modules, and left-right, right-left, and right-right Long dimodules.

Proposition 105. *[161] Let H be Hopf algebra, and assume that H is flat as a k-module. We have pairs of inverse equivalences between the following categories:*

$$\mathcal{M}_H \quad \text{and} \quad {}^H_H\mathcal{M}\left(H\,{}^k_H\,H\right)^k_H$$

$$\mathcal{M}^H \quad \text{and} \quad {}^H_H\mathcal{M}\left(H\,{}^H_k\,k\right)^H_k$$

$$\mathcal{YD}(H,H)^H_H \quad \text{and} \quad {}^H_H\mathcal{M}\left(H\,{}^H_H\,H\right)^H_H$$

Proof. The Fundamental Theorem for Hopf modules (see Proposition 59) tells us that we have an equivalence between the categories $\mathcal{M}$ and ${}^H_H\mathcal{M}(H)$. Recall that $F = H \otimes \bullet : \mathcal{M} \to {}^H_H\mathcal{M}(H)$ and $G = {}^{\mathrm{co}H}(\bullet)$. The left structure on $F(M) = H \otimes M$ is induced by the structure on H.

1. Take $M \in \mathcal{M}_H$. On $H \otimes M$, we then define a right H-action as follows:

$$(h \otimes m)k = hk_{(1)} \otimes mk_{(2)}$$

Easy computations show that $H \otimes M$ is now an H-bimodule, and also an object of ${}^H\mathcal{M}(H)_H$, since

$$\rho^l((h \otimes m)k) = \rho^l(hk_{(1)} \otimes mk_{(2)}) = h_{(1)}k_{(1)} \otimes h_{(2)}k_{(2)} \otimes mk_{(3)}$$
$$= h_{(1)}k_{(1)} \otimes (h_{(2)} \otimes m)k_{(2)}$$

and this shows that

$$H \otimes M \in {}^H_H\mathcal{M}\left(H\,{}^k_H\,H\right)^k_H$$

Conversely, for $N \in {}^H_H\mathcal{M}\left(H\,{}^k_H\,H\right)^k_H$, we define a right H-action on $G(N) = {}^{\mathrm{co}H}N$ by

$$n \cdot h = S(h_{(1)})nh_{(2)}$$

We have to show that $\rho^l(n \cdot h) = 1 \otimes n \cdot h$:

$$\rho^l(n \cdot h) = \rho^l(S(h_{(1)})nh_{(2)})$$
$$= S(h_{(2)})n_{[-1]}h_{(3)} \otimes S(h_{(1)})n_{[0]}h_{(4)}$$
$$= 1 \otimes S(h_{(1)})nh_{(2)} = 1 \otimes n \cdot h$$

2. For $M \in \mathcal{M}^H$, we define a right H-coaction on $H \otimes M$ as follows:

$$\rho^r(h \otimes m) = (h_{(1)} \otimes m_{[0]}) \otimes h_{(2)}m_{[1]}$$

It can be verified easily that $H \otimes M$ is an (H,H)-bicomodule, and a left-right Hopf module, and consequently

$$H \otimes M \in {}_H^H\mathcal{M}\left(H\,{}_k^H\,k\right)_k^H$$

Conversely, for $N \in {}_H^H\mathcal{M}\left(H\,{}_k^H\,k\right)_k^H$, the right H-coaction on N restricts to a right H-coaction on $G(N)$.

3. Let N be a right-right Yetter-Drinfeld module. We already know that $H \otimes M$ is an object of ${}_H^H\mathcal{M}(H)$, ${}^H\mathcal{M}(H)_H$, ${}_H\mathcal{M}(H)^H$, and that is an (H, H)-bimodule and bicomodule. We are left to show that it is a right-right Hopf module.

$$\begin{aligned}
\rho^r((h \otimes m)k) &= \rho^r(hk_{(1)} \otimes mk_{(2)}) \\
&= (hk_{(1)})_{(1)} \otimes (mk_{(2)})_{[0]} \otimes (hk_{(1)})_{(2)}(mk_{(2)})_{[1]} \\
&= h_{(1)}k_{(1)} \otimes m_{[0]}k_{(4)} \otimes h_{(2)}k_{(2)}S(k_{(3)})m_{[1]}k_{(5)} \\
&= h_{(1)}k_{(1)} \otimes m_{[0]}k_{(2)} \otimes h_{(2)}m_{[1]}k_{(3)} \\
&= (h \otimes m)_{[0]}h_{(1)} \otimes (h \otimes m)_{[1]}h_{(2)}
\end{aligned}$$

as needed. Conversely, let $N \in {}_H^H\mathcal{M}\left(H\,{}_H^H\,H\right)_H^H$. We have seen above that $G(N)$ is a right H-module and a right H-comodule. Let us show that it is a Yetter-Drinfeld module. For $n \in {}^{coH}N$, we have

$$\begin{aligned}
\rho^r(n \cdot h) &= \rho^r(S(h_{(1)})nh_{(2)}) \\
&= S(h_{(2)})n_{[0]}h_{(3)} \otimes S(h_{(1)})n_{[1]}h_{(4)} = n_{[0]} \cdot h_{(2)} \otimes S(h_{(1)})n_{[1]}h_{(3)}
\end{aligned}$$

as needed.

We have similar results for the categories of Hopf modules and Long dimodules: they are equivalent to categories of two-sided Hopf modules, but with different compatibility conditions.

Proposition 106. *Let H be Hopf algebra, and assume that H is flat as a k-module. We have pairs of inverse equivalences between the following categories:*

$$\mathcal{M}(H)_H^H \text{ and } {}_H^H\mathcal{M}\left(H\,{}_k^k\,H\right)_H^H$$

$$\mathcal{L}_H^H \text{ and } {}_H^H\mathcal{M}\left(H\,{}_k^k\,k\right)_H^H$$

Proof. We proceed as in Proposition 105. For $M \in \mathcal{M}_H$, we now define the following action on $H \otimes M$:

$$(h \otimes m)k = h \otimes mk$$

It is clear that this action makes $H \otimes M$ into an (H, H)-bimodule, and a right-left H-dimodule. Similarly, if M is a right H-comodule, then $H \otimes M$

is a right H-comodule, with coaction $I_H \otimes \rho_M$. This makes $H \otimes M$ into an (H, H)-bicomodule, and a left-right H-dimodule. If M is a right-right Hopf module, then we find that $H \otimes M$ is an object of ${}^H_H\mathcal{M}\left(H{}^k_k H\right)^H_H$, and if M is a right-right Long dimodule, then we find that $H \otimes M$ is an object of ${}^H_H\mathcal{M}\left(H{}^k_k k\right)^H_H$. Further details are left to the reader.

4.8 Corings and descent theory

Effective descent morphisms Let $i : B \to A$ be a ring homomorphism. It can be verified easily that $\mathcal{C} = A \otimes_B A$, with structure maps

$$\Delta_{\mathcal{C}} : A \otimes_B A \to (A \otimes_B A) \otimes_A (A \otimes_B A) \cong A \otimes_B A \otimes_B A \text{ and } \varepsilon_{\mathcal{C}} : A \otimes_B A \to A$$

given by

$$\Delta_{\mathcal{C}}(a \otimes_B b) = (a \otimes_B 1) \otimes_A (1 \otimes_B b) = a \otimes_B 1 \otimes_B b$$

$$\varepsilon_{\mathcal{C}}(a \otimes_B b) = ab$$

is an A-coring. $\mathcal{C}$ is called the *canonical coring* associated to the ring morphism i. A right $\mathcal{C}$-comodule consists of a right A-module M together with a right A-module map

$$\rho_M : M \to M \otimes_A (A \otimes_B A) \cong M \otimes_B A$$

We will use the Sweedler-Heyneman notation

$$\rho_M(m) = m_{[0]} \otimes_B m_{[1]} \in M \otimes_B A$$

The coassociativity condition and the counit condition then take the form

$$m_{[0][0]} \otimes_B m_{[0][1]} \otimes_B m_{[1]} = m_{[0]} \otimes_B 1 \otimes_B m_{[1]} \tag{4.53}$$

and

$$m_{[0]} m_{[1]} = m \tag{4.54}$$

We have a functor, called the *comparison functor*

$$K = \bullet \otimes_B A : \mathcal{M}_B \to \mathcal{M}^{\mathcal{C}}$$

where the $\mathcal{C}$-comodule structure on $N \otimes_B A$ is the following:

$$\rho_{N \otimes_B A}(n \otimes_B a) = n \otimes_B 1 \otimes_B a$$

i is called an *effective descent morphism* if K is an equivalence of categories. In this situation, a right A-module M is isomorphic to $N \otimes_B A$ for some right B-module N if and only if we can define a right $\mathcal{C}$-comodule structure on M.

In the situation where A and B are commutative, there is an isomorphism between the category of comodules over the canonical coring, and the category of *descent data*, as introduced by Knus and Ojanguren in [109]. Recall from [109, II.3.1] that a *descent datum* consists of a pair (M, g), with $M \in \mathcal{M}_A$, and $g : A \otimes_B M \to M \otimes_B A$ an $A \otimes_B A$-module homorphism such that

$$g_2 = g_3 \circ g_1 : A \otimes_B A \otimes_B M \to M \otimes_B A \otimes_B A \tag{4.55}$$

and

$$\mu_M(g(1 \otimes_B m)) = m \tag{4.56}$$

for all $m \in M$. Here g_i is defined by applying I_A to the i-th tensor position, and g to the two other ones. A morphism of two descent data (M, g) and (M', g') consists of an A-module homomorphism $f : M \to M'$ such that the diagram

$$
\begin{array}{ccc}
A \otimes_B M & \xrightarrow{g} & M \otimes_B A \\
\downarrow{\scriptstyle I_A \otimes_B f} & & \downarrow{\scriptstyle f \otimes_B I_A} \\
A \otimes_B M' & \xrightarrow{g'} & M' \otimes_B A
\end{array}
$$

commutes. $\underline{\mathrm{Desc}}(A/B)$ will be the category of descent data.

Theorem 55. *Let $i : B \to A$ be a morphism of commutative rings. We have an isomorphism of categories*

$$\underline{\mathrm{Desc}}(A/B) \cong \mathcal{M}^{A \otimes_B A}$$

Proof. For a right C-comodule (M, ρ_M), we define $g : A \otimes_B M \to M \otimes_B A$ by

$$g(a \otimes_B m) = m_{[0]} a \otimes_B m_{[1]}$$

It is easy to see that g is an $A \otimes_B A$-module map, and that

$$
\begin{aligned}
(g_3 \circ g_1)(a \otimes_B b \otimes_B m) &= g_3(a \otimes_B m_{[0]} b \otimes_B m_{[1]}) \\
&= m_{[0]} a \otimes_B m_{[1]} b \otimes_B m_{[2]} \\
(4.53) \quad &= m_{[0]} a \otimes_B b \otimes_B m_{[1]} \\
&= g_2(a \otimes_B b \otimes_B m)
\end{aligned}
$$

From (4.54), it follows that

$$\mu_M(g(1 \otimes_B m)) = m_{[0]} \varepsilon_C(1 \otimes_B m_{[1]}) = m_{[0]} m_{[1]} = m$$

and we see that (M, g) is a descent datum. Conversely, if (M, g) is a descent datum, then the map

$$\rho_M : M \to M \otimes_B A \; ; \; \rho_M(m) = g(1 \otimes_B m)$$

makes M into a right C-comodule. All other details are left to the reader.

Corollary 34. *Let* $i : B \to A$ *be a morphism of commutative rings. For* $M \in \mathcal{M}_A$ *and* $g : A \otimes_B M \to M \otimes_B A$, (M, g) *is a descent datum if and only if (4.53) holds and* g *is an isomorphism.*

Proof. First assume that (M, g) is a descent datum, and let ρ_M be the associated $\mathcal{C}$-comodule structure on M. For all $a \in A$ and $m \in M$, we compute

$$(\tau \circ g \circ \tau \circ g)(a \otimes m) = (\tau \circ g)(m_{[1]} \otimes_B m_{[0]}a)$$
$$= \tau\big(g(m_{[1]} \otimes_B m_{[0]})(1 \otimes_B a)\big)$$
$$= \tau\big(m_{[0]}m_{[2]} \otimes_B m_{[1]}a\big)$$
$$= m_{[1]}a \otimes_B m_{[0]}m_{[2]} = a \otimes_B m$$

and it follows that $\tau \circ g \circ \tau$ is a (left and right) inverse of g. Conversely, assume that g is bijective. We can still consider the associated map ρ_M, and we know that ρ_M satisfies (4.53). We are done if we can show that it also satisfies (4.54). Multiplying the second and third factor in (4.53), we see that

$$m_{[0]} \otimes_B m_{[1]}m_{[2]} = m_{[0]} \otimes_B m_{[1]}$$

or

$$g(1 \otimes_B m_{[0]}m_{[1]}) = g(1 \otimes_B m_{[0]})m_{[1]} = g(1 \otimes_B m)$$

Applying g^{-1} to both sides, and then multiplying the two tensor factors, we see that $m_{[0]}m_{[1]} = m$, as needed.

In the situation where A and B are not necessarily commutative, descent data have been introduced by Cipolla [59]. Cipolla's descent data are exactly comodules over the canonical coring. Nuss [149] has proposed alternative descriptions of the category of descent data.

The "faithfully flat descent theorem" states that a sufficient condition for a morphism $i : B \to A$ to be an effective descent morphism is that A is faithfully flat as a B-module.

Proposition 107. *Let* $i : B \to A$ *be a morphism of rings. Then the comparison functor* $K : \mathcal{M}_B \to \mathcal{M}^{\mathcal{C}}$ *has a right adjoint* R.

Proof. R is defined as follows

$$R : \mathcal{M}^{\mathcal{C}} \to \mathcal{M}_B \; ; \; R(M) = M^{co\mathcal{C}} = \{m \in M \mid \rho_M(m) = m \otimes_B 1\}$$

The unit and counit of the adunction are given by

$$\eta_N : N \to (N \otimes_B A)^{co\mathcal{C}} \; ; \; \eta_N(n) = n \otimes_B 1$$

$$\varepsilon_M : M^{co\mathcal{C}} \otimes_B A \to M \; ; \; \varepsilon_M(m \otimes_B a) = ma$$

for all $N \in \mathcal{M}_A$ and $M \in \mathcal{M}^{\mathcal{C}}$.

Proposition 108. *Let* $i : B \to A$ *be a morphism of rings, and assume that* A *is flat as a left B-module. Then the right adjoint R of K is fully faithful.*

Proof. M^{coC} is the coequalizer of the maps

$$0 \longrightarrow M^{coC} \longrightarrow M \underset{I_M \otimes_B i}{\overset{\rho_M}{\rightrightarrows}} M \otimes_B A$$

A is flat as a left B-module, so we have an exact sequence

$$0 \longrightarrow M^{coC} \otimes_B A \longrightarrow M \otimes_B A \underset{I_M \otimes_B i \otimes_B I_A}{\overset{\rho_M \otimes_B I_A}{\rightrightarrows}} M \otimes_B A \otimes_B A \qquad (4.57)$$

From the coassociativity of ρ_M, it now follows that $\rho^r(m) \in M^{coC} \otimes_B A \subset M \otimes_B A \cong M \otimes_A (A \otimes_B A)$, for all $m \in M$, and we have a map $\rho_M : M \to M^{coC} \otimes_B A$. From the counit property, it follows that $\varepsilon_M \circ \rho_M = I_M$. For $m \in M^{coC}$ and $a \in A$, we have

$$\rho_M(\varepsilon_M(m \otimes_B a)) = \rho_M(ma) = \rho_M(m)a = (m \otimes_B 1)a = m \otimes_B a$$

Thus the counit ε_M has an inverse, for all M, and R is fully faithful.

Proposition 109. *Let* $i : B \to A$ *be a morphism of rings, and assume that A is faithfully flat as a left B-module. Then i is an effective descent morphism.*

Proof. We allready know that K has a fully faithful right adjoint R. It remains to be shown that K itself is fully faithful, i.e. η_N is an isomorphism, for all $N \in \mathcal{M}_B$. We first show that the sequence

$$0 \longrightarrow N \otimes_B A \overset{I_N \otimes_B i \otimes_B I_A}{\longrightarrow} N \otimes_B A \otimes_B A \underset{I_N \otimes_B A \otimes_B i \otimes_B I_A}{\overset{\rho_N \otimes_B A \otimes_B I_A}{\rightrightarrows}} N \otimes_B A \otimes_B A \otimes_B A$$

is exact. Indeed, if $\sum_i n_i \otimes_B a_i \otimes_B b_i \in \mathrm{Ker}\,(\rho_{N \otimes_B A} \otimes_B I_A - I_{N \otimes_B A} \otimes_B i \otimes_B I_A)$, then

$$\sum_i n_i \otimes_B a_i \otimes_B 1 \otimes_B b_i = \sum_i n_i \otimes_B 1 \otimes_B a_i \otimes_B b_i$$

Multiplying the third and fourth tensor factor, we find

$$\sum_i n_i \otimes_B a_i \otimes_B b_i = \sum_i n_i \otimes_B 1 \otimes_B a_i b_i \in \mathrm{Im}\,(I_N \otimes_B i \otimes_B I_A)$$

Now A/B is faithfully flat, and we find that the sequence

$$0 \longrightarrow N \overset{I_N \otimes_B i}{\longrightarrow} N \otimes_B A \underset{I_N \otimes_B A \otimes_B i}{\overset{\rho_{N \otimes_B A}}{\rightrightarrows}} N \otimes_B A \otimes_B A$$

is exact, and this means exactly that $\eta_N = I_N \otimes_B i : N \to (N \otimes_B A)^{coC}$ is an isomorphism.

It is somewhat surprising that the converse of Proposition 109 does not hold. Let B be a commutative ring. A morphism $f : M \to M'$ of B-modules is called *pure* if for any B-module N the morphism

$$f \otimes_B I_N : M \otimes_B N \to M' \otimes_B N$$

is monic. The following result is due to Joyal and Tierney (unpublished). For detail, we refer the reader to [132] and [97].

Theorem 56. *Let $i : B \to A$ be a morphism of commutative rings. The following assertions are equivalent:*

1. *i is an effective descent morphism;*
2. *K is fully faithful;*
3. *i is pure as a morphism of B-modules.*

Galois type corings As before let $i : B \to A$ be a morphism of rings, and let $C = A \otimes_B A$ be the canonical coring. We also consider a second A-coring D. $\text{Hom}_{\text{coring}}(C, D)$ will be the k-module consisting of all A-coring maps from C to D. Also

$$G(D) = \{d \in D \mid \Delta_D(d) = d \otimes_A d \text{ and } \varepsilon_D(d) = 1_A\}$$

is the set of grouplike elements of D. Clearly $1_A \otimes_B 1_A \in G(C)$.

Proposition 110. *With notation as above, we have an isomorphism of sets*

$$\text{Hom}_{\text{coring}}(C, D) \cong G(D)^B = \{d \in G(D) \mid bd = db, \text{ for all } b \in B\}$$

Proof. Take a coring homomorphism can : $C \to D$. Then $d = \text{can}(1 \otimes_B 1) \in G(D)^B$. can is completely determined by d, because can is an A-bimodule map:

$$\text{can}(a \otimes_B b) = a\text{can}(1 \otimes_B 1)b = adb \tag{4.58}$$

For $d \in G(D)$, we define can : $C \to D$ using (4.58). Then can is an A-bimodule map, and a coring morphism since

$$\Delta_D(\text{can}(a \otimes_B b)) = \Delta_D(adb) = ad \otimes_A db = (\text{can} \otimes_A \text{can})((a \otimes_B 1) \otimes_A (1 \otimes_B b))$$

and

$$\varepsilon_D(\text{can}(a \otimes_B b)) = \varepsilon_D(adb) = ab = \varepsilon_C(a \otimes_B b)$$

Proposition 111. *Let D be an A-coring. Then*

$$G(D) \cong \{\rho : A \to A \otimes_A D \cong D \mid \rho \text{ makes } A \text{ into a } D\text{-comodule}\}$$

Proof. For $d \in G(\mathcal{D})$, we define $\rho : A \to \mathcal{D}$ by $\rho(a) = 1 \otimes_A da = da$. It is then clear that ρ is right A-linear. (A, ρ) is a right $\mathcal{D}$-comodule since

$$(\rho \otimes_A I_{\mathcal{D}})\rho(a) = 1 \otimes_A d \otimes_A da = 1 \otimes_A \Delta_{\mathcal{D}}(d)a = (I_A \otimes_A \Delta_{\mathcal{D}})\rho(a)$$

and

$$(I_A \otimes_A \varepsilon_{\mathcal{D}})\rho(a) = 1 \otimes_A \varepsilon_{\mathcal{D}}(da) = 1 \otimes_A a = a$$

Conversely, if A is a $\mathcal{D}$-comodule, then $\rho(1_A) = d$ is grouplike.

Corollary 35. *An A-coring $\mathcal{D}$ is isomorphic to the canonical coring $\mathcal{C}$ if and only if there exists a grouplike element $d \in G(\mathcal{D})^B$ such that the map can : $\mathcal{C} \to \mathcal{D}$, $\mathrm{can}(a \otimes_B b) = adb$ is bijective,*

Take $d \in G(\mathcal{D})$, and let ρ be the corresponding $\mathcal{D}$-comodule structure on A. We define

$$A^{\mathrm{co}\mathcal{D}} = \{a \in A \mid \rho(a) = a \otimes_A d = ad\}$$
$$= \{a \in A \mid ad = da\}$$

For $\mathcal{D}$ equal to the canonical coring $\mathcal{C}$ and $d = 1 \otimes_B 1$, we find

$$A^{\mathrm{co}\mathcal{C}} = \{a \in A \mid a \otimes_B 1 = 1 \otimes_B a\}$$

Lemma 24. *Let $\mathcal{D}$ be an A-coring, $d \in G(\mathcal{D})^B$, and ρ the corresponding $\mathcal{D}$-comodule structure on A. Then $i(B) \subset A^{\mathrm{co}\mathcal{C}} \subset A^{\mathrm{co}\mathcal{D}}$. If $\mathcal{C}$ and $\mathcal{D}$ are isomorphic as corings, then $A^{\mathrm{co}\mathcal{C}} = A^{\mathrm{co}\mathcal{D}}$.*

Proof. $i(B) \subset A$ is clear. If $a \in A^{\mathrm{co}\mathcal{C}}$, then $a \otimes_B 1 = 1 \otimes_B a$, and consequently

$$ad = \mathrm{can}(a \otimes_B 1) = \mathrm{can}(1 \otimes_B a) = da$$

If can is bijective, then the converse also holds.

Let $\mathcal{C}$ be the canonical coring associated to a ring morphism $i : B \to A$. An A-coring is said to be of *Galois type* if there exists a grouplike $d \in G(\mathcal{D})^B$ such that the corresponding map can : $\mathcal{C} \to \mathcal{D}$ is an isomorphism, and

$$B \cong i(B) = A^{\mathrm{co}\mathcal{C}} = A^{\mathrm{co}\mathcal{D}}$$

Now assume that $i : B \to A$ is a morphism of k-algebras, and that (A, C, ψ) is a right-right entwining structure. As we have seen in Section 2.7, we can associate an A-coring $\mathcal{D} = A \otimes C$ to it. Assume that this coring is isomorphic to the canonical coring $\mathcal{C}$:

$$\mathcal{C} = A \otimes_B A \cong \mathcal{D} = A \otimes C$$

We will write

$$d = \sum_i a_i \otimes c_i \in G(A \otimes C)^B$$

From Proposition 111, we know that A is a right $\mathcal{D}$-comodule, and therefore an entwined module (see Theorem 17), and a fortiori a right C-comodule. The right C-coaction is

$$\rho(a) = a_{[0]} \otimes a_{[1]} = da = \sum_i a_i a_\psi \otimes c_i^\psi$$

and can can be rewritten in terms of the coaction:

$$\mathrm{can}(a \otimes_B b) = adb = ab_{[0]} \otimes b_{[1]}$$

and we conclude that

1. A is a right C-comodule;
2. can : $A \otimes_B A \to A \otimes C$, $\mathrm{can}(a \otimes_B b) = ab_{[0]} \otimes b_{[1]}$, is an isomorphism;
3. for all $b \in B$: $\rho(i(b)) = i(b)\rho(1)$.

Conversely, assume that $i : B \to A$ is a morphism of k-algebras, and that C is a k-coalgebra such that the three above conditions hold. can is bijective, so the coring structure on $A \otimes_B A$ induces a coring structure on $A \otimes C$. We will show that this coring structure comes from an entwining structure (A, C, ψ). To this end, we apply Theorem 16. We have to verify (2.87-2.89).
It is clear that the natural left A-module structure on $A \otimes C$ makes can into a left A-linear map, so (2.87) holds. The right A-module structure on $A \otimes C$ induced by can is given by

$$(b \otimes c)a = \mathrm{can}(\mathrm{can}^{-1}(b \otimes c)a)$$

Since $\mathrm{can}^{-1}(1_{[0]} \otimes 1_{[1]}) = 1 \otimes_B 1$, we have

$$(1_{[0]} \otimes 1_{[1]})a = \mathrm{can}(1 \otimes a) = a_{[0]} \otimes a_{[1]} \tag{4.59}$$

The comultiplication Δ on $A \otimes C$ is given by

$$\Delta(a \otimes c) = (\mathrm{can} \otimes_A \mathrm{can})\Delta_C(\mathrm{can}^{-1}(a \otimes c)) \in (A \otimes C) \otimes_A (A \otimes C),$$

for all $a \in A$ and $c \in C$. can is bijective, so we can find $a_i, b_i \in A$ such that

$$\mathrm{can}(\sum_i a_i \otimes_B b_i) = \sum_i a_i b_{i[0]} \otimes_B b_{i[1]} = a \otimes c$$

and we compute that

$$\Delta(a \otimes c) = (\mathrm{can} \otimes_A \mathrm{can})\Delta_C(\sum_i a_i \otimes_B b_i)$$

$$= \sum_i \mathrm{can}(a_i \otimes_B 1) \otimes_A \mathrm{can}(1 \otimes_B b_i)$$

$$= \sum_i (a_i 1_{[0]} \otimes 1_{[1]}) \otimes_A (b_{i[0]} \otimes b_{i[1]})$$

$$= \sum_i (a_i 1_{[0]} \otimes 1_{[1]}) b_{i[0]} \otimes_A (1 \otimes b_{i[1]})$$

$$(4.59) \quad = \sum_i (a_i b_{i[0]} \otimes b_{i[1]}) \otimes_A (1 \otimes b_{i[2]})$$

$$= (a \otimes c_{(1)}) \otimes_A (1 \otimes c_{(2)})$$

proving (2.88). (2.89) can be proved as follows:

$$\varepsilon(a \otimes c) = \varepsilon_C(\sum_i a_i \otimes b_i) = \sum_i a_i b_i$$

$$= \sum_i a_i b_{i[0]} \varepsilon_C(b_{i[1]}) = a \varepsilon_C(c)$$

Coalgebra Galois extensions have been introduced in [30] (see also [25] and [32]). Let $i : B \to A$ be a morphism of k-algebras, and C a k-coalgebra. A is called a C-Galois extension of B if the following conditions hold:

1. A is a right C-comodule;
2. can : $A \otimes_B A \to A \otimes C$, $\mathrm{can}(a \otimes_B a') = aa'_{[0]} \otimes_B a'_{[1]}$ is an isomorphism;
3. $B = \{a \in A \mid \rho(a) = a\rho(1)\}$.

Collecting the arguments above, we find

Proposition 112. *Let $i : A \to B$ be a morphism of k-algebras, and C a k-coalgebra. A is C-Galois extension of B for some right C-coaction on A if and only if there exists a right-right entwining structure (A, C, ψ) such that $A \otimes C$ is an A-coring of Galois type.*

Now consider the special case where $C = H$ is a bialgebra, A is a right H-comodule algebra, $d = 1_A \otimes 1_H$ and

$$\psi : H \otimes A \to A \otimes H, \quad \psi(h \otimes a) = a_{[0]} \otimes ha_{[1]}$$

i.e. (A, C, ψ) comes from a Doi-Koppinen datum (H, A, H). Now

$$\mathrm{can} : A \otimes_B A \to A \otimes H, \quad \mathrm{can}(a \otimes_B a') = aa'_{[0]} \otimes_B a'_{[1]}$$

and $A^{coH} = \{a \in A \mid \rho(a) = a \otimes 1\}$. A is then an H-Galois extension of B if and only if can is an isomorphism, and $A^{coH} = B$, which means that A is a Hopf-Galois extension of B in the sense of Section 4.2.

Corings and comonads Let $\mathcal{D}$ be a category. A comonad on $\mathcal{D}$ is a three-tuple $\mathbb{T} = (T, \varepsilon, \Delta)$, where $T : \mathcal{D} \to \mathcal{D}$ is a functor, and $\varepsilon : T \to 1_{\mathcal{D}}$ and $\Delta : T \to T \circ T$ are natural transformations, such that

$$T(\Delta_M) \circ \Delta_M = \Delta_{T(M)} \circ \Delta_M$$

and

$$T(\varepsilon_M) \circ \Delta_M = \varepsilon_{T(M)} \circ \Delta_M = I_{T(M)}$$

for all $M \in \mathcal{D}$. A morphism between two $\mathcal{D}$-comonads $\mathbb{T} = (T, \varepsilon, \Delta)$ and $\mathbb{T}' = (T', \varepsilon', \Delta')$ consists of a natural transformation $\alpha : T \to T'$ such that

$$\varepsilon' \circ \alpha = \varepsilon \quad \text{and} \quad (\alpha * \alpha) \circ \Delta = \Delta' \circ \alpha$$

Here $*$ is the Godement product:

$$(\alpha * \alpha)_M = \alpha_{T'(M)} \circ T(\alpha_M)$$

for all $M \in \mathcal{D}$. $\underline{\text{Comonad}}(\mathcal{D})$ will be the category of comonads on $\mathcal{D}$. For $\mathbb{T} \in \underline{\text{Comonad}}(\mathcal{D})$, a $\mathbb{T}$-coalgebra is a pair (M, ξ), with $M \in \mathcal{D}$, and $\xi : M \to \overline{T(M)}$ a morphism in $\mathcal{D}$ such that

$$\varepsilon_M \circ \xi = I_M \quad \text{and} \quad \Delta_M \circ \xi = T(\xi) \circ \xi$$

A morphism between (M, ξ) and $(M'\xi')$ consists of a morphism $f : M \to M'$ in $\mathcal{D}$ such that

$$T(f) \circ \xi = \xi' \circ f$$

The category of $\mathbb{T}$-coalgebras is denoted by $\mathcal{D}^{\mathbb{T}}$. Monads and algebras over a monad are defined in a similar way, in fact a monad on a category is a comonad on the dual category.

Now let $F : \mathcal{C} \to \mathcal{D}$ be a functor having a right adjoint G, and denote the unit and counit of the adjunction by

$$\eta : 1_{\mathcal{C}} \to GF \quad \text{and} \quad \varepsilon : FG \to 1_{\mathcal{D}}$$

We can associate a monad on $\mathcal{C}$ and a comonad on $\mathcal{D}$ to this adjunction. The comonad on $\mathcal{D}$ can be described as follows:

$$T = FG : \mathcal{D} \to \mathcal{D} \quad ; \quad \Delta = 1_F * \eta * 1_G : FG \to FGFG$$

i.e. $\Delta_M = F(\eta_{G(M)})$, and ε is the counit of the adjunction. Monads and comonads are the right tools to develop categorical descent theory, see [22, Ch. 4] or [123, Ch. 6] for a detailed discussion. Let us explain how comonads are related to corings.

Proposition 113. *For a ring A, we have a full and faithful functor*

$$i : A\text{-}\underline{\text{Coring}} \to \underline{\text{Comonad}}(\mathcal{M}_A)$$

If $B \to A$ is a morphism of rings, then the comonad on $\mathcal{M}_A$ associated to the restriction of scalars functor and its adjoint corresponds to the canonical coring $A \otimes_B A$.

For any A-coring $\mathcal{C}$, we have an isomorphism of categories

$$\mathcal{M}_A^{i(\mathcal{C})} \cong \mathcal{M}^{\mathcal{C}}$$

Proof. An A-coring $\mathcal{C}$ is an A-bimodule, so we have a functor $T: \bullet \otimes_A \mathcal{C}: \mathcal{M}_A \to \mathcal{M}_A$. We define $i(\mathcal{C}) = (T, \varepsilon, \Delta)$ with

$$\varepsilon_M = I_M \otimes_A \varepsilon_{\mathcal{C}}: \; T(M) = M \otimes_A \mathcal{C} \to M$$

$$\Delta_M = I_M \otimes_A \Delta_{\mathcal{C}}: \; T(M) = M \otimes_A \mathcal{C} \to T(T(M)) = M \otimes_A \mathcal{C} \otimes_A \mathcal{C}$$

for all $M \in \mathcal{M}_A$. It is straightforward to verify that (T, ε, Δ) is a comonad, and all other verifications are left to the reader.

Part II

Nonlinear equations

5 Yetter-Drinfeld modules and the quantum Yang-Baxter equation

he study of the quantum Yang-Baxter equation (QYBE) $R^{12}R^{13}R^{23} = R^{23}R^{13}R^{12}$ was one of the stimuli for the development of the theory of quantum groups ([108]). We will prove that special types of Hopf algebras (quasitriangular or co-quasitriangular) play a major role in solving this equation. The main result of this Chapter is the famous FRT theorem due to Faddeev, Reshetikhin, and Takhtajan ([85]). The alternative version of the FRT theorem (using Yetter-Drinfeld modules) proven by Radford in [155] is included, and will be the key in the unification schedule as described in the Preface. On the other hand, some recent results and new directions for studying the QYBE (for instance at set-theoretical level) are also included.

5.1 Notation

In this Section, we introduce some notation that will be used in the subsequent Chapters. Throughout part II, k will be a commutative field and vector spaces are taken over k.

Let A be a k-algebra, and $R \in A \otimes A$. We write

$$R = R^1 \otimes R^2$$

where the summation is implicitly understood. We write

$$R^{12} = R^1 \otimes R^2 \otimes 1_A, \ R^{13} = R^1 \otimes 1_A \otimes R^2, \ R^{23} = 1_A \otimes R^1 \otimes R^2 \ \in A^{\otimes 3}$$

We will use this in particular in the situation where $A = \mathrm{End}_k(M)$, with M a finite dimensional vector space. For $R \in \mathrm{End}_k(M) \otimes \mathrm{End}_k(M) \cong \mathrm{End}_k(M^{\otimes 2})$, we then obtain R^{12}, R^{13}, $R^{23} \in \mathrm{End}_k(M^{\otimes 3})$. If M is infinite dimensional, then $\mathrm{End}_k(M) \otimes \mathrm{End}_k(M)$ is no longer isomorphic to $\mathrm{End}_k(M^{\otimes 2})$, but the above notation still makes sense for $R \in \mathrm{End}_k(M^{\otimes 2})$.

Let M be finite dimensional, fix a basis $\{m_1, \cdots, m_n\}$ of M, and $\{p^1, \cdots, p^n\}$ the corresponding dual basis of the dual module M^*, such that

$$\langle p^i, m_j \rangle = \delta_j^i \tag{5.1}$$

for all $i, j \in \{1, \cdots, n\}$. δ_j^i is the Kronecker symbol. Then $\{e_j^i = p^i \otimes m_j \mid i, j = 1, \cdots, n\}$ and $\{c_j^i = m_j \otimes p^i \mid i, j = 1, \cdots, n\}$ are free bases for respectively

$\text{End}_k(M) \cong M^* \otimes M$ and $\text{End}_k(M^*) \cong M \otimes M^*$. The isomorphisms are given by the rules

$$e_j^i(m_k) = \delta_k^i m_j \quad \text{and} \quad c_j^i(p^k) = \delta_j^k p^i \tag{5.2}$$

$\text{End}_k(M) \cong M^* \otimes M$ is isomorphic to the $n \times n$-matrix algebra $\mathcal{M}_n(k)$, and the algebra structure $M^* \otimes M$ is given by the rules

$$e_j^i e_l^k = \delta_l^i e_j^k \quad \text{and} \quad \sum_{i=1}^n e_i^i = 1 \tag{5.3}$$

$\text{End}_k(M^*) \cong M \otimes M^*$ is isomorphic to the $n \times n$-comatrix coalgebra $\mathcal{M}^n(k)$, and the comultiplication on $M \otimes M^*$ is given by

$$\Delta(c_j^i) = \sum_k c_k^i \otimes c_j^k \quad \text{and} \quad \varepsilon(c_j^i) = \delta_j^i \tag{5.4}$$

We obtain the matrix algebra $\mathcal{M}_n(k)$ and the comatrix coalgebra $\mathcal{M}^n(k)$ after we take $M = k^n$ and the canonical basis. e_j^i is then identified with the elementary matrix having 1 in the (j, i)-position and 0 elsewhere.

A linear map $R : M \otimes M \to M \otimes M$ can be described by its matrix X, with n^4 entries $x_{uv}^{ij} \in k$, where i, j, u, v range from 1 to n. This means

$$R(m_k \otimes m_l) = \sum_{ij} x_{kl}^{ij} m_i \otimes m_j \tag{5.5}$$

or

$$R = \sum_{ijkl} x_{kl}^{ij} e_i^k \otimes e_j^l \tag{5.6}$$

We will often identify R and its matrix, i.e., we will write

$$R = \left(x_{kl}^{ij} \right)$$

Also we will use the *Einstein summation convention*. In a summation, all indices run from 1 to n. If in an expression an index occurs twice, namely once as an upper index and once as a lower index, then it is understood implicitly that we take the sum where this index runs from 1 to n. Indices that occur only once are not summation indices, and an index is not allowed to occur more than twice in one expression. For example, (5.6) is rewritten as

$$R = x_{kl}^{ij} e_i^k \otimes e_j^l$$

5.2 The quantum Yang-Baxter equation and the braid equation

Definition 5. *Let M be a vector space and $R \in \text{End}_k(M \otimes M)$.*

1. R is called a solution of the quantum Yang-Baxter equation (QYBE) if

$$R^{12}R^{13}R^{23} = R^{23}R^{13}R^{12} \qquad (5.7)$$

in $\text{End}_k(M^{\otimes 3})$.
2. R is a solution of the braid equation (BE) if

$$R^{12}R^{23}R^{12} = R^{23}R^{12}R^{23} \qquad (5.8)$$

in $\text{End}_k(M^{\otimes 3})$.

Some authors call (5.8) the quantum Yang-Baxter equation. We will see in Proposition 114 that the two equations are equivalent. Both equations can be stated in a more general context: let A be a k-algebra. $R = \sum R^1 \otimes R^2 \in A \otimes A$ is a solution of the quantum Yang-Baxter equation (resp. the braid equation) if the equation (5.7) (resp. (5.8)) holds in $A \otimes A \otimes A$.

Proposition 114. Let $R \in \text{End}_k(M \otimes M)$. The following statements are equivalent:

1. R is a solution of the BE;
2. $R\tau$ is a solution of the QYBE;
3. τR is a solution of the QYBE;
4. $\tau R\tau$ is a solution of the BE.

Proof. 1. $\Leftrightarrow$ 2. Let $T = R\tau$, and observe that

$$T^{12}T^{13}T^{23} = \tau^{13}R^{23}R^{12}R^{23}, \quad T^{23}T^{13}T^{12} = \tau^{13}R^{12}R^{23}R^{12}$$

The equivalence of 1. and 2. now follows from the fact that τ^{13} is an automorphism of $M \otimes M \otimes M$. 1. $\Leftrightarrow$ 3. can be proved in a similar way, and 1. $\Leftrightarrow$ 4. follows from the equivalence of 1. and 2. and the fact that $\tau R = \tau R\tau\tau$.

If M is finite dimensional, then we can write the QYBE in matrix form. Using the notation introduced in Section 5.1, we find

Proposition 115. Let M be a n-dimensional vector space. $R = \left(x_{kl}^{ij} \right) \in \text{End}_k(M \otimes M)$ is a solution of the QYBE if and only if

$$x_{lk}^{vu}x_{rv}^{pi}x_{gu}^{rj} = x_{vu}^{ij}x_{rk}^{pu}x_{ql}^{rv} \qquad (5.9)$$

for all $i, j, k, l, p, q = 1, \cdots, n$.

Proof. A direct computation that is left to the reader. Observe that (5.9) is a system of n^6 simultaneous equations in n^4 indeterminates.

We will now give some standard solutions of the QYBE. In the finite dimensional case, we keep on using the notation introduced in Section 5.1.

Examples 8. 1. The identity map $I_{M \otimes M}$ and the switch map τ_M are solutions of both QYBE and BE. More generally, if $\{a_{ij} \mid i, j = 1, \cdots, n\}$ is a family of scalars and $\dim(M) = n$, then

$$R : \; M \otimes M \to M \otimes M, \quad R(v_i \otimes v_j) = a_{ij} v_j \otimes v_i$$

is a solution of the QYBE.

2. Suppose that $R \in \mathrm{End}_k(M \otimes M)$ is bijective. Then R is a solution of the QYBE if and only if R^{-1} is a solution of the QYBE.

3. Let M be a finite dimensional vector space and u an automorphism of M. If R is a solution of the QYBE then $^u R = (u \otimes u) R (u \otimes u)^{-1}$ is also a solution of the QYBE.

4. Let $f, g \in \mathrm{End}_k(V)$ such that $fg = gf$. Then $R = f \otimes g$ is a solution of the QYBE since

$$R^{12} R^{13} R^{23} = f^2 \otimes gf \otimes g^2 \quad \text{and} \quad R^{23} R^{13} R^{12} = f^2 \otimes fg \otimes g^2.$$

For example, take $\dim(M) = 2$, and let $f, g \in \mathrm{End}_k(M)$ be given by their matrices with respect to the fixed basis $\{m_1, m_2\}$:

$$f = \begin{pmatrix} a & 1 \\ 0 & a \end{pmatrix} \qquad g = \begin{pmatrix} b & c \\ 0 & b \end{pmatrix}$$

where $a, b, c \in k$. The matrix of $R = f \otimes g$ with respect to the basis $\{m_1 \otimes m_1, m_1 \otimes m_2, m_2 \otimes m_1, m_2 \otimes m_2\}$ is

$$R = \begin{pmatrix} ab & ac & b & c \\ 0 & ab & 0 & b \\ 0 & 0 & ab & ac \\ 0 & 0 & 0 & ab \end{pmatrix} \tag{5.10}$$

As $fg = gf$, R is a solution for the QYBE.

5. Let a, b two non-zero scalars of k. Then

$$R = \begin{pmatrix} a & 0 & 0 & b \\ 0 & -a & 2a & 0 \\ 0 & 0 & a & 0 \\ 0 & 0 & 0 & a \end{pmatrix}$$

is a solution of the QYBE.

6. Let q be a scalar of k, $q \neq 0$, and M a n-dimensional vector space. Then $R_q : \; M \otimes M \to M \otimes M$ given by

$$R_q(v_i \otimes v_j) = \begin{cases} qv_i \otimes v_i & \text{if } i = j \\ v_i \otimes v_j & \text{if } i < j \\ v_i \otimes v_j + (q - q^{-1})v_j \otimes v_i & \text{if } i < j \end{cases} \quad (5.11)$$

is a solution of the QYBE. R_q is called the *classical Yang-Baxter operator* on M.

7. Let A be a k-algebra acting on M. Solutions of QYBE on M can be constructed in a natural way from solutions of the QYBE in $A^{\otimes 3}$: if $R = R^1 \otimes R^2 \in A \otimes A$, then

$$\mathcal{R} = \mathcal{R}_{(R,M,\cdot)} : M \otimes M \to M \otimes M, \quad \mathcal{R}(v \otimes w) = R \cdot (v \otimes w) = R^1 \cdot v \otimes R^2 \cdot w$$

is a solution of the QYBE in $\mathrm{End}(M \otimes M \otimes M)$ since

$$\mathcal{R}^{12}\mathcal{R}^{13}\mathcal{R}^{23}(x \otimes y \otimes z) = R^{12}R^{13}R^{23} \cdot (x \otimes y \otimes z)$$
$$= R^{23}R^{13}R^{12} \cdot (x \otimes y \otimes z) = \mathcal{R}^{23}\mathcal{R}^{13}\mathcal{R}^{12}(x \otimes y \otimes z)$$

for all x, y, $z \in V$.

Of course we can take $A = \mathcal{M}_n(k)$. A acts on $M = k^n$, and then we obtain the previous identification of solutions on $M \otimes M$ and in $A \otimes A$. We give a particular example. For a fixed index $i \in \{1, \cdots, n\}$,

$$R_i = \sum_{j=1}^{n} e_j^i \otimes e_i^j$$

is a solution of the braid equation in $\mathcal{M}_n(k) \otimes \mathcal{M}_n(k) \otimes \mathcal{M}_n(k)$. $V = M^n$ is a left $\mathcal{M}_n(k)$-module via $f \cdot v = f(v)$, for all $f \in \mathrm{End}_k(M)$ and $v \in M$. It follows that

$$\mathcal{R}_i : M \otimes M \to M \otimes M, \quad \mathcal{R}_i(v \otimes w) = \sum_{j=1}^{n} e_j^i \cdot v \otimes e_i^j \cdot w$$

is a solution of the braid equation and hence $\tau\mathcal{R}_i$ and $\mathcal{R}_i\tau$ are solution of the QYBE.

8. The above example can be generalized as follows. Let A be an algebra and $R = R^1 \otimes R^2 \in A \otimes A$ be an A-central element, that is

$$aR^1 \otimes R^2 = R^1 \otimes R^2a$$

for all $a \in A$. This means that $R \in W_1$ in the terminology of Section 3.2. We will prove in Proposition 144 that R is a solution of the FS-equation

$$R^{12}R^{23} = R^{23}R^{13} = R^{13}R^{12}$$

in $A \otimes A \otimes A$ and a fortiori of the braid equation. We have seen in Theorem 27 that any separable or Frobenius algebra gives us an $R \in W_1$, and therefore a solution of the braid equation. We will come back to this in Chapter 8.

9. If A is an algebra, then we have the following interesting solution of the braid equation (cf. [148], [149]):

$$R = R_{(A,\cdot,1_A)} : A \otimes A \to A \otimes A, \quad R(x \otimes y) = xy \otimes 1 + 1 \otimes xy - x \otimes y \quad (5.12)$$

for all $x, y \in A$, is a solution of the braid equation and hence $R\tau$ is a solution of QYBE. A generalization of this solution was recently given in [64]: let A be a k-algebra of dimension greater than one and $a, b, c \in k$. Then

$$R = R_{a,b,c} : A \otimes A \to A \otimes A, \quad R(x \otimes y) = axy \otimes 1 + b1 \otimes xy - cx \otimes y \quad (5.13)$$

is a bijective solution of the braid equation if and only if one of the following conditions holds:

1. $a = c \neq 0$ and $b \neq 0$
2. $b = c \neq 0$ and $a \neq 0$
3. $a = b = 0$ and $c \neq 0$

The inverse of $R_{a,b,c}$ is given by $R_{a,b,c}^{-1} = R_{b^{-1},a^{-1},c^{-1}}$ (in case 1) and 2)), and $R_{0,0,c}^{-1} = R_{0,0,c^{-1}}$ (in case 3)).

Indeed, for $x, y, z \in A$, we find that

$$R^{12} R^{23} R^{12}(x \otimes y \otimes z) = R^{23} R^{12} R^{23}(x \otimes y \otimes z)$$

if and only if

$$a(a-c)(b-c)(x \otimes yz \otimes 1 - xy \otimes 1 \otimes z) = b(a-c)(b-c)(1 \otimes xy \otimes z - x \otimes 1 \otimes yz)$$

and the "if part" follows. It remains to prove the "only if" part. As $\dim_k(A) > 1$, we can take $y = 1$ and $x, z \in A \backslash k1_A$. Then the last equation is equivalent to the following two equations

$$a(a-c)(b-c) = 0, \quad b(a-c)(b-c) = 0$$

and the conclusion follows.

Dually, solutions of the QYBE can be constructed using a coalgebra structure on the vector space M. Let C be a coalgebra. Then

$$R : C \otimes C \to C \otimes C, \quad R(c \otimes d) = \varepsilon(c)d_{(1)} \otimes d_{(2)} + \varepsilon(d)c_{(1)} \otimes c_{(2)} - d \otimes c$$

for all $c, d \in C$ is a bijective solution of the QYBE.

10. Let H be a Hopf algebra with a bijective antipode. Then the maps

$$R : H \otimes H \to H \otimes H, \quad R(g \otimes h) = h_{(3)}gS^{-1}(h_{(2)}) \otimes h_{(1)}$$

and

$$R' : H \otimes H \to H \otimes H, \quad R'(g \otimes h) = h_{(3)}S^{-1}(h_{(1)})g \otimes h_{(2)}$$

for all $g, h \in H$ are solutions of the QYBE. This can be proved by a long but routine computation but a more efficient approach is to use Theorem 59 for the Verma Yetter-Drinfeld structures given in Remark 15.

We have seen that for an algebra A the map

$$R_{(A,\cdot,1_A)} : A \otimes A \to A \otimes A, \quad R_{(A,\cdot,1_A)}(x \otimes y) = xy \otimes 1 + 1 \otimes xy - x \otimes y \quad (5.14)$$

is a solution of the braid equation; moreover $R_{(A,\cdot,1_A)}$ is involutory, i.e. $R^2_{(A,\cdot,1_A)} = I_{A \otimes A}$. In Theorem 57 we will describe all involutory solutions of the braid equation that can be obtained in this way.

Theorem 57. (Nichita [148]) *Let k be a field such that $char(k) \neq 2$, M an n-dimensional vector space and $R \in End(M \otimes M)$ an involutory solution of the braid equation. The following statements are equivalent:*

1. *there exists an algebra structure $(M, \cdot, 1_M)$ on M such that $R = R_{(M,\cdot,1_M)}$;*
2. *the following two conditions holds:*
   a. *$rank_k(R + Id) \leq n$;*
   b. *there exists $0 \neq x_0 \in M$ such that $R(x_0 \otimes x) = x \otimes x_0$ for all $x \in M$.*

Proof. $\underline{2. \Rightarrow 1.}$ As R is involutory it follows from 2b) that

$$R(x \otimes x_0) = x_0 \otimes x \quad (5.15)$$

for all $x \in M$. Hence, for $x \in M$

$$(R + Id)(x_0 \otimes x) = x \otimes x_0 + x_0 \otimes x \quad (5.16)$$

Consider the n-dimensional vector space $W = \{x \otimes x_0 + x_0 \otimes x \mid x \in M\}$. Then

$$W \subseteq (R + Id)(x_0 \otimes M) \subseteq (R + Id)(M \otimes M)$$

It follows from here and 2a) that

$$(R + Id)(M \otimes M) = W.$$

Hence there exists a k-linear map $m_M : M \otimes M \to M$ (this will be the multiplication on M) such that

$$(R + Id)(v \otimes w) = m_M(v \otimes w) \otimes x_0 + x_0 \otimes m_M(v \otimes w) \quad (5.17)$$

for all v, $w \in M$. We denote $m_M(v \otimes w) = vw$. From (5.17) we obtain that $R = R_{(M,m_M,1_M=x_0)}$. We still have to prove that $(M, m_M, 1_M = x_0)$ is an algebra structure on M. Using (5.16) and (5.17) we obtain that

$$(x_0 x) \otimes x_0 + x_0 \otimes (x_0 x) = x \otimes x_0 + x_0 \otimes x$$

for all $x \in M$, and $x_0 x = x$ for any $x \in M$. In a similar way, we obtain that $x x_0 = x$ for all $x \in M$, using (5.15) and (5.17), and it follows that x_0 is a unit of M. The associativity of m_M will follow from the braid condition. For a, b, $c \in M$ the braid equation

$$R^{12}R^{23}R^{12}(a \otimes b \otimes c) = R^{23}R^{12}R^{23}(a \otimes b \otimes c)$$

takes the form

$$2(ab)c \otimes x_0 \otimes x_0 + 2x_0 \otimes (ab)c \otimes x_0 + 2x_0 \otimes x_0 \otimes (ab)c$$
$$= 2a(bc) \otimes x_0 \otimes x_0 + 2x_0 \otimes a(bc) \otimes x_0 + 2x_0 \otimes x_0 \otimes a(bc)$$

As $\mathrm{char}(k) \neq 2$ we obtain that $(ab)c = a(bc)$, hence $(M, m_M, 1_M = x_0)$ is an algebra structure on M.

The proof of the other implication 1. ⇒ 2. is straightforward.

Let A be a k-algebra and $R \in A \otimes A$ a solution of the QYBE. Let $F \in A \otimes A$ be invertible and $\tilde{R} = \tau(F^{-1})RF$, where τ switches the two tensor components. If $F = u \otimes u$, where u is an invertible element of A, then $\tau(F^{-1})RF = (u^{-1} \otimes u^{-1})R(u \otimes u)$ is again a solution of the QYBE, but for general F this is no longer true. If $\tilde{R}$ is a solution of the QYBE then we call it a *twist* of R. To find out a necessary and sufficient condition such that $\tilde{R}$ remains a solution of QYBE is not easy. We give a sufficient condition in Theorem 58. We will use the following notation: for $x = x^1 \otimes x^2 \otimes x^3 \in A \otimes A \otimes A$, we write $x^{123} = x$, $x^{231} = x^2 \otimes x^3 \otimes x^1$, etc.

Theorem 58. (Kulish, Mudrov [112]) *Let A be a k-algebra, $R \in A \otimes A$ an invertible solution of the QYBE in $A \otimes A \otimes A$ and F an invertible element of $A \otimes A$. Assume that there exists invertible elements U, $V \in A \otimes A \otimes A$ such that*

$$UF^{12} = VF^{23} \tag{5.18}$$
$$R^{12}U = U^{213}R^{12} \tag{5.19}$$
$$R^{23}V = V^{132}R^{23} \tag{5.20}$$

Then $\tilde{R} := \tau(F^{-1})RF$ is a solution on the QYBE.

Proof. In the computations presented below, bars are used to denote the inverse, for example $\overline{F} = F^{-1}$. Using (5.18), (5.19) and (5.20) we find

$$\overline{F}^{12}\overline{V}^{312}R^{13}R^{23}U^{123}F^{12}$$
$$= \overline{F}^{12}\overline{V}^{312}R^{13}R^{23}V^{123}F^{23}$$
$$= \overline{F}^{12}\overline{V}^{312}R^{13}V^{132}R^{23}F^{23} = \overline{F}^{12}\overline{V}^{312}R^{13}V^{132}F^{32}\tilde{R}^{23}$$
$$= \overline{F}^{12}\overline{V}^{312}R^{13}U^{132}F^{13}\tilde{R}^{23} = \overline{F}^{12}\overline{V}^{312}U^{312}R^{13}F^{13}\tilde{R}^{23}$$
$$= \overline{F}^{12}\overline{V}^{312}U^{312}F^{31}\tilde{R}^{13}\tilde{R}^{23} = \overline{F}^{12}\overline{V}^{312}V^{312}F^{12}\tilde{R}^{13}\tilde{R}^{23}$$
$$= \tilde{R}^{13}\tilde{R}^{23}$$

Using this formula we obtain

$$\tilde{R}^{12}\tilde{R}^{13}\tilde{R}^{23}(\tilde{R}^{12})^{-1}$$
$$= \overline{F}^{21}R^{12}F^{12}\overline{F}^{12}\overline{V}^{312}R^{13}R^{23}U^{123}F^{12}\overline{F}^{12}\overline{R}^{12}F^{21}$$
$$= \overline{F}^{21}R^{12}\overline{V}^{312}R^{13}R^{23}U^{123}\overline{R}^{12}F^{21} = \overline{F}^{21}\overline{V}^{321}R^{12}R^{13}R^{23}\overline{R}^{12}U^{213}F^{21}$$
$$= \overline{F}^{21}\overline{V}^{321}R^{23}R^{13}U^{213}F^{21} = \tau^{12}(\overline{F}^{12}\overline{V}^{312}R^{13}R^{23}U^{123}F^{12})$$
$$= \tau^{12}(\tilde{R}^{13}\tilde{R}^{23}) = \tilde{R}^{23}\tilde{R}^{13}$$

and $\tilde{R}$ is a solution of the QYBE.

Remarks 9. 1. In fact, the pair (U, V) satisfying conditions (5.18-5.20) is determined by an invertible $G \in A \otimes A \otimes A$ such that

$$R^{12}G = G^{213}\overline{F}^{21}R^{12}F^{12}$$

and

$$R^{23}G = G^{132}\overline{F}^{32}R^{23}F^{23}.$$

Indeed, let us denote $G = UF^{12} = VF^{23}$. Then $U = G\overline{F}^{12}$, $V = G\overline{F}^{23}$ and the equations (5.19)-(5.20) are written in terms of G as above.

2. An important example of pairs (F, G) can be obtained using Drinfeld's construction [76]. Let H be a Hopf algebra and $F \in H \otimes H$ an invertible element. Now, we deform the comultiplication on H using F and define $\Delta_F(h) := F^{-1}\Delta(h)F$, for all $h \in H$. Drinfeld proved that if F satisfy the *twist equation*

$$(\Delta \otimes Id)(F)F^{12} = (Id \otimes \Delta)(F)F^{23}$$

then H_F, which is equal to H as an algebra, and with comultiplication Δ_F, is a Hopf algebra.

Now it is easy to see that $(F, G := (\Delta \otimes Id)(F)F^{12})$ satisfy the conditions of Theorem 58. Consequently, if F is a twist in the sense of Drinfeld and $R \in H \otimes H$ is a solution of the QYBE, then $\tilde{R} = \tau(F^{-1})RF$ is a solution on the QYBE.

A method of constructing twists for $H = kG$, where G is a nonabelian group was given in [81].

5.3 Hopf algebras versus the QYBE

In this Section, we discuss how we can construct solutions of the QYBE starting from a Hopf algebra.

Quasitriangular bialgebras

Definition 6. *A quasitriangular bialgebra (resp. Hopf algebra) is a pair* (H, R), *where* H *is a biagebra (resp. a Hopf algebra) and* $R = R^1 \otimes R^2 \in H \otimes H$ *such that*
(QT1) $\Delta(R^1) \otimes R^2 = R^{13}R^{23};$

(QT2) $\varepsilon(R^1)R^2 = 1;$
(QT3) $R^1 \otimes \Delta(R^2) = R^{13}R^{12};$
(QT4) $R^1\varepsilon(R^2) = 1;$
(QT5) $\Delta^{\mathrm{cop}}(h)R = R\Delta(h),$ for all $h \in H.$

Remarks 10. 1. If (H, R) is a quasitriangular Hopf algebra, then R is invertible and $R^{-1} = S(R^1) \otimes R^2$. In this case, the condition (QT5) tell us that Δ^{cop} is a twist of Δ in the sense of Drinfeld. A quasitriangular Hopf algebra (H, R) is called *triangular* if $R^{-1} = \tau(R)$.
2. Let H be a finite dimensional bialgebra and $R = R^1 \otimes R^2 \in H \otimes H$. We define

$$\lambda_R : H^* \to H^{\mathrm{cop}}, \quad \lambda_R(f) = \langle f, R^1 \rangle R^2$$

for all $f \in H^*$. Then R satisfies (QT1)-(QT4) if and only if λ_R is a bialgebra map.
3. Let (H, R) be a quasitriangular Hopf algebra and M, N left H-modules. Then the map

$$\sigma_{M,N} : M \otimes N \to N \otimes M, \quad \sigma_{M,N}(m \otimes n) = R^2 n \otimes R^1 m$$

for all $m \in M$, $n \in N$, is an isomorphism of left H-modules. The inverse of it is given by

$$\sigma_{M,N}^{-1} : N \otimes M \to M \otimes N, \quad \sigma_{M,N}^{-1}(n \otimes m) = S(R^1)m \otimes R^2 n$$

for all $m \in M$, $n \in N$.

Examples 9. 1. Let H be a cocommutative Hopf algebra. Then $(H, R = 1 \otimes 1)$ is a quasitriangular Hopf algebra.
Conversely, assume that (H, R) is a quasitriangular Hopf algebra such that $R \in \mathrm{Im}(\Delta)$. Then H is cocommutative.
Indeed, let $h \in H$ such that $R = \Delta(h)$. Using (QT1) we obtain $(h' = h)$:

$$h_{(1)} \otimes h_{(2)} \otimes h_{(3)} = h_{(1)} \otimes h'_{(1)} \otimes h_{(2)}h'_{(2)}$$

and

$$h = h_{(1)}h_{(2)}S(h_{(3)}) = h_{(1)}h'_{(1)}S(h_{(2)}h'_{(2)}) = \varepsilon(h)\varepsilon(h')1_H.$$

and we conclude that $h = a1_H$ for some $a \in k$. From (QT2), we get that $a = 1$ and therefore, $R = \Delta(1) = 1 \otimes 1$. Now, (QT5) gives us $\Delta^{\mathrm{cop}} = \Delta$, i.e. H is cocommutative.
2. The most important example of quasitriangular Hopf algebra is probably the Drinfeld double $D(H)$ of a finite dimensional Hopf algebra. We have discussed the Drinfeld double already in Section 4.4. Here we will use the representation $D(H) = H \bowtie H^*$. The bialgebra structure is then given by the formulas

$$(h \bowtie f)(h' \bowtie f') = \sum h_{(2)}h' \bowtie f * \langle f', S^{-1}h_{(3)}?h_{(1)} \rangle$$

$$\Delta_{D(H)}(h \bowtie f) = \sum (h_{(1)} \bowtie f_{(2)}) \otimes (h_{(2)} \bowtie f_{(1)})$$

$$\varepsilon_{D(H)} = \varepsilon_H \otimes \varepsilon_{H^* \text{cop}}$$

for $h, h' \in H$ and $f, f' \in H^*$. Let $\{e_i, e_i^*\}$ be a finite dual basis of H and R be the canonical element

$$R = \sum_i (e_i \bowtie \varepsilon) \otimes (1 \bowtie e_i^*).$$

Then, by a routine computation using the multiplication rule of $D(H)$ and the fact that $\{e_i, e_i^*\}$ is a dual basis, we can prove that $(D(H), R)$ is a quasitriangular Hopf algebra ([125]). Moreover, as $H \to D(H)$, $h \to h \bowtie \varepsilon$, is a Hopf algebra map we conclude that any finite dimensional Hopf algebra can be embedded into a quasitriangular one. Now, using Proposition 118 we obtain that the above canonical element R is a solution of the QYBE in $D(H) \otimes D(H) \otimes D(H)$.

Sometimes it is possible to describe all quasitriangular structures on a given Hopf algebra. Consider the Hopf algebra $E(n)$ over a field k with characteristic different from 2, with generators g and $x_1, \cdots, x_n$ and relations

$$g^2 = 1, \quad x_i^2 = 0, \quad x_i g = -g x_i, \quad x_i x_j = -x_j x_i$$

and the coalgebra structure is given by

$$\Delta(g) = g \otimes g, \quad \Delta(x_i) = x_i \otimes g + 1 \otimes x_i,$$

$$\varepsilon(g) = 1, \quad \varepsilon(x_i) = 0$$

for all $i, j = 1, \cdots n$. This Hopf algebra can be obtained using the Ore extension method, as described in [13], and plays a role in the classification theory of pointed Hopf algebras (cf. [36]). For $n = 1$, we recover Sweedler's four dimensional Hopf algebra, which was the first example of a noncommutative noncocommutative Hopf algebra. The following result is due to Panaite and Van Oystaeyen [150].

Proposition 116. *Let k be a field of characteristic different from 2. There is a bijective correspondence between all quasitriangular structures on $E(n)$ and the $n \times n$-matrices with entries in k.*

Proof. We refer to [150] for full detail, and restrict ourselves to giving the description of the quasitriangular structure R_A corresponding to a given $A \in \mathcal{M}_n(k)$. For two subsets P and F of $\{1, \cdots, n\}$ with the same cardinality, we consider the matrix

$$A_{P,F} = (a_{ij})_{i \in P, j \in F}$$

If $P = \{i_1, \cdots, i_s\} \subset \{1, \cdots, n\}$ such that $i_1 < i_2 < \cdots < i_s$, we denote $x_P = x_{i_1} \cdots x_{i_s}$. For a matrix $A \in M_n(k)$ let

$$R_A = 2^{-1}\left(1 \otimes 1 + g \otimes 1 + 1 \otimes g - g \otimes g\right) + 2^{-1} \sum_{|P|=|F|} (-1)^{\frac{|P|(|P|-1)}{2}} \det(A_{P,F}) \times$$

$$\times \left(x_P \otimes g^{|P|}x_F + gx_P \otimes g^{|P|}x_F + x_P \otimes g^{|P|+1}x_F - gx_P \otimes g^{|P|+1}x_F\right)$$

Then, taking into account that $\{g^j x_P | P \subset \{1, \cdots, n\}, j = 0, 1\}$ is a basis of $E(n)$, we can prove that $(E(n), R_A)$ is a quasitriangular Hopf algebra.

As we observed above, $E(1) = H_4$, Sweedler's Hopf algebra. The quasitriangular structures on H_4 are given by

$$R_a = 2^{-1}\left(1 \otimes 1 + g \otimes 1 + 1 \otimes g - g \otimes g\right) + 2^{-1}a\left(x \otimes x + x \otimes gx + gx \otimes gx - gx \otimes x\right)$$

where a ranges over k.

There exist Hopf algebras without quasitriangular structure. The easiest example is the dual of a nonabelian finite group ring.

Proposition 117. *For any finite nonabelian group G, $(kG)^*$ has no quasitriangular structure.*

Proof. Fist we recall the coalgebra structure of $(kG)^*$: let $\{p_g \mid g \in G\}$ be the dual basis of $\{g \mid g \in G\}$, that is $p_g(h) = \delta_{g,h}$ for all $h \in G$. Then

$$\Delta(p_g) = \sum_{x \in G} p_x \otimes p_{x^{-1}g}, \quad \text{and} \quad p_g p_h = \delta_{g,h} p_g$$

for all $g, h \in G$. As G is non-abelian we can pick $g, h \in G$ such that $gh \neq hg$; let $M = kp_h$ and $N = kp_g$. If we can show that $M \otimes N$ and $N \otimes M$ are not isomorphic as $(kG)^*$-modules, then it follows from Remark 10 3) that there is no R such that $((kG)^*, R)$ is quasitriangular. We have

$$p_{gh} \cdot (M \otimes N) = p_{gh} \cdot (p_h \otimes p_g) = \sum_{x \in G} p_x p_h \otimes p_{x^{-1}gh} p_g = 0,$$

and, on the other hand

$$p_{gh} \cdot (N \otimes M) = p_{gh} \cdot (p_g \otimes p_h) = \sum_{x \in G} p_x p_g \otimes p_{x^{-1}gh} p_h = p_g \otimes p_h.$$

Hence, $p_{gh} \in \text{Ann}_{kG^*}(M \otimes N) \setminus \text{Ann}_{kG^*}(N \otimes M)$.

Proposition 118. *Let (H, R) be a quasitriangular bialgebra. Then R is a solution of the QYBE in $H \otimes H \otimes H$. In particular, for any left H-module M, the homotety*

$$\mathcal{R} = \mathcal{R}_{R,V} : M \otimes M \to M \otimes M, \quad \mathcal{R}(v \otimes w) = R \cdot (v \otimes w) = R^1 \cdot v \otimes R^2 \cdot w$$

is a solution of the QYBE in $\text{End}(M \otimes M \otimes M)$.

Proof. It follows from (QT1) and (QT5) that

$$\Delta^{\mathrm{cop}}(R^1) \otimes R^2 = R^1_{(2)} \otimes R^1_{(1)} \otimes R^2 = (\tau \otimes Id)(\Delta(R^1) \otimes R^2)$$
$$= (\tau \otimes Id)(R^{13} R^{23}) = R^{23} R^{13}$$

and we have proved the formula
$(QT1')$ $\quad R^{23} R^{13} = \Delta^{\mathrm{cop}}(R^1) \otimes R^2.$
Using this formula and then (QT5) and (QT1) we obtain $(r = R)$:

$$R^{23} R^{13} R^{12} = (\Delta^{\mathrm{cop}}(r^1) \otimes r^2) R^{12} = \Delta^{\mathrm{cop}}(r^1) R \otimes r^2$$
$$= R\Delta(r^1) \otimes r^2 = R^{12}(\Delta(r^1) \otimes r^2) = R^{12} R^{13} R^{23}$$

i.e. R is a solution of the QYBE.

Remark 13. Using Proposition 118 we obtain a rich class of solutions of the QYBE. However, not every solution is of this form: Radford [155] constructed a counterexample, namely the classical Yang-Baxter operator R_q constructed in (5.11), in the case where q is not a root of unity.

Coquasitriangular bialgebras Let H be a bialgebra. Recall that a k-linear map $\sigma : H \otimes H \to k$ is called *convolution invertible* if it has an inverse σ^{-1} in the convolution algebra $\mathrm{Hom}(H \otimes H, k)$, that is, $\sigma^{-1} : H \otimes H \to k$ satisfies

$$\sigma(x_{(1)} \otimes y_{(1)})\sigma^{-1}(x_{(2)} \otimes y_{(2)}) = \sigma^{-1}(x_{(1)} \otimes y_{(1)})\sigma(x_{(2)} \otimes y_{(2)}) = \varepsilon(x)\varepsilon(y)1_H$$

for all $x, y \in H$. In the sequel, $\sigma_{12}, \sigma_{13}, \sigma_{23} : H \otimes H \otimes H \to k$ will be the k-linear maps defined by

$$\sigma_{12}(x \otimes y \otimes z) = \varepsilon(z)\sigma(x \otimes y), \quad \sigma_{13}(x \otimes y \otimes z) = \varepsilon(y)\sigma(x \otimes z),$$

$$\sigma_{23}(x \otimes y \otimes z) = \varepsilon(x)\sigma(y \otimes z),$$

for all $x, y, z \in H$.

Definition 7. *A coquasitriangular (or braided) bialgebra (resp. Hopf algebra) is a pair (H, σ), where H is a biagebra (resp. Hopf algebra) and $\sigma : H \otimes H \to k$ is a k-linear map such that*
(B1) $\quad \sigma(x_{(1)} \otimes y_{(1)})y_{(2)}x_{(2)} = \sigma(x_{(2)} \otimes y_{(2)})x_{(1)}y_{(1)};$
(B2) $\quad \sigma(x \otimes 1) = \varepsilon(x);$
(B3) $\quad \sigma(x \otimes yz) = \sigma(x_{(1)} \otimes y)\sigma(x_{(2)} \otimes z);$
(B4) $\quad \sigma(1 \otimes x) = \varepsilon(x);$
(B5) $\quad \sigma(xy \otimes z) = \sigma(y \otimes z_{(1)})\sigma(x \otimes z_{(2)}),$
for all $x, y, z \in H$.

Remarks 11. 1. The two notions of quasitriangular bialgebra and coquasitriangular bialgebra are dual: a finite dimensional bialgebra H is coquasitriangular if and only if H^* a quasitriangular. The correspondence between R and

σ is the following: assume that $(H^*, R = \sum_i f_i \otimes g_i)$ is quasitriangular, and define $\sigma = \sigma_R$ by the formula

$$\sigma(h \otimes g) = \sum_i f_i(g)g_i(h)$$

for all g, $h \in H$. Then (H, σ) is coquasitriangular.

Conversely, assume that (H, σ) is coquasitriangular. As H is finite dimensional, we can identify $H^* \otimes H^*$ and $(H \otimes H)^*$, and we take $R \in H^* \otimes H^*$ equal to $\sigma \in (H \otimes H)^*$. Then (H^*, R) is a quasitriangular bialgebra.

In particular, the duals of all the finite dimensional quasitriangular bialgebras discussed in Example 9 are coquasitriangular.

2. Coquasitriangular bialgebras can be viewed as generalizations of commutative bialgebras. In particular, if H is commutative, then $(H, \varepsilon_H \otimes \varepsilon_H)$ is coquasitriangular. We remark that the condition (B1) is then noting else then the comutativity condition $yx = xy$.

3. If (H, σ) is a coquasitriangular Hopf algebra then σ is convolution invertible with inverse $\sigma^{-1}(x \otimes y) = \sigma(x \otimes S(y))$.

Sometimes it is possible to describe all possible coquasitriangular structures on a Hopf algebra. An elementary example is given in Proposition 119

Proposition 119. [53, Lemma 1.2] *Let H be a finite dimensional commutative and cocommutative bialgebra. Then we have a bijective correspondence between coquasitriangular structures σ on H and bialgebra maps $\theta : H \to H^*$.*

Proof. If H is commutative and cocommutative, then (B1) holds for every σ. The canonical isomorphism

$$(H \otimes H)^* \cong \mathrm{Hom}(H, H^*)$$

sending $\sigma \in (H \otimes H)^*$ to $\theta : H \to H^*$ given by

$$\langle \theta(h), k \rangle = \sigma(h \otimes k)$$

restricts to an isomorphism between the vector space consisting of all σ satisfying (B2)-(B5) and bialgebra maps $H \to H^*$.

Example 20. Assume that $H = kC_n$, and let g be a generator of the cyclic group C_n. A Hopf algebra map $\theta : H \to H^*$ is completely determined by $\langle \theta(g), g \rangle = \lambda$. Since $\theta(g^n) = \varepsilon$, it follows that $\lambda^n = 1$, and we find that coquasitriangular structures on kC_n are in bijective correspondence with the n-th roots of 1 in k.

Our next result is the dual version of Proposition 118.

Proposition 120. *Let (H, σ) be a coquasitriangular bialgebra. Then*

$$\sigma_{12} * \sigma_{13} * \sigma_{23} = \sigma_{23} * \sigma_{13} * \sigma_{12}$$

in the convolution algebra $\text{Hom}(H^{\otimes 3}, k)$.
In particular, for any right H-comodule (M, ρ), *the k-linear map*

$$R = R_{(\sigma, M, \rho)} : M \otimes M \to M \otimes M, \quad R(u \otimes v) = \sigma(u_{[1]} \otimes v_{[1]}) u_{[0]} \otimes v_{[0]} \quad (5.21)$$

is a solution of the QYBE in $\text{End}(M \otimes M \otimes M)$.

Proof. Let x, y, $z \in H$. A direct calculation shows that

$$(\sigma_{23} * \sigma_{13} * \sigma_{12})(x \otimes y \otimes z) = \sigma(y_{(1)} \otimes z_{(1)}) \sigma(x_{(1)} \otimes z_{(2)}) \sigma(x_{(2)} \otimes y_{(2)})$$

On the other hand

$$(\sigma_{12} * \sigma_{13} * \sigma_{23})(x \otimes y \otimes z) = \sigma(x_{(1)} \otimes y_{(1)}) \sigma(x_{(2)} \otimes z_{(1)}) \sigma(y_{(2)} \otimes z_{(2)})$$

$$(B3) = \sigma\left(x \otimes \sigma(y_{(2)} \otimes z_{(2)}) y_{(1)} z_{(1)}\right)$$

$$(B1) = \sigma\left(x \otimes \sigma(y_{(1)} \otimes z_{(1)}) z_{(2)} y_{(2)}\right)$$

$$= \sigma(y_{(1)} \otimes z_{(1)}) \sigma(x \otimes z_{(2)} y_{(2)})$$

$$(B3) = \sigma(y_{(1)} \otimes z_{(1)}) \sigma(x_{(1)} \otimes z_{(2)}) \sigma(x_{(2)} \otimes y_{(2)})$$

hence we have proved that σ is a solution of the QYBE $\sigma_{12} * \sigma_{13} * \sigma_{23} = \sigma_{23} * \sigma_{13} * \sigma_{12}$ in the convolution algebra $\text{Hom}(H \otimes H \otimes H, k)$.
The second statement then follows from the formulas

$$R^{12} R^{13} R^{23}(u \otimes v \otimes w) = \sigma_{12} * \sigma_{13} * \sigma_{23}(u_{[1]} \otimes v_{[1]} \otimes w_{[1]}) u_{[0]} \otimes v_{[0]} \otimes w_{[0]}$$

$$R^{23} R^{13} R^{12}(u \otimes v \otimes w) = \sigma_{23} * \sigma_{13} * \sigma_{12}(u_{[1]} \otimes v_{[1]} \otimes w_{[1]}) u_{[0]} \otimes v_{[0]} \otimes w_{[0]}$$

for all u, M, $w \in V$.

Remarks 12. 1. Let (H, σ) be a finite dimensional coquasitriangular bialgebra and (M, ρ) a right H-comodule. Then (H^*, R_σ) is a quasitriangular bialgebra, M has a structure $\cdot$ of left H^*-module and $R_{(\sigma, M, \rho)} = \mathcal{R}_{(R_\sigma, M, \cdot)}$.
2. We have a remarkable difference between quasitriangular and coquasitriangular bialgebras. In the next Section, we will prove the FRT-Theorem: if $R : M \otimes M \to M \otimes M$ is a solution of the QYBE, with M finite dimensional, then we can find a coquasitriangular bialgebra H coacting on M in such a way that $R = R_{(\sigma, M, \rho)}$. We do not have a similar result for quasitriangular bialgebras, as is shown by Radford's example, see Remark 13. Apparently this is in contradiction with the duality between Propositions 120 and 118. This is explained by the fact that the coquasitriangular bialgebra H constructed in the FRT Theorem is not necessarily finite dimensional (even if M is finite dimensional).

Yetter-Drinfeld modules Let H be a bialgebra. We recall from Section 4.4 that a (left-right) Yetter-Drinfeld module over H is a threetuple $(M, \cdot, \rho)$, where $(M, \cdot)$ is a left H-module, (M, ρ) is a right H-comodule and

$$h_{(1)} \cdot m_{[0]} \otimes h_{(2)} m_{[1]} = (h_{(2)} \cdot m)_{[0]} \otimes (h_{(2)} \cdot m)_{[1]} h_{(1)} \qquad (5.22)$$

for all $h \in H$, $m \in M$. $_H \mathcal{YD}^H$ is the category of Yetter-Drinfeld module over H and H-linear H-colinear maps. As we have seen in Section 4.4, the compatibility relation (5.22) can be rewritten in the case where H has a twisted antipode $\overline{S}$, and then it takes the form

$$\rho(h \cdot m) = h_{(2)} \cdot m_{[0]} \otimes h_{(3)} m_{[1]} \overline{S}(h_{(1)}) \qquad (5.23)$$

Example 21. Let G be a group and $H = kG$. Using (5.23) we easily obtain that $M \in _{kG}\mathcal{YD}^{kG}$ if and only if M is a *crossed G-module* , that is M is a left $k[G]$-module and there exists $\{M_\sigma \mid \sigma \in G\}$ a family of k-subspaces of M such that

$$M = \oplus_{\sigma \in G} M_\sigma \quad \text{and} \quad g \cdot M_\sigma \subseteq M_{g\sigma g^{-1}}$$

for all g, $\sigma \in G$.

The following Theorem generalizes and unifies Propositions 118 and 120.

Theorem 59. (Yetter [189]) *Let H be a bialgebra and $(M, \cdot, \rho) \in _H\mathcal{YD}^H$ be a left-right Yetter-Drinfeld module. Then the map*

$$R = R_{(M,\cdot,\rho)} : \ M \otimes M \to M \otimes M, \quad R(m \otimes n) = n_{[1]} \cdot m \otimes n_{[0]}$$

for all m, $n \in M$, is a solution of the QYBE. Futhermore, if H is a Hopf algebra then R is bijective.

Proof. For l, m, $n \in M$ we have

$$R^{23}R^{13}R^{12}(l \otimes m \otimes n) = R^{23}R^{13}\Big(m_{[1]} \cdot l \otimes m_{[0]} \otimes n\Big)$$

$$= R^{23}\Big(n_{[1]}m_{[1]} \cdot l \otimes m_{[0]} \otimes n_{[0]}\Big)$$

$$= n_{[2]}m_{[1]} \cdot l \otimes n_{[1]} \cdot m_{[0]} \otimes n_{[0]}$$

and

$$R^{12}R^{13}R^{23}(l \otimes m \otimes n) = R^{12}R^{13}\Big(l \otimes n_{[1]} \cdot m \otimes n_{[0]}\Big)$$

$$= R^{12}\Big(n_{[1]} \cdot l \otimes n_{[2]} \cdot m \otimes n_{[0]}\Big)$$

$$= (n_{[2]} \cdot m)_{[1]} \cdot (n_{[1]} \cdot l) \otimes (n_{[2]} \cdot m)_{[0]} \otimes n_{[0]}$$

$$= \Big((n_{[1](2)} \cdot m)_{[1]} n_{[1](1)}\Big) \cdot l \otimes (n_{[1](2)} \cdot m)_{[0]} \otimes n_{[0]}$$

$$(5.22) = n_{[2]}m_{[1]} \cdot l \otimes n_{[1]} \cdot m_{[0]} \otimes n_{[0]}$$

i.e. R is a solution of the QYBE. Now, if S is the antipode of H then the map

$$R^{-1} : M \otimes M \to M \otimes M, \quad R^{-1}(m \otimes n) = S(n_{[1]}) \cdot m \otimes n_{[0]}$$

for all m, $n \in M$ is the inverse of R.

Remark 14. If $(M, \cdot, \rho) \in {}_H \mathcal{YD}^H$ then the map $R'_{(M,\cdot,\rho)} := \tau R \tau$ given by

$$R'_{(M,\cdot,\rho)} : M \otimes M \to M \otimes M, \quad R'_{(M,\cdot,\rho)}(m \otimes n) = m_{[0]} \otimes m_{[1]} \cdot n$$

is also a solution of the QYBE.

We will prove in Section 5.4 that Theorem 59 has a converse in the situation where M is finite dimensional. The problem will be to find a Yetter-Drinfeld module structure on M. We end this Section presenting two different ways to construct Yetter-Drinfeld modules. First, any representation (resp. corepresentation) of a quasitriangular (resp. coquasitriangular) bialgebra gives us a Yetter-Drinfeld module. This also explains why Propositions 118 and 120 are special cases of Theorem 59.

Proposition 121. *Let (H, R) be a quasitriangular bialgebra and $(M, \cdot)$ be a left H-module. Then M is an object in ${}_H \mathcal{YD}^H$ via the right H-coaction*

$$\rho = \rho_R : \ M \to M \otimes H, \quad \rho(m) = R^2 \cdot m \otimes R^1$$

for all $m \in M$. Furthermore, $\mathcal{R}_{(R,M,\cdot)} = R_{(M,\cdot,\rho)}$.

Proof. First we show that $(M, \rho = \rho_R)$ is a right H-comodule. A direct calculation gives us that $(r = R)$:

$$(\rho \otimes Id)\rho(m) = r^2 R^2 \cdot m \otimes r^1 \otimes R^1, \quad \text{and} \quad (Id \otimes \Delta)\rho(m) = R^2 \cdot m \otimes \Delta(R^1)$$

for all $m \in M$. Applying τ to (QT1), we obtain $R^2 \otimes \Delta(R^1) = r^2 R^2 \otimes r^1 \otimes R^1$ and $(\rho \otimes Id)\rho = (Id \otimes \Delta)\rho$. Taking (QT2) into account, we have

$$\varepsilon(m_{[1]})m_{[0]} = \varepsilon(R^1)R^2 \cdot m = m$$

i.e. (M, ρ) is a right H-comodule. Let us prove now the compatibility condition (5.22). For $h \in H$ and $m \in M$ we have

$$h_{(1)} \cdot m_{[0]} \otimes h_{(2)} m_{[1]} = h_{(1)} R^2 \cdot m \otimes h_{(2)} R^1$$

and

$$(h_{(2)} \cdot m)_{[0]} \otimes (h_{(2)} \cdot m)_{[1]} h_{(1)} = R^2 h_{(2)} \cdot m \otimes R^1 h_{(1)}.$$

Applying τ to (QT5) we find

$$h_{(1)} R^2 \otimes h_{(2)} R^1 = R^2 h_{(2)} \otimes R^1 h_{(1)}$$

hence $(M, \cdot, \rho_R) \in {}_H \mathcal{YD}^H$. Finally

$$R_{(M,\cdot,\rho_R)}(m \otimes n) = n_{[1]} \cdot m \otimes n_{[0]} = R^1 \cdot m \otimes R^2 \cdot n = \mathcal{R}_{(R,M,\cdot)}(m \otimes n)$$

which finishes our proof.

The dual result is the following; we leave the details as an exercise to the reader.

Proposition 122. *Let (H, σ) be a coquasitriangular bialgebra and (M, ρ) be a right H-comodule. Then M is an object in $_H\mathcal{YD}^H$ via the left H-action*

$$h \cdot m = \sigma(m_{[1]} \otimes h)m_{[0]}$$

for all $h \in H$ and $m \in M$. Furthermore $R_{(\sigma, M, \rho)} = R_{(M, \cdot, \rho)}$.

Recall from Chapters 2 and 4 that the forgetful functors $_H\mathcal{YD}^H \to {}_H\mathcal{M}$ and $_H\mathcal{YD}^H \to \mathcal{M}^H$ respectively have a right and left adjoint. This gives us a second method of constructing Yetter-Drinfeld modules: start with an H-module or an H-comodule, and apply the appropriate adjoint functors. We describe this explicitly in the next Proposition.

Proposition 123. *Let H be a bialgebra with a twisted antipode.*

1. *For any left H-module M, $M \otimes H \in {}_H\mathcal{YD}^H$ via the following structures*

$$h \cdot (m \otimes k) = h_{(2)}m \otimes h_{(3)}kS^{-1}(h_{(1)}) \quad \text{and} \quad \rho(m \otimes h) = m \otimes h_{(1)} \otimes h_{(2)}$$

   for all $h, k \in H$ and $m \in M$. The functor $\bullet \otimes H : {}_H\mathcal{M} \to {}_H\mathcal{YD}^H$ is a right adjoint of the forgetful functor $_H\mathcal{YD}^H \to {}_H\mathcal{M}$.
2. *For any right H-comodule N, $H \otimes N \in {}_H\mathcal{YD}^H$ via the following structures*

$$h \cdot (k \otimes n) = hk \otimes n \quad \text{and} \quad \rho(h \otimes n) = h_{(2)} \otimes n_{[0]} \otimes h_{(3)}n_{[1]}S^{-1}(h_{(1)})$$

   for all $h, k \in H$ and $n \in N$. The functor $H \otimes \bullet : \mathcal{M}^H \to {}_H\mathcal{YD}^H$ is a left adjoint of the forgetful functor $_H\mathcal{YD}^H \to \mathcal{M}^H$.

Proof. This is a special case of Theorem 15, Example 13 and Example 14.

Remark 15. k is a left H-module and a right H-comodule, the structure maps are the trivial ones. It follows that H can be viewed as an Yetter-Drinfeld module in two ways, namely

$$h \cdot k = h_{(2)}kS^{-1}(h_{(1)}) \quad \text{and} \quad \rho = \Delta$$

and

$$h \cdot k = hk \quad \text{and} \quad \rho(h) = h_{(2)} \otimes h_{(3)}S^{-1}(h_{(1)})$$

for all $h, k \in H$. These Yetter-Drinfeld modules are sometimes called the *Verma Yetter-Drinfeld modules* [83].

5.4 The FRT Theorem

First, we recall some well-known Lemmas.

Lemma 25. *Let M be a finite dimensional vector space with basis $\{m_1, \cdots, m_n\}$, and let C be a coalgebra. Let $\{c_l^v \mid v, l = 1, \cdots, n\}$ be a family of elements in C, and consider the k-linear map*

$$\rho: \; M \to M \otimes C, \quad \rho(m_l) = m_v \otimes c_l^v$$

(M, ρ) is a right C-comodule if and only if the matrix (c_l^v) is comultiplicative, i.e.

$$\Delta(c_k^j) = c_u^j \otimes c_k^u \quad \text{and} \quad \varepsilon(c_k^j) = \delta_k^j \tag{5.24}$$

for all $j, k = 1, \cdots, n$.

Writing $B = (c_l^v)$, we can formally rewrite (5.24) as

$$\Delta(B) = B \otimes B \quad \text{and} \quad \varepsilon(B) = I_n$$

Lemma 26. *Let (C, Δ, ε) be a coalgebra. On the tensor algebra $(T(C), M, u)$, there exists a unique bialgebra structure $(T(C), M, u, \overline{\Delta}, \overline{\varepsilon})$ such that $\overline{\Delta}(c) = \Delta(c)$ and $\overline{\varepsilon}(c) = \varepsilon(c)$ for all $c \in C$. In addition, the inclusion $i: \; C \to T(C)$ is a coalgebra map.*
Furthermore, if M is a vector space and $\mu: \; C \otimes M \to M$, $\mu(c \otimes m) = c \cdot m$ is a linear map, then there exists a unique left $T(C)$-module structure on M, $\overline{\mu}: T(C) \otimes M \to M$, such that $\overline{\mu}(c \otimes m) = c \cdot m$, for all $c \in C$, $m \in M$.

Proof. This is an immediate application of the universal property of the tensor algebra $T(C)$.

Lemma 27. *Let H be a bialgebra generated as an algebra by $\mathcal{G} = (g_{ij})$ and let $\sigma: \; H \otimes H \to k$ be a k-linear map satisfying (B3) and (B5). If σ satisfies (B1) for any x, $y \in \mathcal{G}$, then (B1) holds for all elements of H.*

Proof. Left to the reader.

Now, we can prove the main results of this chapter.

Theorem 60. *Let M be a finite dimensional vector space and $R \in \mathrm{End}_k(M \otimes M)$ a solution of the QYBE. Then there exists a bialgebra $A(R)$ and a unique k-linear map $\sigma: \; A(R) \otimes A(R) \to k$ such that $(A(R), \sigma)$ is a coquasitriangular bialgebra, M has a structure of right $A(R)$-comodule (M, ρ), and $R = R_{(\sigma, M, \rho)}$.*

Proof. First we construct the bialgebra $A(R)$. Let $\{m_1, \cdots, m_n\}$ be a basis of M and (x_{uv}^{ij}) a family of scalars of k such that

$$R(m_u \otimes m_v) = x_{uv}^{ij} m_i \otimes m_j \tag{5.25}$$

for all u, $v = 1, \cdots, n$.

Let $(C, \Delta, \varepsilon) = \mathcal{M}^n(k)$ be the comatrix coalgebra (see Section 5.1), and let $\rho : M \to M \otimes C$ given by

$$\rho(m_l) = m_v \otimes c_l^v \qquad (5.26)$$

for all $l = 1, \cdots, n$. Then, by Lemma 25, M is a right C-comodule. We know from Lemma 26 that $T(C)$ is a bialgebra. $i : C \to T(C)$ is a coalgebra map, so M is a right $T(C)$-comodule via

$$M \xrightarrow{\rho} M \otimes C \xrightarrow{I \otimes i} M \otimes T(C)$$

This right $T(C)$-coaction will also be denoted by ρ.

Now, we define the quantum Yang-Baxter obstructions yb_{kl}^{ij} by the following formula

$$yb_{kl}^{ij} = x_{vu}^{ij} c_k^u c_l^v - x_{lk}^{vu} c_v^i c_u^j \qquad (5.27)$$

for all i, j, k, $l = 1, \cdots, n$. We can easily prove that

$$\Delta(yb_{kl}^{ij}) = yb_{uv}^{ij} \otimes c_k^u c_l^v + c_u^i c_v^j \otimes yb_{kl}^{uv} \quad \text{and} \quad \varepsilon(yb_{kl}^{ij}) = 0$$

for all i, j, k, $l = 1, \cdots, n$. Let I be the two-sided ideal of $T(C)$ generated by all yb_{kl}^{ij}. It follows from the above formulas that I is a biideal of $T(C)$; hence

$$A(R) = T(C)/I$$

is a bialgebra and M has a right $A(R)$-comodule structure via the canonical projection $T(C) \to A(R)$. We conclude $A(R)$ is the free algebra with generators c_i^j, i, $j = 1, \cdots, n$, and relations

$$x_{vu}^{ij} c_k^u c_l^v = x_{lk}^{vu} c_v^i c_u^j$$

for all i, j, k, $l = 1, \cdots, n$. The comultiplications and the counit are given in such a way that (c_j^i) is a comultiplicative matrix.

Now we construct $\sigma : A(R) \otimes A(R) \to k$ such that $(A(R), \sigma)$ is coquasitriangular and $R = R_{(\sigma, M, \rho)}$. First, we define

$$\sigma : C \otimes C \to k, \quad \sigma(c_v^i \otimes c_u^j) = x_{vu}^{ij} \qquad (5.28)$$

for all i, j, u, $v = 1, \cdots, n$. Then we extend the map σ to the whole of $T(C)$ using the relations (B2-B5). The new map $T(C) \otimes T(C) \to k$ is also denoted by σ. We claim that this map factorizes to a map $A(R) \otimes A(R) \to k$; we have to show that

$$\sigma(T(C) \otimes I) = \sigma(I \otimes T(C)) = 0$$

Using (B3) and (B5), this comes down to proving

$$\sigma(c_q^p \otimes yb_{kl}^{ij}) = 0 \quad \text{and} \quad \sigma(yb_{kl}^{ij} \otimes c_q^p) = 0$$

We prove only first formula and leave the second one to the reader.

$$\sigma(c_q^p \otimes yb_{kl}^{ij}) = x_{vu}^{ij}\sigma(c_q^p \otimes c_k^u c_l^v) - x_{lk}^{vu}\sigma(c_q^p \otimes c_v^i c_u^j)$$

$$\text{(B3)} \qquad = x_{vu}^{ij}\sigma(c_r^p \otimes c_k^u)\sigma(c_q^r \otimes c_l^v) - x_{lk}^{vu}\sigma(c_r^p \otimes c_v^i)\sigma(c_q^r \otimes c_u^j)$$

$$\qquad = x_{vu}^{ij} x_{rk}^{pu} x_{ql}^{rv} - x_{lk}^{vu} x_{rv}^{pi} x_{qu}^{rj}$$

$$\text{(5.9)} \qquad = 0$$

We have constructed $\sigma : A(R) \otimes A(R) \to k$ such that (B2)-(B5) holds. From Lemma 27 we know that it suffices to verify (B1) on the algebra generators c_j^i. For $x = c_{pq}$, $y = c_{rs}$, we have

$$\sigma(x_{(1)} \otimes y_{(1)})y_{(2)}x_{(2)} = \sigma(c_v^p \otimes c_u^r)c_s^u c_q^v = x_{vu}^{pr} c_s^u c_q^v$$

and

$$\sigma(x_{(2)} \otimes y_{(2)})x_{(1)}y_{(1)} = \sigma(c_q^v \otimes c_s^u)c_v^p c_u^r = x_{qs}^{vu} c_p^p c_u^r$$

These are equal in $A(R)$ since

$$yb_{sq}^{pr} = \sum_{u,v} x_{vu}^{pr} c_s^u c_q^v - x_{qs}^{vu} c_p^p c_u^r = 0$$

in $A(R)$. We conclude that σ satisfies (B1) and that $(A(R), \sigma)$ is a coquasi-triangular bialgebra.

Finally $R = R_{(\sigma, M, \rho)}$: for any two elements m_u, m_v, in the basis of M, we have

$$R_{(\sigma, M, \rho)}(m_v \otimes m_u) = \sigma(c_v^i \otimes c_u^j)m_i \otimes m_j$$

$$= x_{vu}^{ij} m_i \otimes m_j$$

$$\text{(5.25)} \qquad = R(m_v \otimes m_u)$$

This formula also shows that σ is unique with the condition $R = R_{(\sigma, M, \rho)}$ and this completes our proof.

Theorem 60 is originally due to Faddeev, Reshetikhin, and Takhtajan (see [85]) and is now usually referred to as the "FRT" theorem. An alternative version, using Yetter-Drinfeld modules instead of coquasitriangular bialgebras, was proved by Radford in [155]. Combining Theorem 60 and Proposition 122, we obtain Radford's version of the FRT Theorem.

Theorem 61. *Let M be a finite dimensional vector space and $R \in \mathrm{End}_k(M \otimes M)$ a solution of the QYBE. Then there exists a bialgebra $A(R)$ and an $A(R)$-Yetter-Drinfeld module structure $(M, \cdot, \rho)$ on M such that $R = R_{(M, \cdot, \rho)}$.*

Remark 16. Recently Di-Ming Lu [119] showed that the bialgebra $A(R)$ can be viewed as a special case of the bialgebras that arise from Tannaka duality.

Example 22. Take a scalar $q \neq 0$ and

$$R_q = \begin{pmatrix} q & 0 & 0 & 0 \\ 0 & 1 & q - q^{-1} & 0 \\ 0 & 0 & 1 & 0 \\ 0 & 0 & 0 & q \end{pmatrix}$$

the classical Yang-Baxter operator in dimension two.

The bialgebra $A(R_q)$ from the FRT theorem is $M_q(2)$, using the notation from quantum group theory. It has the following description: as an algebra, $A(R_q)$ has generators x, y, z, t, and relations

$$xy = qyx, \quad xz = qzx, \quad yz = zy$$

$$yt = qty, \quad zt = qtz, \quad xt - tx = (q - q^{-1})yz$$

The comultiplication Δ and the counit ε of $A(R_q)$ are given by

$$\Delta \begin{pmatrix} x & y \\ z & t \end{pmatrix} = \begin{pmatrix} x & y \\ z & t \end{pmatrix} \otimes \begin{pmatrix} x & y \\ z & t \end{pmatrix}, \quad \varepsilon \begin{pmatrix} x & y \\ z & t \end{pmatrix} = I_2$$

To see this, we consider the sixteen relations

$$\sum_{u,v} x_{vu}^{ij} c_k^u c_l^v = \sum_{u,v} x_{lk}^{vu} c_v^i c_u^j$$

for all i, j, k, $l = 1, 2$. Writing $c_1^1 = x$, $c_2^1 = y$, $c_1^2 = z$, $c_2^2 = t$, we find after a lengthy calculation that the six commutation rules given above are the only independent ones among the sixteen relations. It is interesting to note that the element $\det_q = xt - qyz$ is a central and grouplike element of $M_q(2)$, called the *quantum determinant* of order two. In particular, taking the quotient $M_q(2)/(\det_q - 1)$ we obtain the Hopf algebra $SL_q(2)$. Its antipode is given by the formula

$$S(x) = t, \; S(y) = -qy, \; S(z) = -q^{-1}z, \; S(t) = x$$

This construction can be generalized for an arbitrary n leading to the bialgebras (resp. Hopf algebras) $M_q(n)$ (resp. $SL_q(n)$).

5.5 The set-theoretic braid equation

The quantum Yang-Baxter equation and the braid equation can be considered on the level of sets. Let S be a set, and $R : S \times S \to S \times S$ a (bijective) map. We can still define the maps

$$R^{12}, \ R^{13}, \ R^{23} : \ S^3 \to S^3$$

and we say that S is a solution of the quantum Yang-Baxter equation (resp. the braid equation) if (5.7) (resp. (5.8)) is an equality for functions $S^3 \to S^3$. As in the vector space case, R is a solution of the QYBE if and only if τR is a solution of the braid equation; now the map $\tau : \ S \times S \to S \times S$ is the transposition $\tau(x, y) = (y, x)$.

Example 23. [92] Let $\triangleright : \ S \times S \to S$, $(x, y) \to x \triangleright y$ be a function. Then

$$R : \ S \times S \to S \times S, \quad R(x, y) = (x \triangleright y, x)$$

is a solution of the braid equation if and only if

$$x \triangleright (y \triangleright z) = (x \triangleright y) \triangleright (x \triangleright z)$$

for all x, y, $z \in S$. A pair $(S, \triangleright)$ is called a *rack* if the above equation holds and the map $x \triangleright \bullet : S \to S$, $y \to x \triangleright y$, is a bijection for all $x \in S$. Racks served as a tool for the construction of invariants of knots and links and recently for the construction of finite dimensional pointed Hopf algebras. For a further study and examples of racks we refer to [6] and the references therein.

The problem now is to find solutions of the set theoretic QYBE. First we will construct solutions in the case that $S = G$ is a group. The study of the braid equation on sets was started recently in [84] and then continued in [121], [160].

Definition 8. *Let G be a group and let*

$$\xi : \ G \times G \to G, \quad \xi(x, y) = {}^{x}y$$

be a left action of G on itself, and

$$\eta : \ G \times G \to G, \quad \eta(x, y) = x^{y}$$

be a right action of G on itself. (ξ, η) is called a compatible pair of actions *if*

$$xy = ({}^{x}y)(x^{y}) \tag{5.29}$$

for all x, $y \in G$.

In a pair of compatible actions, the left action is completely determined by the right action, and vice versa. So we could reformulate the definition in terms of one of the two actions, but this is not very appropriate since the conditions for the remaining action is not very transparent. An immediate example of a pair of compatible actions is (ξ, η) where ξ is trivial and η is the conjugation, i.e.

$${}^{x}y = y \quad \text{and} \quad y^{x} = x^{-1}yx$$

Theorem 62. (Lu, Yan, Zhu [121]) *Let (ξ, η) be a compatible pair of actions of a group G on itself. Then the map*

$$R = R_{(\xi,\eta)} : \; G \times G \to G \times G, \quad R(x,y) = ({}^x y, x^y) \tag{5.30}$$

is a bijective solution of the braid equation on the set G.

Proof. We write

$$R^{12} R^{23} R^{12}(x,y,z) = (x_1, y_1, z_1), \qquad R^{23} R^{12} R^{23}(x,y,z) = (x_2, y_2, z_2).$$

It follows from the compatibility condition (5.29) that $x_1 y_1 z_1 = x_2 y_2 z_2$, so it suffices to show that $x_1 = x_2$ and $z_1 = z_2$.
A direct calculation tells us that $x_1 = {}^{({}^x y)(x^y)} z$ and $x_2 = {}^{(xy)} z$ and using (5.29) once more, we find that $x_1 = x_2$. In a similar way we can prove that $z_1 = z_2$.
We still have to prove that R is bijective. First we claim that

$$y^{-1} = {}^{(x^y)^{-1}} \left(({}^x y)^{-1} \right), \quad x^{-1} = \left((x^y)^{-1} \right)^{({}^x y)^{-1}} \tag{5.31}$$

for all $x, y \in G$. Indeed, we have

$$\left({}^{x^y} (y^{-1}) \right) x = \left({}^{x^y} (y^{-1}) \right) \left((x^y)^{y^{-1}} \right) \overset{(5.29)}{=} x^y y^{-1} \overset{(5.29)}{=} ({}^x y)^{-1} x,$$

proving the first formula follows; the proof of the second formula is similar. Let i be the bijection $i : \; G \times G \to G \times G$, $i(x,y) = (y^{-1}, x^{-1})$. For all x, $y \in G$, we compute that

$$(i \circ R)^2(x,y) = i\left({}^{(x^y)^{-1}} \left(({}^x y)^{-1} \right), \left((x^y)^{-1} \right)^{({}^x y)^{-1}} \right) \overset{(5.31)}{=} i(y^{-1}, x^{-1}) = (x, y)$$

This means that $(i \circ R)^2 = \mathrm{Id}$, and R is bijective.

In Theorem 63, we will provide two alternative descriptions of the class of solutions given by compatible pairs. First we introduce braiding operators on groups, which can be viewed as set theoretic analogs of coquasitriangular Hopf algebras.

Definition 9. *Let G be a group with multiplication m. A braiding operator on G is a bijective map $\sigma : \; G \times G \to G \times G$ satisfying the following conditions:*
(BO1) $\sigma(m \times \mathrm{Id}) = (\mathrm{Id} \times m)\sigma_{12}\sigma_{23}$;
(BO2) $\sigma(\mathrm{Id} \times m) = (m \times \mathrm{Id})\sigma_{23}\sigma_{12}$;
(BO3) $\sigma(1, x) = (x, 1)$ *and* $\sigma(x, 1) = (1, x)$, *for all* $x \in G$;
(BO4) $m\sigma = m$.

Remarks 13. 1. The map $\sigma(x, y) = (y, y^{-1}xy)$ is a braiding operator on any group G, called the *conjugate braiding*.

2. (BO4) can be viewed as a generalized commutativity condition.

3. In Section 2.3, we have discussed the smash product of algebras. A similar construction can be carried out in the category of groups: given two groups G and H, consider the product set $G \times H$, a map $\sigma : H \times G \to G \times H$, and the following multiplication on $G \times H$:

$$m_{G \times_\sigma H} = (m \times m)(Id \times \sigma \times Id)$$

The group theoretic analog of Theorem 7 tells us that $(G \times H, m_{G \times_\sigma G}, (1_G, 1_H))$ is a group if and only if (BO1)-(BO3) are satisfied (in Definition 9 these conditions are stated in the case $G = H$, but they still make sense in the case $G \neq H$. This can also restated in terms of the factorization problem for groups. A groups X factorizes through G and H if and only if there exist morphisms of groups $i : G \to X$ and $j : H \to X$ such that $m_X \circ (i \circ j) : G \times H \to X$ is a group isomorphism. We can show that there is a one-one correspondence between group structures on $G \times H$ for which the canonical inclusions are group homomorphisms and maps $\sigma : H \times G \to G \times H$ satisfying (BO1-BO3).

Next we recall the notion of 1-cocycle.

Definition 10. *Let G, A be groups, and assume that G acts as a group of automorphisms on A. A bijective 1-cocycle of G with coefficients in A is a bijection $\pi : G \to A$ such that*

$$\pi(xy) = \pi(x)(x \cdot \pi(y)) \tag{5.32}$$

for all x, $y \in G$.

We will next show that there is a close relation between the three concepts introduced above, namely compatible pairs of actions, braiding operators, and 1-cocycles.

Theorem 63. (Lu, Yan, Zhu [121]) *Let G be a group. The following three sets are isomorphic as sets:*

1. *$\mathcal{CA}$ consisting of all compatible pairs of actions (ξ, η) of G on itself;*
2. *$\mathcal{BO}$ consisting of all braiding operators $\sigma : G \times G \to G \times G$;*
3. *$\mathcal{BC}$ consisting of all triple $(A, \cdot, \pi)$, where A is a group on which G acts as a group of automorphisms, $\pi : G \to A$ is a bijective 1-cocycle.*

Proof. First, let $(\xi, \eta) \in \mathcal{CA}$ be a a compatible pair of actions and define $\sigma = \sigma_{(\xi, \eta)}$ by the formula

$$\sigma(x, y) = (^x y, x^y). \tag{5.33}$$

We have to prove that σ is a braiding operator. We denote

$$\sigma(xy, z) = (x_1, y_1), \qquad (id \times m)\sigma_{12}\sigma_{23}(x, y, z) = (x_2, y_2).$$

It follows from the compatibility condition (5.29) that $x_1 y_1 = xyz = x_2 y_2$. So, it is enough to prove that $x_1 = x_2$. A direct computation gives

$$x_1 = {}^{(xy)}z, \quad \text{and} \quad x_2 = {}^{x}({}^{y}z)$$

which are equal because ξ is a left action, and (BO1) follows. (BO2) can be proved in a similar way. Taking $x = 1$ in (5.29), we find that $1^y = 1$ for any $y \in G$, and $\sigma(1, y) = ({}^1 y, 1^y) = (y, 1)$ and the first equality of (BO3) follows; the second one is obtained in a similar way. Finally, the equality (BO4) is exactly the compatibility condition (5.29), and we can conclude that σ is a braiding operator.

Now let σ is a braiding operator and define $(\xi, \eta) = (\xi, \eta)_\sigma$ using (5.33). Then looking at the first component of (BO1), we obtain ${}^{(xy)}z = {}^{x}({}^{y}z)$; the first equality in (BO3) implies ${}^1 x = x$. This proves ξ is a left action. Similarly, we can show that η is a right action. The compatibility condition (5.29) then follows directly from (BO4).

We now have well-defined maps $\mathcal{CA} \to \mathcal{BO}$ and $\mathcal{BO} \to \mathcal{CA}$, and it is clear that they are each others inverses.

Take a compatible pair $(\xi, \eta) \in \mathcal{CA}$, and define $(A, \cdot, \pi) = (A, \cdot, \pi)_{(\xi, \eta)}$ as follows: $A = G$ as a set, with new multiplication

$$x \star y = x({}^{x^{-1}}y). \tag{5.34}$$

for all $x, y \in G$. Replacing y by ${}^x y$ in (5.34), we find

$$xy = x \star {}^x y$$

and $\pi = I : G \to A$ is a bijective 1-cocycle. We still have to show that G is a group and that G acts as a group of automorphisms on A. First, it is obvious that 1 is a left unit, and ${}^x(x^{-1})$ is a right inverse of x with respect to the multiplication $\star$. Using (5.29), we find

$$x \star y = x({}^{x^{-1}}y) = y(({x^{-1}})^y)^{-1} \tag{5.35}$$

and it follows that 1 is also a right unit, and that $(x^{({}^{x^{-1}})})^{-1}$ is a left inverse of x.

Now let us show that G acts on A. Using (5.29), we find

$$\left({}^x(yz)\right)\left(x^{(yz)}\right) = xyz = ({}^x y)(x^y)z = ({}^x y)({}^{x^y}z)((x^y)^z) \tag{5.36}$$

Now η is a right action, so

$${}^x(yz) = ({}^x y)({}^{x^y}z) \tag{5.37}$$

Let $t \in G$ and put $z = {}^{y^{-1}}t$ in (5.37). We obtain

$$
{}^{x}(y \star t) = {}^{x}(yz) = ({}^{x}y)({}^{x^{y}}z) = ({}^{x}y)({}^{(x^{y}y^{-1})}t)
$$

$$
(5.29) \quad = ({}^{x}y)({}^{({}^{x}y)^{-1}x}t) = {}^{x}y \star {}^{x}t, \tag{5.38}
$$

and we have shown that G acts on A. Finally A is associative, since

$$
x \star (y \star t) = x \bigl({}^{x^{-1}}(y \star t) \bigr) \overset{(5.38)}{=} x \bigl({}^{x^{-1}}y \bigr) \bigl({}^{({}^{x^{-1}}y)^{-1}x^{-1}}t \bigr)
$$

$$
= (x \star y) \bigl({}^{(x\star y)^{-1}}t \bigr) = (x \star y) \star t.
$$

Our next step is to the inverse of the map $(\xi, \eta) \to (A, \cdot, \pi)_{(\xi,\eta)}$. Take $(A, \cdot, \pi) \in \mathcal{BC}$. we use the bijection π to identify G and A, and use $\star$ to denote the product on G induced by the product on A, and ξ to denote the left action of G on G induced by the action $\cdot$ on G on A. Then (5.32) is equivalent to (5.34). Finally, we define η using (5.29), so that (5.35) also holds. It remains to be shown that η is a right action. Since ξ is a left action, we have $1 \star 1 = 1({}^{1^{-1}}1) = 1$. It follows that 1 is also a unit with respect to $\star$. Taking $x = 1$ in (5.35), we obtain $y^{1} = y$, for all $y \in G$. On the other hand, by the compatibility condition (5.29), we still have the equality (5.36). Moreover, since ξ acts as automorphisms of $(G, \star)$, we have

$$
{}^{x}(yz) = {}^{x}(y \star t) = {}^{x}y \star {}^{x}t = ({}^{x}y)({}^{x^{y}}z),
$$

where, in the last equality, we used part of the computations in (5.38). It follows from (5.36) that

$$
x^{(yz)} = (x^{y})^{z},
$$

i.e. η is a right action.

Combining Theorem 62 and Theorem 63, we obtain the following Corollary.

Corollary 36. *Any braiding operator* $\sigma : G \times G \to G \times G$ *is a bijective solution of the braid equation.*

Recall that a map $R : S \times S \to S \times S$ is called non-degenerate if the maps

$$
R(x, -), \ R(-, x) : \ S \to S
$$

are bijective, for all $x \in S$. A remarkable result, due to Lu, Yan, and Zhu [121], is the following set-theoretic version of the FRT Theorem.

Theorem 64. *Let* $R : \ S \times S \to S \times S$ *be a bijective and non-degenerate solution of the braid equation. Then there exists a group* $G = G(S, R)$, *a map* $i : \ S \to G$ *and a braiding operator* $\sigma : \ G \times G \to G \times G$ *on* G *such that* $\sigma(i \times i) = (i \times i)R$. *Furthermore* (G, σ) *is universal with this property.*

Proof. (Sketch - we refer to [121] for full detail) The group G is the group generated by the set S and subject to the relations $xy = uv$ whenever $R(x, y) = (u, v)$. The hard part of the proof is then to extend R to σ.

An open problem is the following: is $i : \ S \to G$ injective. If R is involutory, then we can show that i is injective, see [84].

6 Hopf modules and the pentagon equation

We study the Hopf equation $R^{12}R^{23} = R^{23}R^{13}R^{12}$; this equation is equivalent to the pentagon equation $R^{12}R^{13}R^{23} = R^{23}R^{12}$. It is older ([122]) than the QYBE, and, in a sense, more basic, as it is an expression of an associativity constraint rather than a commutativity constraint. The pentagon equation plays a key role in the theory of duality for von Neumann algebras ([174], [10]).

In this Chapter we study these equivalent equations from a purely algebraic point of view. This brings us in the framework of Chapter 5, where we study the quantum Yang-Baxter equation. We will see that the category of Hopf modules plays (via an FRT type theorem) the same role in the study of the Hopf equation and the pentagon equation, as the category of Yetter-Drinfeld modules in the study of the QYBE. We will apply our results to constructing new examples of noncommutative noncocommutative bialgebras which are different from those arising from the FRT theorem for the QYBE. One of the main applications is that our theory leads to a structure Theorem and the classification of finite dimensional Hopf algebras ([137]).

6.1 The Hopf equation and the pentagon equation

We will start with the following

Definition 11. *Let M be a vector space and $R \in \mathrm{End}(M \otimes M)$.*

1. *R is called a solution of the Hopf equation if*

$$R^{23}R^{13}R^{12} = R^{12}R^{23} \qquad (6.1)$$

2. *R is called a solution of the pentagon equation if*

$$R^{12}R^{13}R^{23} = R^{23}R^{12} \qquad (6.2)$$

In the literature (cf. [170]), the pentagon equation is sometimes called the *fusion equation*. The Hopf equation (resp. the pentagon equation) can be obtained from the QYBE

$$R^{23}R^{13}R^{12} = R^{12}R^{13}R^{23}$$

by deleting the middle term from the right (resp. left) hand side. The Hopf (resp. pentagon) equation can be defined more generally for elements $R \in A \otimes A$, where A is a k-algebra. In this case the Equations 6.1 and 6.2 should hold in $A \otimes A \otimes A$.

Let us now show that the Hopf equation is equivalent to the pentagon equation.

Proposition 124. *Let M be a vector space and $R \in \mathrm{End}(M \otimes M)$.*

1. *R is a solution of the Hopf equation if and only if $W = \tau R \tau$ is a solution of the pentagon equation.*
2. *If $R \in \mathrm{End}(M \otimes M)$ is bijective, then R is a solution of the Hopf equation if and only if R^{-1} is a solution of the pentagon equation.*

Proof. 1. follows from the formulas:

$$W^{12}W^{13}W^{23} = R^{23}R^{13}R^{12}\tau^{13}, \quad W^{23}W^{12} = R^{12}R^{23}\tau^{13}$$

2. is straightforward.

Remark 17. Let $R \in \mathrm{End}(M \otimes M)$ and put $T = R\tau$. Then

$$T^{23}T^{12}T^{23} = R^{23}R^{13}R^{12}\tau^{13}, \quad T^{12}\tau^{23}T^{12} = R^{12}R^{23}\tau^{13}.$$

and it follows that R is a solution of the Hopf equation if and only if T is a solution of the equation

$$T^{12}\tau^{23}T^{12} = T^{23}T^{12}T^{23} \tag{6.3}$$

In [170], this equation is called the *3-cocycle equation*. Similar arguments show that R is a solution of the pentagon equation if and only if τR is a solution of the 3-cocycle equation.

In the next Proposition, we present the Hopf equation in matrix form.

Proposition 125. *We fix a basis $\{v_1, \cdots, v_n\}$ of a finite dimensional vector space M. Let $R, S \in \mathrm{End}(M \otimes M)$ given by their matrices*

$$R(v_k \otimes v_l) = x^{ij}_{kl} v_i \otimes v_j, \quad S(v_k \otimes v_l) = y^{ij}_{kl} v_i \otimes v_j,$$

for all $k, l = 1, \cdots, n$, where (x^{ji}_{uv}), (y^{ji}_{uv}) are two families of scalars in k. Then

$$R^{23}S^{13}S^{12} = S^{12}R^{23}$$

if and only if

$$x^{ij}_{vu}y^{pu}_{\beta k}y^{\beta v}_{ql} = x^{\alpha j}_{lk}y^{pi}_{q\alpha}$$

for all $i, j, k, l, p, q = 1, \cdots, n$. In particular, R is a solution of the Hopf equation if and only if

$$x^{ij}_{vu}x^{pu}_{\beta k}x^{\beta v}_{ql} = x^{\alpha j}_{lk}x^{pi}_{q\alpha} \tag{6.4}$$

for all $i, j, k, l, p, q = 1, \cdots, n$.

Proof. For $k, l, q = 1, \cdots, n$ we have:

$$R^{23}S^{13}S^{12}(v_q \otimes v_l \otimes v_k) = R^{23}S^{13}\left(y_{ql}^{\beta v}v_\beta \otimes v_v \otimes v_k\right)$$
$$= R^{23}\left(y_{\beta k}^{pu}y_{ql}^{\beta v}v_p \otimes v_v \otimes v_u\right)$$
$$= \left(x_{vu}^{ij}y_{\beta k}^{pu}y_{ql}^{\beta v}\right)v_p \otimes v_i \otimes v_j$$

and

$$S^{12}R^{23}(v_q \otimes v_l \otimes v_k) = S^{12}\left(x_{lk}^{\alpha j}v_q \otimes v_\alpha \otimes v_j\right)$$
$$= x_{lk}^{\alpha j}y_{q\alpha}^{pi}v_p \otimes v_i \otimes v_j$$
$$= \sum_{i,j,p}\left(x_{lk}^{\alpha j}y_{q\alpha}^{pi}\right)v_p \otimes v_i \otimes v_j$$

and the result follows.

Now we will see how solutions of the Hopf equation can be used to construct bialgebras without counit.

Proposition 126. *Let A be an algebra and $R \in A \otimes A$ an invertible solution of the Hopf equation. The comultiplication*

$$\Delta_l : \ A \to A \otimes A, \quad \Delta_l(a) = R(1 \otimes a)R^{-1}$$

is coassociative and an algebra map, i.e. $(A, \cdot, 1_A, \Delta_l)$ is a bialgebra without counit. The same result holds for the comultiplication

$$\Delta_r : \ A \to A \otimes A, \quad \Delta_r(a) = R^{-1}(a \otimes 1)R$$

Proof. For $a \in A$, we compute

$$(I \otimes \Delta_l)\Delta_l(a) = R^{23}R^{13}(1 \otimes 1 \otimes a)(R^{23}R^{13})^{-1}$$
$$(\Delta_l \otimes I)\Delta_l(a) = R^{12}R^{23}(1 \otimes 1 \otimes a)(R^{12}R^{23})^{-1}$$

R is a solution of the Hopf equation, so $(R^{13})^{-1}(R^{23})^{-1}R^{12}R^{23} = R_{12}$, and it follows that Δ_l is coassociative if and only if

$$(1 \otimes 1 \otimes a)R_{12} = R_{12}(1 \otimes 1 \otimes a)$$

both sides are equal to $R_{12} \otimes a$, and it follows that Δ_l is coassociative. It is obvious that Δ_l is an algebra map.

Solutions of the pentagon equation in operator algebras are given in [10]. We mention one example, playing a key role in the classification of multiplicative commutative operators, and then we turn to purely algebraic examples. Let G be a locally compact group and dg a right Haar measure on G. Then $V_G(\xi)(s, t) = \xi(st, t)$ is a solution of the pentagon equation.

Examples 10. 1. The identity map $\mathrm{Id}_{M \otimes M}$ is a solution of the Hopf equation.

2. Let M be a finite dimensional vector space and u an automorphism of M. If R is a solution of the Hopf equation then $^u R = (u \otimes u) R (u \otimes u)^{-1}$ is also a solution of the Hopf equation.

Indeed, as $\mathrm{End}(M \otimes M) \cong \mathrm{End}(M) \otimes \mathrm{End}(M)$, we can view $R = \sum f_i \otimes g_i$, where $f_i, g_i \in \mathrm{End}(M)$. Then $^u R = \sum u f_i u^{-1} \otimes u g_i u^{-1}$ and

$$(^u R)^{12} (^u R)^{23} = (u \otimes u \otimes u) R^{12} R^{23} (u \otimes u \otimes u)^{-1},$$

$$(^u R)^{23} (^u R)^{13} (^u R)^{12} = (u \otimes u \otimes u) R^{23} R^{13} R^{12} (u \otimes u \otimes u)^{-1},$$

hence $^u R$ is also a solution of the Hopf equation.

3. Let $f, g \in \mathrm{End}(M)$ such that $f^2 = f$, $g^2 = g$ and $fg = gf$. Then, $R = f \otimes g$ is a solution of the Hopf equation. A direct computation shows that

$$R^{23} R^{13} R^{12} = f^2 \otimes fg \otimes g^2, \quad R^{12} R^{23} = f \otimes gf \otimes g$$

so the above conclusion follows. With this example in mind, we can view solutions of the Hopf equation as generalizations of idempotent endomorphisms of M, since $R = f \otimes I$ (or $R = I \otimes f$) is a solution of the Hopf equation if and only if $f^2 = f$.

Let M be a two dimensional vector space with basis $\{m_1, m_2\}$, and identify $\mathrm{End}(M)$ with $\mathcal{M}_2(k)$. Consider

$$f_q = \begin{pmatrix} 1 & q \\ 0 & 0 \end{pmatrix} \tag{6.5}$$

where q is a scalar of k. Then $f_q^2 = f_q$. $g_q = I_V - f_q$ is also an idempotent endomorphism of M and $g_q f_q = f_q g_q$. Thus, we obtain that $R_q = f_q \otimes g_q$ is a solution of the Hopf equation. The matrix of R_q with respect to the basis $\{m_1 \otimes m_1, m_1 \otimes m_2, m_2 \otimes m_1, m_2 \otimes m_2\}$ is

$$R_q = \begin{pmatrix} 0 & -q & 0 & -q^2 \\ 0 & 1 & 0 & q \\ 0 & 0 & 0 & 0 \\ 0 & 0 & 0 & 0 \end{pmatrix}$$

Now let $g = I_M$ and $R_q' = f_q \otimes I_M$. The matrix of R_q' is

$$R_q' = \begin{pmatrix} 1 & 0 & q & 0 \\ 0 & 1 & 0 & q \\ 0 & 0 & 0 & 0 \\ 0 & 0 & 0 & 0 \end{pmatrix}$$

and R'_q is also a solution of the Hopf equation.

4. Let A be a k-algebra a, $b \in A$ and $R \in A \otimes A$ given by

$$R = 1 \otimes 1 + a \otimes b$$

Then R is a solution of the pentagon equation if and only if

$$a \otimes (ab - ba - 1) \otimes b = a \otimes a \otimes b^2 + a^2 \otimes b \otimes b + a^2 \otimes ba \otimes b^2 \quad (6.6)$$

Moreover, if a or b are nilpotent elements then R is invertible. Now if we take a and b such that

$$\begin{cases} a^2 = b^2 = 0 \\ ab - ba = 1 \end{cases} \quad (6.7)$$

or

$$\begin{cases} a^2 = 0 \\ b^2 = b \\ ab - ba = a + 1 \end{cases} \quad (6.8)$$

then (6.6) holds and R is a solution of the pentagon equation.

We give some explicit examples of this type in the case where $A = \mathcal{M}_n(k)$, with $\mathrm{char}(k)|n$.

Assume that $\mathrm{char}(k) = 2$ and let $n = 2q$ where q is a positive integer. Then

$$a = e_1^2 + e_3^4 + \cdots + e_{2q-1}^{2q}, \quad b = e_2^1 + e_4^3 + \cdots + e_{2q}^{2q-1}$$

is a solution of the (6.7), and hence

$$R = I_n \otimes I_n + \sum_{i,j=1}^{q} e_{2i-1}^{2i} \otimes e_{2j}^{2j-1}$$

is an invertible solution of the pentagon equation. On the other hand, for an arbitrary invertible matrix $X \in \mathcal{M}_q(k)$, the matrices a, $b \in \mathcal{M}_n(k)$ given by

$$b = e_1^1 + e_2^2 + \cdots e_q^q, \quad a = \begin{pmatrix} I_q & X^{-1} \\ X & I_q \end{pmatrix}$$

form a solution of the (6.8) and hence

$$R = R_X = I_n \otimes I_n + \sum_{i=1}^{q} a \otimes e_i^i$$

is an invertible solution of the pentagon equation.

5. R_q and R'_q from Example 3. are also solutions of the QYBE, because they are of the form $f \otimes g$ with $fg = gf$. Let us now construct a solution of the Hopf equation which is not a solution of the QYBE.

Let G be a group and M a left G-graded $k[G]$-module. The map

$$R: \; M \otimes M \to M \otimes M, \quad R(u \otimes v) = \sum_\sigma \sigma \cdot u \otimes v_\sigma \qquad (6.9)$$

is a solution of the Hopf equation and is not a solution of the quantum Yang-Baxter equation.

To prove that R is a solution of the Hopf equation, it suffices to show that it (6.1) holds for homogeneous elements of $M^{\otimes 3}$. For $u_\sigma \in M_\sigma$, $u_\tau \in M_\tau$ and $u_\theta \in M_\theta$, we compute

$$
\begin{aligned}
R^{23} R^{13} R^{12} (u_\tau \otimes u_\tau \otimes u_\theta) &= R^{23} R^{13} (\tau \cdot u_\sigma \otimes u_\tau \otimes u_\theta) \\
&= R^{23} (\theta\tau \cdot u_\sigma \otimes u_\tau \otimes u_\theta) \\
&= \theta\tau \cdot u_\sigma \otimes \theta \cdot u_\tau \otimes u_\theta \\
&= R^{12} (u_\sigma \otimes \theta \cdot u_\tau \otimes u_\theta) \\
&= R^{12} R^{23} (u_\sigma \otimes u_\tau \otimes u_\theta)
\end{aligned}
$$

and R is a solution of the Hopf equation. A similar calculation shows that

$$R^{12} R^{13} R^{23} (u_\sigma \otimes u_\tau \otimes u_\theta) = \theta\tau\theta \cdot u_\sigma \otimes \theta \cdot u_\tau \otimes u_\theta$$

and we see that R is not a solution of the QYBE.

6. Let G be a group and M be a G-crossed module, that is a kG-Yetter-Drinfeld module. These means that M is G-graded and a left $k[G]$-module such that

$$g \cdot M_\sigma \subseteq M_{g\sigma g^{-1}}$$

for all $g, \sigma \in G$. Then R given by (6.9) is a solution of the QYBE but not a solution of the Hopf equation.

7. Take $0 \neq q \neq 1 \in k$, and consider the classical two dimensional Yang-Baxter operator

$$
R = \begin{pmatrix}
q & 0 & 0 & 0 \\
0 & 1 & q - q^{-1} & 0 \\
0 & 0 & 1 & 0 \\
0 & 0 & 0 & q
\end{pmatrix}
$$

R is a solution of the QYBE and is not a solution for the Hopf equation. Indeed, the element in the $(1,1)$-position of $R^{23} R^{13} R^{12}$ is q^3, while the element in the $(1,1)$-position of $R^{12} R^{23}$ is q^2, i.e. R is not a solution of the Hopf equation.

8. For a bialgebra H,

$$R: \; H \otimes H \to H \otimes H, \quad R(g \otimes h) = \sum h_{(1)} g \otimes h_{(2)}$$

for all $g, h \in H$, is a solution of the Hopf equation. This operator was introduced by Takesaki in [174] for a Hopf-von Neumann algebra $(\mathcal{A}, \Delta)$. On the other hand the map

$$W : \; H \otimes H \to H \otimes H, \quad W(g \otimes h) = \sum g_{(1)} \otimes g_{(2)}h$$

for all g, $h \in H$, is a solution of the pentagon equation.

9. Let H be a Hopf algebra with an antipode S. Then H/k is a Hopf-Galois extension ([140]), i.e. the canonical map

$$\beta : \; H \otimes H \to H \otimes H, \quad \beta(g \otimes h) = \sum gh_{(1)} \otimes h_{(2)}$$

is bijective. β is a solution of the Hopf equation. Furthermore,

$$R' : \; H \otimes H \to H \otimes H, \quad R'(g \otimes h) = \sum g_{(1)} \otimes S(g_{(2)})h$$

is also a solution of the Hopf equation.

We have already mentioned that the pentagon equation is in a certain sense more basic than the QYBE. By this we mean that the pentagon equation expresses an associativity constraint (cf. [122], [170]), while the QYBE expresses a commutativity constraint. The following two results explain this point of view. Proposition 127 is in fact an elementary observation. Theorem 65 is a more recent result due to Davydov [65].

Proposition 127. *Let M be a vector space, $m : \; M \otimes M \to M$, $m(v \otimes w) = v \cdot w$ be a k-linear map and*

$$R = R_m : \; M \otimes M \to M \otimes M, \quad R(v \otimes w) = v \otimes w \cdot v$$

for all v, $w \in M$. Then R_m is a solution of the pentagon equation if and only if (M, m) is an associative algebra.

Proof. For u, v, $w \in M$ we have

$$R^{12}R^{13}R^{23}(u \otimes v \otimes w) = u \otimes v \cdot u \otimes (w \cdot v) \cdot u$$

and

$$R^{23}R^{12}(u \otimes v \otimes w) = u \otimes v \cdot u \otimes w \cdot (v \cdot u)$$

Hence R is a solution of the pentagon equation iff (V, m) is an algebra structure.

Let $R \in \mathrm{End}(M \otimes M)$ be an invertible solution of the pentagon equation. We introduce the following product $\boxtimes_M$ on $\mathcal{M}_k$:

$$U_1 \boxtimes_M U_2 = U_1 \otimes M \otimes U_2$$

For any U_1, U_2, $U_3 \in \mathcal{M}_k$, we have an isomorphism

$$\varphi_{U_1, U_2, U_3} = R^{24} : \; U_1 \otimes M \otimes U_2 \otimes M \otimes U_3 \to U_1 \otimes M \otimes U_2 \otimes M \otimes U_3$$

One can verify that this makes $(\mathcal{M}_k, \boxtimes, \varphi)$ into a monoidal category (without identity object). In fact, we have

Theorem 65. *There exists a bijective correspondence between all monoidal structures on the category of k-vector spaces (without identity object) and all bijective solutions of pentagon equation on various k-vector spaces.*

Proof. Let $\mathcal{M}_k$ be the category of k-vector spaces. By the functoriality any tensor product functor $\boxtimes : \mathcal{M}_k \times \mathcal{M}_k \to \mathcal{M}_k$ is defined by the vector space $M := k \boxtimes k$. Let $\boxtimes_M$ be the tensor product functor corresponding to a given vector space M. For arbitrary vector spaces U_1, U_2, we have that

$$U_1 \boxtimes_M U_2 = U_1 \otimes M \otimes U_2.$$

We have to explain what the associativity constraint means for the tensor product $\boxtimes_M$. Using the naturality, any associativity constraint φ for $\boxtimes_M$ is given by the automorphism $R = R_\varphi = \varphi_{k,k,k} \in \operatorname{Aut}(M^{\otimes 2})$ that arises from the sequences of isomorphisms:

$$M \otimes M = k \boxtimes_M M = k \boxtimes_M (k \boxtimes_M k) \xrightarrow{\varphi_{k,k,k}} (k \boxtimes_M k) \boxtimes_M k = M \boxtimes_M k = M \otimes M$$

For arbitrary vector spaces U_1, U_2, U_3 the associativity constraint $\varphi_{U_1,U_2,U_3} : U_1 \boxtimes_M (U_2 \boxtimes_M U_3) \to (U_1 \boxtimes_M U_2) \boxtimes_M U_3$ is given in terms of R via the following diagram

i.e. $\varphi_{U_1,U_2,U_3} = R^{24}$. In particular we obtain

$$\varphi_{k\boxtimes_M k,k,k} = \varphi_{M,k,k} = R^{23}, \quad \varphi_{k,k\boxtimes_M k,k} = \varphi_{k,M,k} = R^{13},$$
$$\varphi_{k,k,k\boxtimes_M k} = \varphi_{k,k,M} = R^{12}$$

On the other hand

$$I_k \boxtimes_M \varphi_{k,k,k} = R^{23}, \quad \varphi_{k,k,k} \boxtimes_M I_k = R^{12}.$$

Now the commutativity of the pentagon diagram for φ from the definition of a monoidal category is equivalent to the pentagon equation for R

$$R^{12} R^{13} R^{23} = R^{23} R^{12}$$

and the proof is complete.

Definition 12. *Let M be a vector space and $R \in \operatorname{End}(M \otimes M)$.*

1. *R is called commutative if $R^{12} R^{13} = R^{13} R^{12}$.*
2. *R is called cocommutative if $R^{13} R^{23} = R^{23} R^{13}$.*

Remarks 14. 1. Let $R \in \operatorname{End}(M \otimes M)$. Then R is a commutative solution of the Hopf equation if and only if $W = \tau R \tau$ is a cocommutative solution of the pentagon equation.
Indeed, $R^{12} R^{13} = R^{13} R^{12}$ if and only if

$$\tau^{12} W^{12} \tau^{12} \tau^{13} W^{13} \tau^{13} = \tau^{13} W^{13} \tau^{13} \tau^{12} W^{12} \tau^{12}. \tag{6.10}$$

Using the formulas

$$\tau^{12} \tau^{13} = \tau^{23} \tau^{12}, \quad \tau^{13} \tau^{12} = \tau^{12} \tau^{23},$$

$$W^{12} \tau^{23} = \tau^{23} W^{13}, \quad \tau^{12} W^{13} = W^{23} \tau^{12},$$

$$W^{13} \tau^{12} = \tau^{12} W^{23}, \quad \tau^{23} W^{12} = W^{13} \tau^{23}$$

we find that the equation (6.10) is equivalent to

$$\tau^{12} \tau^{23} W^{13} W^{23} \tau^{12} \tau^{13} = \tau^{13} \tau^{12} W^{23} W^{13} \tau^{23} \tau^{12}.$$

The conclusion follows after we observe that

$$\tau^{12} \tau^{13} \tau^{12} \tau^{23} = \tau^{23} \tau^{12} \tau^{13} \tau^{12} = \operatorname{Id}.$$

2. Suppose that $R \in \operatorname{End}(M \otimes M)$ is bijective. Then, R is a cocommutative solution of the Hopf equation if and only if $\tau R^{-1} \tau$ is a commutative solution of the Hopf equation.

6.2 The FRT Theorem for the Hopf equation

We begin with the following Proposition, which can be viewed as an analog of Theorem 59. It explains the role of Hopf modules in the theory of the Hopf equation.

Proposition 128. *Let H be a bialgebra and $(M, \cdot, \rho) \in {}_H\mathcal{M}^H$ an H-Hopf module. Then the natural map*

$$R_{(M, \cdot, \rho)}(m \otimes n) = n_{[1]} \cdot m \otimes n_{[0]}$$

is a solution of the Hopf equation. If H is commutative then $R_{(M, \cdot, \rho)}$ is a commutative solution of the Hopf equation.

Proof. Let $R = R_{(M, \cdot, \rho)}$. For l, m, $n \in M$ we have

$$
\begin{aligned}
R^{23}R^{13}R^{12}(l \otimes m \otimes n) &= R^{23}R^{13}\left(m_{[1]} \cdot l \otimes m_{[0]} \otimes n\right) \\
&= R^{23}\left(n_{[1]}m_{[1]} \cdot l \otimes m_{[0]} \otimes n_{[0]}\right) \\
&= n_{[2]}m_{[1]} \cdot l \otimes n_{[1]} \cdot m_{[0]} \otimes n_{[0]} \\
&= n_{[1](2)}m_{[1]} \cdot l \otimes n_{[1](1)} \cdot m_{[0]} \otimes n_{[0]} \\
&= (n_{[1]} \cdot m)_{[1]} \cdot l \otimes (n_{[1]} \cdot m)_{[0]} \otimes n_{[0]} \\
&= R^{12}\left(l \otimes n_{[1]} \cdot m \otimes n_{[0]}\right) \\
&= R^{12}R^{23}(l \otimes m \otimes n)
\end{aligned}
$$

and R is a solution of the Hopf equation. We have

$$
R^{12}R^{13}(l \otimes m \otimes n) = m_{[1]}n_{[1]} \cdot l \otimes m_{[0]} \otimes n_{[0]}
$$

and

$$
R^{13}R^{12}(l \otimes m \otimes n) = n_{[1]}m_{[1]} \cdot l \otimes m_{[0]} \otimes n_{[0]}
$$

If H is commutative, then it follows that $R^{12}R^{13} = R^{13}R^{12}$.

Remark 18. If $(M, \cdot, \rho)$ is an H-Hopf module then the map

$$
R'_{(M, \cdot, \rho)} = \tau R_{(M, \cdot, \rho)}\tau : \ M \otimes M \to M \otimes M, \quad R'_{(M, \cdot, \rho)}(m \otimes n) = m_{[0]} \otimes m_{[1]} \cdot n
$$

is a solution of the pentagon equation.

Before stating the FRT Theorem, we state two Lemmas; the proofs are straighforward.

Lemma 28. *Let H be a bialgebra, $(M, \cdot)$ a left H-module and (M, ρ) a right H-comodule. Then the set*

$$
\{h \in H \mid \rho(h \cdot m) = h_{(1)} \cdot m_{[0]} \otimes h_{(2)}m_{[1]}, \forall m \in M\}
$$

is a subalgebra of H.

Lemma 29. *Let H be a bialgebra, $(M, \cdot)$ a left H-module and (M, ρ) a right H-comodule. If I is a biideal of H such that $I \cdot M = 0$, then, with the natural structures, $(M, \cdot')$ is a left H/I-module, (M, ρ') a right H/I-comodule and $R_{(M, \cdot', \rho')} = R_{(M, \cdot, \rho)}$*

Theorem 66. *Let M be a finite dimensional vector space and $R \in \mathrm{End}(M \otimes M)$ be a solution of the Hopf equation.*

1. *There exists a bialgebra $B(R)$ such that M has a structure of $B(R)$-Hopf module $(M, \cdot, \rho)$ and $R = R_{(M, \cdot, \rho)}$.*

2. *The bialgebra $B(R)$ is universal with respect to this property: if H is a bialgebra such that $(M, \cdot', \rho') \in {}_H\mathcal{M}^H$ and $R = R_{(M, \cdot', \rho')}$ then there exists a unique bialgebra map $f : B(R) \to H$ such that $\rho' = (I \otimes f)\rho$. Furthermore, $a \cdot m = f(a) \cdot' m$, for all $a \in B(R)$, $m \in M$.*

3. *If R is commutative, then there exists a commutative bialgebra $\overline{B}(R)$ such that M has a structure of $\overline{B}(R)$-Hopf module $(M, \cdot', \rho')$ and $R = R_{(M, \cdot', \rho')}$.*

Proof. 1. The proof will be given in several steps. Let $\{m_1, \cdots, m_n\}$ be a basis for M and (x_{uv}^{ji}) be the matrix of R, i.e.

$$R(m_v \otimes m_u) = x_{vu}^{ij} m_i \otimes m_j \tag{6.11}$$

for all $u, v = 1, \cdots, n$.

Let $(C, \Delta, \varepsilon) = \mathcal{M}^n(k)$, be the comatrix coalgebra of order n. As in Section 5.1, $\{c_j^i \mid i, j = 1, \cdots, n\}$ is the canonical basis of C. Recall that

$$\Delta(c_j^i) = c_u^i \otimes c_j^u, \quad \varepsilon(c_j^i) = \delta_j^i \tag{6.12}$$

for all $j, k = 1, \cdots, n$. Let $\rho : M \to M \otimes C$ given by

$$\rho(m_l) = m_v \otimes c_l^v \tag{6.13}$$

for all $l = 1, \cdots, n$. It follows from Lemma 25 that M is a right C-comodule. Let $T(C)$ be the bialgebra structure on the tensor algebra $T(C)$ which extends Δ and ε (using Lemma 26). As the inclusion $i : C \to T(C)$ is a coalgebra map, M has a right $T(C)$-comodule structure via

$$M \xrightarrow{\;\rho\;} M \otimes C \xrightarrow{\;I \otimes i\;} M \otimes T(C)$$

The right $T(C)$-comodule structure on M will also be denoted by ρ. We will now define a left $T(C)$-module structure on M in such a way that $R = R_{(M, \cdot, \rho)}$. First we define

$$\mu : C \otimes M \to M, \quad \mu(c_u^j \otimes m_v) = x_{vu}^{ij} m_i$$

for all $j, u, v = 1, \cdots n$. From lemma 26, there exists a unique left $T(C)$-module structure on $(M, \cdot)$ such that

$$c_u^j \cdot m_v = \mu(c_u^j \otimes m_v) = x_{vu}^{ij} m_i$$

for all $j, u, v = 1, \cdots, n$. We then have for all $u, v = 1, \cdots, n$ that

$$\begin{aligned}
R_{(M, \cdot, \rho)}(m_v \otimes m_u) &= c_u^j \cdot m_v \otimes m_j \\
&= x_{vu}^{ij} m_i \otimes m_j \\
&= R(m_v \otimes m_u)
\end{aligned}$$

and $(M, \cdot, \rho)$ has a structure of left $T(C)$-module and right $T(C)$-comodule such that $R = R_{(M, \cdot, \rho)}$.

Now, we define the *obstructions* $\chi(i, j, k, l)$, measuring how far away M is from being a $T(C)$-Hopf module. Keeping in mind that $T(C)$ is generated as an algebra by (c_j^i) and using Lemma 28 we compute

$$h_{(1)} \cdot m_{[0]} \otimes h_{(2)} m_{[1]} - \rho(h \cdot m)$$

for $h = c_k^j$, and $m = m_l$, for $j, k, l = 1, \cdots, n$. We have

$$h_{(1)} \cdot m_{[0]} \otimes h_{(2)} m_{[1]} = c_u^j \cdot m_v \otimes c_k^u c_l^v = m_i \otimes \left(x_{vu}^{ij} c_k^u c_l^v \right)$$

and

$$\rho(h \cdot m) = \rho(c_k^j \cdot m_l) = x_{lk}^{\alpha j} (m_\alpha)_{[0]} \otimes (m_\alpha)_{[1]}$$
$$= x_{lk}^{\alpha j} m_i \otimes c_\alpha^i = m_i \otimes \left(x_{lk}^{\alpha j} c_\alpha^i \right)$$

Let

$$\chi_{kl}^{ij} = x_{vu}^{ij} c_k^u c_l^v - x_{lk}^{\alpha j} c_\alpha^i \tag{6.14}$$

for all $i, j, k, l = 1, \cdots, n$. Then

$$h_{(1)} \cdot m_{[0]} \otimes h_{(2)} m_{[1]} - \rho(h \cdot m) = m_i \otimes \chi(i, j, k, l) \tag{6.15}$$

Let I be the two-sided ideal of $T(C)$ generated by all χ_{kl}^{ij}, $i, j, k, l = 1, \cdots, n$. The following assertion is the key point of our proof:

I is a bi-ideal of $T(C)$ and $I \cdot M = 0$.

We first prove that I is a coideal; this will result from the following formula:

$$\Delta(\chi_{kl}^{ij}) = \chi_{ab}^{ij} \otimes c_k^a c_l^b + c_p^i \otimes \chi_{kl}^{pj} \tag{6.16}$$

Indeed, we have:

$$\Delta(\chi_{kl}^{ij}) = x_{vu}^{ij} \Delta(c_k^u) \Delta(c_l^v) - x_{lk}^{\alpha j} \Delta(c_\alpha^i)$$
$$= x_{vu}^{ij} c_a^u c_b^v \otimes c_k^a c_l^b - x_{lk}^{\alpha j} c_p^i \otimes c_\alpha^p$$
$$= \left(x_{vu}^{ij} c_a^u c_b^v \right) \otimes c_k^a c_l^b - c_p^i \otimes \left(x_{lk}^{\alpha j} c_\alpha^p \right)$$
$$= \left(\chi_{ab}^{ij} + x_{ba}^{\gamma j} c_\gamma^i \right) \otimes c_k^a c_l^b$$
$$\quad - c_p^i \otimes \left(-\chi_{kl}^{pj} + x_{sr}^{pj} c_k^r c_l^s \right)$$
$$= \chi_{ab}^{ij} \otimes c_k^a c_l^b + c_p^i \otimes \chi_{kl}^{pj}$$

where in the last equality we use the fact that

$$x_{ba}^{\gamma j} c_\gamma^i \otimes c_k^a c_l^b = x_{sr}^{pj} c_p^i \otimes c_k^r c_l^s$$

Hence, the formula (6.16) holds. On the other hand

$$\varepsilon\left(\chi_{kl}^{ij}\right)= x_{lk}^{ij} - x_{lk}^{ij} = 0$$

so we proved that I is a coideal of $T(C)$.

The proof of the fact that $I \cdot M = 0$ relies on the fact that R is a solution of the Hopf equation. For $z \in M$, j, $k = 1, \cdots, n$, we have the following formula:

$$\left(R^{23} R^{13} R^{12} - R^{12} R^{23}\right)(z \otimes m_k \otimes m_j) = \chi(r, s, j, k) \cdot z \otimes m_r \otimes m_s \quad (6.17)$$

Indeed:

$$\left(R^{23} R^{13} R^{12}\right)(z \otimes m_k \otimes m_j) = \left(R^{23} R^{13}\right)(c_k^\alpha \cdot z \otimes m_\alpha \otimes m_j)$$
$$= R^{23}\left(c_j^\beta c_k^\alpha \cdot z \otimes m_\alpha \otimes m_\beta)\right)$$
$$= x_{\alpha\beta}^{rs} c_j^\beta c_k^\alpha \cdot z \otimes m_r \otimes m_s$$

and

$$\left(R^{12} R^{23}\right)(z \otimes m_k \otimes m_j) = R^{12}(z \otimes c_j^s \cdot m_k \otimes m_s)$$
$$= R^{12}(z \otimes x_{kj}^{\alpha s} m_\alpha \otimes m_s)$$
$$= x_{kj}^{\alpha s} c_\alpha^r \cdot z \otimes m_r \otimes m_s$$

so we obtain

$$\left(R^{23} R^{13} R^{12} - R^{12} R^{23}\right)(z \otimes m_k \otimes m_j)$$
$$= \left(x_{\alpha\beta}^{rs} c_j^\beta c_k^\alpha - x_{kj}^{\alpha s} c_\alpha^r\right) \cdot z \otimes m_r \otimes m_s = \chi_{jk}^{rs} \cdot z \otimes m_r \otimes m_s$$

proving (6.17). But R is a solution of the Hopf equation, hence $\chi_{jk}^{rs} \cdot z = 0$, for all $z \in M$, j, k, r, $s = 1, \cdots, n$, and we conclude that $I \cdot M = 0$. Now we define

$$B(R) = T(C)/I$$

M has a right $B(R)$-comodule structure via the canonical projection $T(C) \to B(R)$ and a left $B(R)$-module structure as $I \cdot M = 0$. The (c_j^i) generate $B(R)$ and $\chi_{kl}^{ij} = 0$ in $B(R)$, so using (6.15) we find that $(M, \cdot, \rho) \in {}_{B(R)}\mathcal{M}^{B(R)}$ and, using Lemma 29, $R = R_{(M, \cdot, \rho)}$. This finishes the proof of Part 1) of the Theorem.

2. Let H be a bialgebra and suppose that $(M, \cdot', \rho') \in {}_H\mathcal{M}^H$ and $R = R_{(M, \cdot', \rho')}$. Let (c_{ij}') be a family of elements in H such that

$$\rho'(m_l) = m_v \otimes c_l'^v$$

Then

$$R(m_v \otimes m_u) = c_u'^j \cdot' m_v \otimes m_j$$

and

$$c_u'^j \cdot' m_v = x_{vu}^{ij} m_i = c_u^j \cdot m_v$$

Let

$$\chi_{kl}'^{ij} = x_{vu}^{ij} c_k'^u c_l'^v - x_{lk}^{\alpha j} c_\alpha'^i$$

The universal property of the tensor algebra $T(C)$ implies that there exists a unique algebra map $f_1 : T(C) \to H$ such that $f_1(c_j^i) = c_j'^i$, for all $i, j = 1, \cdots, n$. As $(M, \cdot', \rho') \in {}_H\mathcal{M}^H$ we get that $\chi_{kl}'^{ij} = 0$, and hence $f_1(\chi_{kl}^{ij}) = 0$, for all $i, j, k, l = 1, \cdots, n$. So the map f_1 factorizes through a map

$$f : B(R) \to H, \quad f(c_j^i) = c_j'^i$$

Of course, for m_l an arbitrary element of the given basis of M, we have

$$(I \otimes f)\rho(m_l) = m_v \otimes f(c_l^v) = m_v \otimes c_l'^v = \rho'(m_l)$$

Conversely, the relation $(I \otimes f)\rho = \rho'$ necessarily implies $f(c_j^i) = c_j'^i$, which proves the uniqueness of f. This completes the proof of part 2) the theorem.
3. For $z \in M$ and $j, k = 1, \cdots, n$ we have the formula:

$$\left(R^{12}R^{13} - R^{13}R^{12}\right)(z \otimes m_k \otimes m_j) = \left(c_k^r c_j^s - c_j^s c_k^r\right) \cdot z \otimes m_r \otimes m_s \quad (6.18)$$

Let $\bar{I}$ be the two-sided ideal of $T(C)$ generated by I and all $[c_k^r, c_j^s]$. It follows from the formula

$$\Delta\left([c_k^r, c_j^s]\right) = \left([c_a^r, c_b^s] \otimes c_j^b c_k^a + c_a^r c_b^s \otimes [c_k^a, c_j^b]\right)$$

that $\bar{I}$ is also a coideal of $T(C)$ and from equation (6.18) we get that $\bar{I} \cdot M = 0$. Define now

$$\bar{B}(R) = T(C)/\bar{I}.$$

Then $\bar{B}(R)$ is a commutative bialgebra, M has a structure of $\bar{B}(R)$-Hopf module $(M, \cdot', \rho')$ and $R = R_{(M, \cdot', \rho')}$.

Combining Proposition 128 and Theorem 66, we obtain the following result.

Corollary 37. *Let M be a finite dimensional vector space and $R \in \text{End}(M \otimes M)$. Then R is a solution of the Hopf equation if and only if there exists a bialgebra $B(R)$ such that M has a structure of left-right $B(R)$-Hopf module $(M, \cdot, \rho)$ and $R = R_{(M, \cdot, \rho)}$.*

Remarks 15. 1. The obstruction elements χ_{kl}^{ij} constructed in Theorem 66 are different from the homogenous elements yb_{kl}^{ij} defined in 5.27 which correspond to the quantum Yang-Baxter equation. Despite that, as in the case of the quantum Yang-Baxter equation, the elements χ_{kl}^{ij} play the same role: the two-sided ideal generated by them is also a coideal which annihilates M, which is the key point of the proof.

2. All commutative bialgebras $\overline{B}(R)$ which come from a commutative solution of the Hopf equation are quotients for various bialgebra structures which can be given on a polynomial algebra $k[Y_1, \cdots, Y_n]$ in commutative variables $Y_1, \cdots, Y_n$.

Let $(M, \cdot, \rho) \in {}_H\mathcal{M}^H$ be a Hopf module over a bialgebra H. In Remark 18, we have seen that the map

$$R' = R'_{(M,\cdot,\rho)} : M \otimes M \to M \otimes M, \quad R'_{(M,\cdot,\rho)}(m \otimes n) = m_{[0]} \otimes m_{[1]} \cdot n \quad (6.19)$$

is a solution of the pentagon equation. Furthermore, if H has an antipode, then R is invertible with inverse

$$R'^{-1}(m \otimes n) = m_{[0]} \otimes S(m_{[1]})n$$

We have seen in Proposition 124 that the Hopf equation and the pentagon equation are equivalent; thus it is no surprise that we have also an FRT Theorem for the pentagon equation.

Theorem 67. *Let M be a finite dimensional vector space and $R \in \text{End}(M \otimes M)$. R is a solution of the pentagon equation if and only if there exists a bialgebra $P(R)$ such that M has a structure of left-right $P(R)$-Hopf module $(M, \cdot, \rho)$ and $R = R'_{(M,\cdot,\rho)}$.*

Proof. The proof is completely similar to the one of Theorem 66. Keeping the same notation, the bialgebra $P(R)$ is defined as follows:

• $P(R)$ is the free algebra generated by $(c_j^i)_{i,j=1,\cdots,n}$ and the relations

$$x_{uv}^{ij} c_k^u c_l^v = x_{kl}^{ia} c_a^j \quad (6.20)$$

for all $i, j, k, l = 1, \cdots, n$.

• the comultiplication and the counit is defined in such a way that the matrix (c_j^i) is comultiplicative.

The structure of M as an object in ${}_{P(R)}\mathcal{M}^{P(R)}$ is the following: M is a right $P(R)$-comodule via

$$\rho : M \to M \otimes P(R), \quad \rho(m_u) = m_j \otimes c_u^j$$

and has a unique left $P(R)$-module structure $\cdot$ such that $R = R'_{(M,\cdot,\rho)}$ given by

$$c_v^j \cdot m_u = x_{vu}^{ji} m_i$$

for all $j, u, v = 1, \cdots, n$.

Furthermore, looking at the obstruction (6.14), we see that $P(R) = B(\tau R \tau)$.

We have seen in Chapter 5 that the FRT Theorem 61 for the QYBE is a special case of Theorem 60, stating that every solution R of the QYBE has the form $R = R_{(\sigma, M, \rho)}$, for a coquasitriangular bialgebra $(A(R), \sigma)$ and a right $A(R)$-comodule structure ρ on M. We will prove a similar result for the Hopf equation; the first step is to introduce a new class of bialgebras, that will play same role for the Hopf equation as coquasitriangular algebras for the QYBE.

Definition 13. *Let H be a bialgebra and C be a subcoalgebra of H. A k-linear map $\sigma : C \otimes H \to k$ is called a Hopf function if:*
(H1) $\sigma(c_{(1)} \otimes h_{(1)})h_{(2)}c_{(2)} = \sigma(c_{(2)} \otimes h)c_{(1)}$
(H2) $\sigma(c \otimes 1) = \varepsilon(c)$
(H3) $\sigma(c \otimes hk) = \sigma(c_{(1)} \otimes h)\sigma(c_{(2)} \otimes k)$
for all $c \in C$, h, $k \in H$. In this case we will say that (H, C, σ) is a bialgebra with a Hopf function.

Remarks 16. 1. An immediate question is the following: why is σ only defined on $C \otimes H$, and not on the whole of $H \otimes H$ (as in Definition 7)? The answer is that the bialgebra with a Hopf function (H, H, σ) is (k, k, I_k), and this is due to the condition (H1). This can be easily seen: taking $c = 1_H$ in (H1), we find

$$\sigma(1_H \otimes h)1_H = \sigma(1_H \otimes h_{(1)})h_{(2)} \qquad (6.21)$$

and $T_\sigma : H \to k$, $T_\sigma(h) := \sigma(1_H \otimes h)$ is a right integral in H^*, and it is also an algebra map, by (H3), implying $H = k$.
2. (H2) and (H3) are exactly (B2) and (B3), of course with the restriction that they hold only on $C \otimes H$. Also the left hand side of (H1) is the same as the left hand side of (B1), but their right hand sides are considerably different.
3. Let (H, C, σ) be a bialgebra with a Hopf function. If σ is right convolution invertible in $\text{Hom}(C \otimes H, k)$, then (H2) follows from (H3). Indeed, for $c \in C$ we have:

$$\begin{aligned}
\sigma(c \otimes 1) &= \sigma(c_{(1)} \otimes 1)\varepsilon(c_{(2)}) \\
&= \sigma(c_{(1)} \otimes 1)\sigma(c_{(2)} \otimes 1)\sigma^{-1}(c_{(3)} \otimes 1) \\
&= \sigma(c_{(1)} \otimes 1)\sigma^{-1}(c_{(2)} \otimes 1) = \varepsilon(c)
\end{aligned}$$

If H has an antipode S, then σ is invertible and $\sigma^{-1}(c \otimes h) = \sigma(c \otimes S(h))$, for all $c \in C$, $h \in H$.

Now we will explain the relation between (H1) and the concept of right integral in H^*. Let H be a bialgebra and C a subcoalgebra of H. If $T \in H^*$ is a right integral then the map

$$\sigma_T : C \otimes H \to k, \qquad \sigma_T(c \otimes h) = \varepsilon(c)T(h), \quad \forall c \in C, h \in H$$

satisfies (H1).

Conversely, if $1_H \in C$ and $\sigma : C \otimes H \to k$ satisfies (H1) then the map

$$T_\sigma : H \to k, \qquad T_\sigma(h) := \sigma(1_H \otimes h), \quad \forall h \in H$$

is a right integral. Assume that H has an antipode and that (H2) holds. Then $T_\sigma(1_H) = 1_k$, and, by Maschke's Theorem for Hopf algebras, H is cosemisimple. As $T_{\sigma_T} = T$, we obtain that the map

$$p: \{\sigma : C \otimes H \to k \mid \sigma \text{ satisfies } (H1)\} \to \int_{H^\bullet}^r, \quad \sigma \to T_\sigma$$

is a projection, with section

$$i: \int_{H^\bullet}^r \to \{\sigma : C \otimes H \to k \mid \sigma \text{ satisfies } (H1)\}, \quad T \to \sigma_T$$

We summarize our results in the next Proposition.

Proposition 129. *Let H be a bialgebra and C a subcoalgebra of H.*

1. *If $T \in H^*$ is a right integral on H, then the map $\sigma_T : C \otimes H \to k$ satisfies (H1).*
2. *if $1_H \in C$ and $\sigma : C \otimes H \to k$ satisfies (H1), then the map $T_\sigma : H \to k$ is a right integral on H. Furthermore, if (H2) holds and H has an antipode, then H is cosemisimple.*

Proposition 130. *Let H be a bialgebra, C a subcoalgebra of H and $\sigma : C \otimes H \to k$ a k-linear map satisfying (H3). Suppose that (H1) holds on a basis of C and a system of algebra generators of H. Then (H1) holds for any $c \in C$ and $h \in H$.*

Proof. Let $c \in C$ be an element of the given basis of C and x, $y \in H$ two of the algebra generators of H. We are done if we can show that (H1) holds for (c, xy). This is a straightforward computation.

$$
\begin{aligned}
\sigma(c_{(2)} \otimes xy)c_{(1)} &= \sigma(c_{(2)(1)} \otimes x)\sigma(c_{(2)(2)} \otimes y)c_{(1)} \\
&= \sigma(c_{(2)} \otimes y)\sigma(c_{(1)(2)} \otimes x)c_{(1)(1)} \\
&= \sigma(c_{(2)} \otimes y)\sigma(c_{(1)(1)} \otimes x_{(1)})x_{(2)}c_{(1)(2)} \\
&= \sigma(c_{(1)} \otimes x_{(1)})x_{(2)}\sigma(c_{(2)(2)} \otimes y)c_{(2)(1)} \\
&= \sigma(c_{(1)} \otimes x_{(1)})x_{(2)}\sigma(c_{(2)(1)} \otimes y_{(1)})y_{(2)}c_{(2)(2)} \\
&= \sigma(c_{(1)} \otimes x_{(1)})\sigma(c_{(2)} \otimes y_{(1)})x_{(2)}y_{(2)}c_{(3)} \\
&= \sigma(c_{(1)} \otimes x_{(1)}y_{(1)})x_{(2)}y_{(2)}c_{(2)} \\
&= \sigma(c_{(1)} \otimes (xy)_{(1)})(xy)_{(2)}c_{(2)}
\end{aligned}
$$

Examples 11. 1. Let $H = k[G]$ be a group algebra. Let C be an arbitrary subcoalgebra of H. Then there exists no Hopf function $\sigma : C \otimes k[G] \to k$. Indeed, any subcoalgebra of $k[G]$ has the form $k[F]$, where F is a subset of G. Suppose that there exists $\sigma : k[F] \otimes k[G] \to k$ a Hopf function. From (H2) we get that $\sigma(f \otimes 1) = 1$ for all $f \in F$. Now let $g \in G$, $g \neq 1$ and $f \in F$. From (H1) we obtain $\sigma(f \otimes g)gf = \sigma(f \otimes g)f$ i.e. $\sigma(f \otimes g) = 0$ for all $f \in F$. But then, using (H3), we find a contradiction:

$$1 = \sigma(f \otimes 1) = \sigma(f \otimes gg^{-1}) = \sigma(f \otimes g)\sigma(f \otimes g^{-1}) = 0,$$

2. Let $H = k_q < x, y \mid xy = qyx >$ be the quantum plane:

$$\Delta(x) = x \otimes x, \quad \Delta(y) = y \otimes 1 + x \otimes y, \quad \varepsilon(x) = 1, \quad \varepsilon(y) = 0.$$

$C = kx$ is a subcoalgebra of H. For $a \in k$ let $\sigma_a : C \otimes H \to k$ be given by

$$\sigma_a(x \otimes 1) = 1, \quad \sigma_a(x \otimes x) = 0, \quad \sigma_a(x \otimes y) = a$$

and extend σ_a to the whole of $C \otimes H$ using (H3). Then, σ_a is a Hopf function. By Proposition 130, it suffices to check that (H1) holds for $h \in \{x, y\}$. For $h = x$, (H1) is

$$\sigma_a(x \otimes x)x^2 = \sigma_a(x \otimes x)x$$

which holds if and only if $\sigma_a(x \otimes x) = 0$. For $h = y$, (H1) has the form

$$\sigma_a(x \otimes y)x + \sigma_a(x \otimes x)yx = \sigma_a(x \otimes y)x,$$

which is true, as $\sigma_a(x \otimes x) = 0$. In fact, we have also proved the converse: if (H, C, σ) is a bialgebra with a Hopf function, then $\sigma = \sigma_a$ for some $a \in k$.
3. Let M be a monoid and $N = \{n \in M \mid xn = n, \text{ for all } x \in M\}$. Let $H = k[M]$ and $C = k[F]$, with $F \subset N$. Let $\sigma : k[F] \otimes k[M] \to k$ be such that $\sigma(f \otimes 1) = 1$ and $\sigma(f, \bullet) : M \to (k, \cdot)$ is a morphism of monoids for all $f \in F$. Then σ is a Hopf function.
We give such a concrete example. Let $a \in k$ and $\mathcal{F}_a(k) = \{u : k \to k \mid u(a) = a\}$. $(\mathcal{F}_a(k), \circ)$ is a monoid, and $F = \{I_k\} \subset N$.

$$\sigma : k[F] \otimes k[\mathcal{F}_a(k)] \to k, \quad \sigma(I_k \otimes u) = 1$$

is a Hopf function.

Let H be a bialgebra, $C \subset H$ a subcoalgebra, and $\sigma : C \otimes H \to k$ a k-linear map. As usual, $\sigma_{12}, \sigma_{13}, \sigma_{23} : C \otimes C \otimes H \to k$ are defined by:

$$\sigma_{12}(c \otimes d \otimes x) = \varepsilon(x)\sigma(c \otimes d), \quad \sigma_{13}(c \otimes d \otimes x) = \varepsilon(d)\sigma(c \otimes x)$$

$$\sigma_{23}(c \otimes d \otimes x) = \varepsilon(c)\sigma(d \otimes x)$$

for all $c, d \in C$, $x \in H$.

Proposition 131. *Let (H, C, σ) be a bialgebra with a Hopf function $\sigma : C \otimes H \to k$. Then*

$$\sigma_{23} * \sigma_{13} * \sigma_{12} = \sigma_{12} * \sigma_{23} \tag{6.22}$$

in $\mathrm{Hom}(C \otimes C \otimes H, k)$.
If (M, ρ) is a right C-comodule, then the map

$$R = R_{(\sigma, M, \rho)} : \ M \otimes M \to M \otimes M, \quad R(m \otimes n) = \sigma(m_{[1]} \otimes n_{[1]}) m_{[0]} \otimes n_{[0]}$$

is a solution of the Hopf equation.

Proof. For $c, d \in C$ and $x \in H$, we have:

$$(\sigma_{12} * \sigma_{23})(c \otimes d \otimes x) = \varepsilon(x_{(1)})\sigma(c_{(1)} \otimes d_{(1)})\varepsilon(c_{(2)})\sigma(d_{(2)} \otimes x_{(2)})$$
$$= \sigma(c \otimes d_{(1)})\sigma(d_{(2)} \otimes x)$$

and

$$(\sigma_{23} * \sigma_{13} * \sigma_{12})(c \otimes d \otimes x)$$
$$= \varepsilon(c_{(1)})\sigma(d_{(1)} \otimes x_{(1)})\varepsilon(d_{(2)})\sigma(c_{(2)} \otimes x_{(2)})\varepsilon(x_{(3)})\sigma(c_{(3)} \otimes d_{(3)})$$
$$= \sigma(c_{(1)} \otimes x_{(2)})\sigma(c_{(2)} \otimes d_{(2)})\sigma(d_{(1)} \otimes x_{(1)})$$
$$= \sigma(c \otimes x_{(2)}d_{(2)})\sigma(d_{(1)} \otimes x_{(1)})$$
$$= \sigma\Big(c \otimes \sigma(d_{(1)} \otimes x_{(1)})x_{(2)}d_{(2)}\Big)$$
$$= \sigma\Big(c \otimes \sigma(d_{(2)} \otimes x)d_{(1)}\Big)$$
$$= \sigma(c \otimes d_{(1)})\sigma(d_{(2)} \otimes x)$$

proving (6.22). The fact that R is a solution of the Hopf equation follows from (6.22) and from the formulas:

$$R^{12}R^{23}(u \otimes v \otimes w) = R^{12}\Big(\sigma(v_{[1]} \otimes w_{[1]})u \otimes v_{[0]} \otimes w_{[0]}\Big)$$
$$= \sigma(u_{[1]} \otimes v_{[0][1]})\sigma(v_{[1]} \otimes w_{[1]})u_{[0]} \otimes v_{[0][0]} \otimes w_{[0]}$$
$$= \sigma(u_{[1]} \otimes v_{[1](1)})\sigma(v_{[1](2)} \otimes w_{[1]})u_{[0]} \otimes v_{[0]} \otimes w_{[0]}$$
$$= \Big(\sigma_{12} * \sigma_{23}\Big)(u_{[1]} \otimes v_{[1]} \otimes w_{[1]})u_{[0]} \otimes v_{[0]} \otimes w_{[0]}$$

and

$$R^{23}R^{13}R^{12}(u \otimes v \otimes w)$$
$$= R^{23}R^{13}\Big(\sigma(u_{[1]} \otimes v_{[1]})u_{[0]} \otimes v_{[0]} \otimes w\Big)$$
$$= R^{23}\Big(\sigma(u_{[0][1]} \otimes w_{[1]})\sigma(u_{[1]} \otimes v_{[1]})u_{[0][0]} \otimes v_{[0]} \otimes w_{[0]}\Big)$$
$$= R^{23}\Big(\sigma(u_{[1]} \otimes w_{[1]})\sigma(u_{[2]} \otimes v_{[1]})u_{[0]} \otimes v_{[0]} \otimes w_{[0]}\Big)$$

$$= \sigma(v_{[0][1]} \otimes w_{[0][1]})\sigma(u_{[1]} \otimes w_{[1]})\sigma(u_{[2]} \otimes v_{[1]})$$
$$u_{[0]} \otimes v_{[0][0]} \otimes w_{[0][0]}$$
$$= \sigma(v_{[1]} \otimes w_{[1]})\sigma(u_{[1]} \otimes w_{[2]})\sigma(u_{[2]} \otimes v_{[2]})u_{[0]} \otimes v_{[0]} \otimes w_{[0]}$$
$$= \sigma(u_{[1](1)} \otimes w_{[1](2)})\sigma(u_{[1](2)} \otimes v_{[1](2)})\sigma(v_{[1](1)} \otimes w_{[1](1)})$$
$$u_{[0]} \otimes v_{[0]} \otimes w_{[0]}$$
$$= \Big(\sigma_{23} * \sigma_{13} * \sigma_{12}\Big)(u_{[1]} \otimes v_{[1]} \otimes w_{[1]})u_{[0]} \otimes v_{[0]} \otimes w_{[0]}$$

We will present the analog of Theorem 60 for the Hopf equation. The bialgebra $B(R)$ and the coaction ρ of $B(R)$ on M are as in Theorem 66.

Theorem 68. *Let M be a finite dimensional vector space and $R \in \mathrm{End}(M \otimes M)$ a solution of the Hopf equation. Let C be the subcoalgebra of $B(R)$ generated by the (c_j^i).*

1. *There exists a unique Hopf function $\sigma : C \otimes B(R) \to k$ such that $R = R_{(\sigma,M,\rho)}$.*
2. *If R is bijective and commutative, then σ is invertible in the convolution algebra $\mathrm{Hom}(C \otimes B(R), k)$.*

Proof. 1. First we prove that σ is unique. Let $\sigma : C \otimes B(R) \to k$ be a Hopf function such that $R = R_\sigma$. Let $u, v = 1, \cdots, n$. Then

$$R_{(\sigma,M,\rho)}(m_v \otimes m_u) = \sigma\Big((m_v)_{[1]} \otimes (m_u)_{[1]}\Big)(m_v)_{[0]} \otimes (m_u)_{[0]}$$
$$= \sigma(c_v^i \otimes c_u^j)m_i \otimes m_j$$

and the fact that $R_\sigma(m_v \otimes m_u) = R(m_v \otimes m_u)$ implies

$$\sigma(c_v^i \otimes c_u^j) = x_{vu}^{ij} \tag{6.23}$$

As $B(R)$ is generated as an algebra by the (c_j^i), the relations (6.23) with (H2) and (H3) imply the uniqueness of σ.

Our next goal is to prove the existence of σ. First we define $\sigma_0 : C \otimes C \to k$ using (6.23). Then we extend σ_0 to a map $\sigma_1 : C \otimes T(C) \to k$ such that (H2) and (H3) hold. In order to prove that σ_1 factorizes trough a map $\sigma : C \otimes B(R) \to k$, we have to show that $\sigma_1(C \otimes I) = 0$, whith I the two-sided ideal of $T(C)$ generated by all χ_{kl}^{ij}. This follows from the fact that

$$\sigma_1(c_q^p \otimes \chi_{kl}^{ij}) = x_{vu}^{ij}\sigma_1(c_q^p \otimes c_k^u c_l^v) - x_{lk}^{\alpha j}\sigma_1(c_q^p \otimes c_\alpha i)$$
$$= x_{vu}^{ij}\sigma_1(c_\beta^p \otimes c_k^u)\sigma_1(c_q^\beta \otimes c_l^v) - x_{lk}^{\alpha j}x_{q\alpha}^{pi}$$
$$= x_{vu}^{ij}x_{\beta k}^{pu}x_{ql}^{\beta v} - x_{lk}^{\alpha j}x_{q\alpha}^{pi} = 0$$

We have constructed $\sigma : C \otimes B(R) \to k$ such that (H2) and (H3) hold and $R = R_{(\sigma,M,\rho)}$. We are left to prove (H1). By proposition 130, it suffices to check (H1) for $c = c_l^i$ and $h = c_k^j$, for all $i, j, k, l = 1, \cdots, n$. We have

$$\sigma(c_{(1)} \otimes h_{(1)})h_{(2)}c_{(2)} = \sum_{u,v} \sigma(c_v^i \otimes c_u^j)c_k^u c_l^v = \sum_{u,v} x_{uv}^{ji} c_k^u c_l^v$$

and

$$\sigma(c_{(2)} \otimes h)hc_{(1)} = \sum_\alpha \sigma(c_l^\alpha \otimes c_k^j)c_\alpha^i = \sum_\alpha x_{kl}^{j\alpha} c_\alpha^i$$

so

$$\sigma(c_{(1)} \otimes h_{(1)})h_{(2)}c_{(2)} - \sigma(c_{(2)} \otimes h)hc_{(1)} = \chi(i,j,k,l) = 0$$

and (H1) also holds, as needed.

2. Assume that R is bijective and let $S = R^{-1}$. Let (y_{vu}^{ij}) be the matrix of S, i.e.

$$S(m_v \otimes m_u) = y_{vu}^{ij} m_i \otimes m_j,$$

As $RS = SR = \mathrm{Id}_{M \otimes M}$ we have

$$x_{\alpha\beta}^{pi} y_{qj}^{\alpha\beta} = \delta_q^p \delta_j^i, \qquad y_{\alpha\beta}^{pi} x_{qj}^{\alpha\beta} = \delta_q^p \delta_j^i$$

We define

$$\sigma_0' : C \otimes C \to k, \quad \sigma_0'(c_v^i \otimes c_u^j) = y_{vu}^{ij}$$

and we extend σ_0' to a map $\sigma_1' : C \otimes T(C) \to k$ in such a way that σ_1' satisfies (H2) and (H3). First we prove that σ_1' is the convolution inverse of σ_1:

$$\sigma_1\big((c_q^p)_{(1)} \otimes (c_j^i)_{(1)}\big)\sigma_1'\big((c_q^p)_{(2)} \otimes (c_j^i)_{(2)}\big) = \sigma_1(c_\alpha^p \otimes c_\beta^i)\sigma_1'(c_q^\alpha \otimes c_j^\beta)$$
$$= x_{\alpha\beta}^{pi} y_{qj}^{\alpha\beta} = \delta_q^p \delta_j^i = \varepsilon(c_q^p)\varepsilon(c_j^i)$$

and

$$\sigma_1'\big((c_q^p)_{(1)} \otimes (c_j^i)_{(1)}\big)\sigma_1\big((c_q^p)_{(2)} \otimes (c_j^i)_{(2)}\big) = \sigma_1'(c_\alpha^p \otimes c_\beta^i)\sigma_1(c_q^\alpha \otimes c_j^\beta)$$
$$= y_{\alpha\beta}^{pi} x_{qj}^{\alpha\beta} = \delta_q^p \delta_j^i = \varepsilon(c_q^p)\varepsilon(c_j^i)$$

To show that $\sigma : C \otimes B(R) \to k$ is also convolution invertible, we have to show that σ_1' factorizes through a map $\sigma' : C \otimes B(R) \to k$. We will prove that this happens if and only if $R^{12}R^{13} = R^{13}R^{12}$, i.e. R is commutative. σ_1' factorizes if and only if

$$\sigma_1'(c_q^p \otimes \chi_{kl}^{ij} = 0,$$

or, equivalently,

$$x_{vu}^{ij} \sigma_1'(c_q^p \otimes c_k^u c_l^v) = x_{lk}^{\alpha j} \sigma_1'(c_q^p \otimes c_\alpha^i)$$

which is equivalent to

$$x_{uv}^{ij} \sigma_1'(c_\beta^p \otimes c_k^u)\sigma_1'(c_q^\beta \otimes c_l^v) = x_{lk}^{\alpha j} y_{q\alpha}^{pi}$$

and

$$x_{vu}^{ij} y_{\beta k}^{pu} y_{ql}^{\beta v} = x_{lk}^{\alpha j} y_{q\alpha}^{pi}$$

From Proposition 125, it follows that this is equivalent to

$$R^{23} S^{13} S^{12} = S^{12} R^{23}$$

Since $S = R^{-1}$, this last equation turns into

$$R^{12} R^{23} = R^{23} R^{12} R^{13} \tag{6.24}$$

Finally R is a bijective solution of the Hopf equation, i.e. $R^{12} R^{23} = R^{23} R^{13} R^{12}$, and we see that (6.24) is equivalent to $R^{12} R^{13} = R^{13} R^{12}$, completing the proof of the Theorem.

Remark 19. There is a major difference between part 2) of Theorem 68 and the corresponding statement for the quantum Yang-Baxter equation. In the QYBE case, the bijectivity of a solution R implies that the map $\sigma : A(R) \otimes A(R) \to k$ making $A(R)$ coquasitriangular is convolution invertible. Behind this is the observation that R is a solution of the QYBE if and only if R^{-1} is a solution of the QYBE. In the Hopf equation situation, things are complicated by the fact that a bijective R is a solution of the Hopf equation if and only if R^{-1} is a solution of the pentagon equation.

Now we introduce the dual notion of a bialgebra with a Hopf function (H, C, σ). This corresponds to the concept of quasitriangular bialgebra.

Definition 14. *Let H be a bialgebra and A a subalgebra of H. An element $R = R^1 \otimes R^2 \in A \otimes H$ is called a Hopf element if*
(HE1) $\Delta(R^1) \otimes R^2 = R^{13} R^{23}$
(HE2) $\varepsilon(R^1) R^2 = 1$
(HE3) $\Delta^{\mathrm{cop}}(a) R = R(1 \otimes a)$
for all $a \in A$. We will say that (H, A, R) is a bialgebra with a Hopf element.

Remarks 17. 1. (HE1) and (HE2) are exactly (QT1) and (QT2), up to the fact that $R \in A \otimes H$ in the Hopf case, while $R \in H \otimes H$ in the quasitriangular case. (HE3) is obtained by modifying the right hand side of (QT5), in such a way that an integral type condition appears. Let us explain this in detail. Take $t = R^1 \varepsilon(R^2) \in A$. (HE3) can be re written as

$$a_{(2)} R^1 \otimes a_{(1)} R^2 = R^1 \otimes R^2 a \tag{6.25}$$

for all $a \in A$. Applying $I \otimes \varepsilon$ to (6.25), we find

$$at = \varepsilon(a) t$$

for all $a \in A$. Hence, if A is a subbialgebra of H, then t is a left integral in A. Applying $I \otimes \varepsilon \otimes \varepsilon$ to (HE1) we get $t^2 = t$. It follows that $t = tt = \varepsilon(t) t$, hence $\varepsilon(t) = 1$. Using the Maschke theorem for Hopf algebras, we conclude that: if (H, A, R) is a bialgebra with a Hopf element and A is a finite dimensional subbialgebra of H with an antipode, then A is semisimple.

Conversely, if t is a left integral in A, then $R = t \otimes 1$ satisfies (HE3).

2. Let H be a bialgebra and A a subalgebra of H. Then $R = 1 \otimes 1$ is a Hopf element if and only if $A = k$.

Indeed, if $R = 1 \otimes 1$ then (HE3) takes the form $\Delta^{\mathrm{cop}}(a) = 1 \otimes a$, for all $a \in A$. Hence, $a = \varepsilon(a)1_H$, for all $a \in A$, i.e. $A = k$.

3. Let (H, A, R) be a bialgebra with a Hopf element. Suppose that H has an antipode S. Then R is invertible and $R^{-1} = S(R^1) \otimes R^2$. Moreover $u = S(R^2)R^1 \in H$ satisfies the condition

$$S(a)u = \varepsilon(a)u,$$

for all $a \in A$. This formula is obtained if we apply $m_H \tau (I \otimes S)$ to (6.25). We observe that if $A \neq k$ then u is not invertible (if u is invertible, then $R^{-1} = S(R^1) \otimes R^2 = \varepsilon(R^1) \otimes R^2 = 1 \otimes 1$, i.e. $A = k$).

Proposition 132. *If (H, A, R) is a bialgebra with a Hopf element, then*

$$R^{23} R^{13} R^{12} = R^{12} R^{23} \tag{6.26}$$

in $A \otimes H \otimes H$. If $(M, \cdot)$ is a left H-module, then the map

$$\mathcal{R} = \mathcal{R}_{(R,M,\cdot)} : \quad M \otimes M \to M \otimes M, \quad \mathcal{R}(m \otimes n) = R^1 \cdot m \otimes R^2 \cdot n$$

is a solution of the Hopf equation.

Proof. (HE1) is equivalent to

(HE1') $\quad \Delta^{\mathrm{cop}}(R^1) \otimes R^2 = R^{23} R^{13}$

Writing $r = R$, we compute

$$R^{23} R^{13} R^{12} = \left(\Delta^{\mathrm{cop}}(R^1) \otimes R^2 \right)(r^1 \otimes r^2 \otimes 1) = \Delta^{\mathrm{cop}}(R^1)R \otimes R^2$$
$$= R(1 \otimes R^1) \otimes R^2 = r^1 \otimes r^2 R^1 \otimes R^2 = R^{12} R^{23}$$

The second statement follows from (6.26) since

$$\mathcal{R}^{23} \mathcal{R}^{13} \mathcal{R}^{12}(l \otimes m \otimes n) = R^{23} R^{13} R^{12} \cdot (l \otimes m \otimes n)$$

and

$$\mathcal{R}^{12} \mathcal{R}^{23}(l \otimes m \otimes n) = R^{12} R^{23} \cdot (l \otimes m \otimes n)$$

for all $l, m, n \in M$.

6.3 New examples of noncommutative noncocommutative bialgebras

Using Theorem 66 we can construct new examples of noncommutative non-cocommutative bialgebras. These are different from the ones that appear in quantum group theory, and this is because the relations that we factor out are not always homogeneous.

If M is a vector space with basis $\{m_1, \cdots, m_n\}$, then the matrix of a k-linear map $R : M \otimes M \to M \otimes M$ is an $n^2 \times n^2$-matrix. We will write this matrix with respect to the basis $\{m_i \otimes m_j \mid i, j = 1, \cdots, n\}$, in the lexicographic order. For example (6.27) is the matrix of R with respect to the basis $\{m_1 \otimes m_1, m_1 \otimes m_2, m_2 \otimes m_1, m_2 \otimes m_2\}$.

Our first example is a solution of the Hopf equation in characteristic two, giving rise to a five dimensional noncommutative noncocommutative bialgebra $B(R)$.

Proposition 133. *Let k be a field and R be the matrix of $\mathcal{M}_4(k)$ given by*

$$R = \begin{pmatrix} 1 & 0 & 0 & 0 \\ 0 & 1 & 1 & 0 \\ 0 & 0 & 1 & 0 \\ 0 & 0 & 0 & 1 \end{pmatrix} \tag{6.27}$$

1. *R is a solution of the Hopf equation if and only if k has the characteristic two. In this case R is commutative.*
2. *If $\operatorname{char}(k) = 2$, then the bialgebra $B(R)$ is the algebra with generators x, y, z and relations*

$$x^2 = x, \quad y^2 = z^2 = yx = yz = 0, \quad xy = y, \quad xz = zx = z$$

The comultiplication Δ and the counit ε are given by:

$$\Delta(x) = x \otimes x + y \otimes z, \quad \Delta(y) = x \otimes y + y \otimes x + y \otimes zy,$$

$$\Delta(z) = z \otimes x + x \otimes z + zy \otimes z,$$

$$\varepsilon(x) = 1, \quad \varepsilon(y) = \varepsilon(z) = 0.$$

Furthermore, $\dim_k(B(R)) = 5$.

Proof. 1. A direct computation tells us that

$$R^{12}R^{23} = \begin{pmatrix} 1 & 0 & 0 & 0 & 0 & 0 & 0 & 0 \\ 0 & 1 & 1 & 0 & 0 & 0 & 0 & 0 \\ 0 & 0 & 1 & 0 & 1 & 0 & 0 & 0 \\ 0 & 0 & 0 & 1 & 0 & 1 & 1 & 0 \\ 0 & 0 & 0 & 0 & 1 & 0 & 0 & 0 \\ 0 & 0 & 0 & 0 & 0 & 1 & 1 & 0 \\ 0 & 0 & 0 & 0 & 0 & 0 & 1 & 0 \\ 0 & 0 & 0 & 0 & 0 & 0 & 0 & 1 \end{pmatrix}$$

and

$$R^{23}R^{13}R^{12} = \begin{pmatrix} 1 & 0 & 0 & 0 & 0 & 0 & 0 & 0 \\ 0 & 1 & 1 & 0 & \alpha & 0 & 0 & 0 \\ 0 & 0 & 1 & 0 & 1 & 0 & 0 & 0 \\ 0 & 0 & 0 & 1 & 0 & 1 & 1 & 0 \\ 0 & 0 & 0 & 0 & 1 & 0 & 0 & 0 \\ 0 & 0 & 0 & 0 & 0 & 1 & 1 & 0 \\ 0 & 0 & 0 & 0 & 0 & 0 & 1 & 0 \\ 0 & 0 & 0 & 0 & 0 & 0 & 0 & 1 \end{pmatrix}$$

where $\alpha = 1 + 1$. Hence, R is a solution of the Hopf equation if and only if $\operatorname{char}(k) = 2$. In this case we also have that

$$R^{12}R^{13} = R^{13}R^{12} = \begin{pmatrix} 1 & 0 & 0 & 0 & 0 & 0 & 0 & 0 \\ 0 & 1 & 0 & 0 & 1 & 0 & 0 & 0 \\ 0 & 0 & 1 & 0 & 1 & 0 & 0 & 0 \\ 0 & 0 & 0 & 1 & 0 & 1 & 1 & 0 \\ 0 & 0 & 0 & 0 & 1 & 0 & 0 & 0 \\ 0 & 0 & 0 & 0 & 0 & 1 & 0 & 0 \\ 0 & 0 & 0 & 0 & 0 & 0 & 1 & 0 \\ 0 & 0 & 0 & 0 & 0 & 0 & 0 & 1 \end{pmatrix}$$

i.e. R is commutative.

2. Suppose that $\operatorname{char}(k) = 2$. Among the scalars (x_{ij}^{kl}) that define $R = (x_{ij}^{kl})$ via the formula (6.11) the only nonzero elements are

$$x_{11}^{11} = x_{22}^{22} = x_{12}^{12} = x_{21}^{21} = x_{21}^{12} = 1.$$

Now, the relations $\chi_{kl}^{ij} = 0$, written in lexicographic order are:

$$c_1^1 c_1^1 = c_1^1, \quad c_1^1 c_2^1 = c_2^1, \quad c_2^1 c_1^1 = 0, \quad c_2^1 c_2^1 = 0,$$

$$c_1^2 c_1^1 + c_1^1 c_1^2 = 0, \quad c_2^1 c_2^1 + c_1^1 c_2^2 = c_1^1,$$

$$c_2^2 c_1^1 + c_2^1 c_1^2 = c_1^1, \quad c_2^2 c_2^1 + c_2^1 c_2^2 = c_2^1,$$

$$c_1^1 c_1^2 = c_1^2, \quad c_1^1 c_2^2 = c_2^2, \quad c_2^1 c_1^2 = 0, \quad c_2^1 c_2^2 = 0,$$

$$c_1^2 c_1^2 = 0, \quad c_1^2 c_2^2 = c_1^2, \quad c_2^2 c_1^2 = c_1^2, \quad c_2^2 c_2^2 = c_2^2.$$

Now, if we write $c_1^1 = x$, $c_2^1 = y$, $c_1^2 = z$, $c_2^2 = t$ and using that char$(k) = 2$ we get the following relations:

$$x^2 = x, \quad y^2 = z^2 = yx = yz = 0, \quad xy = y, \quad xz = zx = z,$$

$$zy = x + t, \quad t^2 = t, \quad xt = t, \quad tx = x,$$

$$yt = 0, \quad ty = y, \quad zt = tz = z.$$

So, t is in the free algebra generated by x, y, z and

$$t = zy - x.$$

If we substitute t in all the relations in which t is involved, then these become identities. The relations given in the statement of the proposition remain valid. The formula for Δ follows, as the matrix

$$\begin{pmatrix} x & y \\ z & t \end{pmatrix}$$

is comultiplicative.

We will now prove that $\dim_k(B(R)) = 5$. More precisely, we will show in an elementary way (without using the Diamond Lemma) that $\{1, x, y, z, zy\}$ is a k-basis for $B(R)$. From the relations which define $B(R)$ we obtain:

$$x(zy) = zy, \quad (zy)^2 = (zy)x = y(zy) = (zy)y = z(zy) = (zy)z = 0.$$

These relations tell us that $\{1, x, y, z, zy\}$ generate $B(R)$ as a vector space over k. We are done if we can show that $\{1, x, y, z, zy\}$ is linearly independent. Let a, b, c, d, $e \in k$ be such that

$$a + bx + cy + dz + e(zy) = 0$$

Left multiplication by y gives $a = 0$. Right multiplication by z gives $b = 0$, and then right multiplication by x gives $d = 0$. Finally left multiplication by z gives $c = 0$, and $e = 0$ follows automatically. Hence $\{1, x, y, z, zy\}$ is a k-basis for $B(R)$.

Remarks 18. 1. The proof also tells us that $B(R)$ can be described as follows:

– As a vector space, $B(R)$ is five dimensional with basis $\{1, x, y, z, t\}$.
– The multiplication is given by:

$$x^2 = x, \quad y^2 = z^2 = 0, \quad t^2 = t,$$

$$xy = y, \quad yx = 0, \quad xz = zx = z, \quad xt = t, \quad tx = x,$$

$$yz = 0, \quad zy = x + t, \quad yt = 0, \quad ty = y, \quad zt = tz = z.$$

– The comutiplcation Δ and the counity ε are defined in such way that the matrix

$$\begin{pmatrix} x & y \\ z & t \end{pmatrix}$$

is comultiplicative.

Now, let C be the four dimensional subcoalgebra of $B(R)$ with k-basis $\{x, y, z, t\}$. The map $\sigma : C \otimes B(R) \to k$ defined by

$$\sigma(x \otimes 1) = 1, \ \sigma(x \otimes x) = 1, \ \sigma(x \otimes y) = 0, \ \sigma(x \otimes z) = 0, \ \sigma(x \otimes t) = 1,$$
$$\sigma(y \otimes 1) = 0, \ \sigma(y \otimes x) = 0, \ \sigma(y \otimes y) = 0, \ \sigma(y \otimes z) = 1, \ \sigma(y \otimes t) = 0,$$
$$\sigma(z \otimes 1) = 0, \ \sigma(z \otimes x) = 0, \ \sigma(z \otimes y) = 0, \ \sigma(z \otimes z) = 1, \ \sigma(z \otimes t) = 0,$$
$$\sigma(t \otimes 1) = 1, \ \ \sigma(t \otimes x) = 1, \ \ \sigma(t \otimes y) = 0, \ \ \sigma(t \otimes z) = 0, \ \ \sigma(t \otimes t) = 0,$$

is a Hopf function. As R is bijective and commutative we obtain that σ is convolution invertible. Since k has characteristic two, $R^{-1} = R$, and $\sigma^{-1} = \sigma$. The bialgebra $B(R)$ is not a Hopf algebra. Localizing $B(R)$ we obtain a Hopf algebra isomorphic to the group algebra kC_2. Indeed, let S be a potential antipode. Then:

$$S(x)x + S(y)z = 1 \quad \text{and} \quad S(x)y + S(y)t = 0$$

If we multiply the second equation to the right with z we get $S(y)z = 0$, so $S(x)x = 1$. But $x^2 = x$, so $x = 1$ and then $y = 0$, $t = 1$. We obtain the Hopf algebra $k < z \mid z^2 = 0 >$, $\Delta(z) = z \otimes 1 + 1 \otimes z$, $\varepsilon(z) = 0$. If we denote $g = z + 1$ then $g^2 = 1$, $\Delta(g) = g \otimes g$, $\varepsilon(g) = 1$, hence $B(R)$ is isomorphic to the Hopf algebra kG, where $G = \{1, g\}$ is a group with two elements.

2. R is commutative, so we can construct the bialgebra

$$\overline{B}(R) = k[X, Z]/(X^2 - X, Z^2, XZ - Z)$$

Its coalgebra structure is given by

$$\Delta(X) = X \otimes X, \quad \Delta(Z) = X \otimes Z + Z \otimes X$$

$$\varepsilon(X) = 1, \quad \varepsilon(Z) = 0.$$

In the next Propositions we will construct the bialgebras that arise from the solutions of the Hopf equation given in Example 10 3). Taking quotients of one of these bialgebras, we will be able to construct a noncommutative noncocommutative bialgebra of dimension $2n + 1$, for any positive integer n and any field k.

First we will construct the bialgebra $B(R_q'')$, which can be associated to the solution $R_q'' = f_q \otimes f_q$, where f_q is defined in Example 10. It is worthwhile to note that these bialgebras do not depend on $q \in k \setminus \{0\}$.

Proposition 134. *Let $q \in k$ and R''_q the solution of the Hopf equation given by*

$$R''_q = \begin{pmatrix} 1 & q & q & q^2 \\ 0 & 0 & 0 & 0 \\ 0 & 0 & 0 & 0 \\ 0 & 0 & 0 & 0 \end{pmatrix}$$

Let $E^2_q(k)$ be the bialgebra $B(R''_q)$.

1. *If $q = 0$, then $E^2_0(k)$ is the free algebra with generators x, y, z and relations*

$$x^2 = x, \qquad xz = zx = z^2 = 0.$$

   The comultiplication Δ and the counit ε are given by

$$\Delta(x) = x \otimes x, \quad \Delta(y) = y \otimes y, \quad \Delta(z) = x \otimes z + z \otimes y$$

$$\varepsilon(x) = \varepsilon(y) = 1, \quad \varepsilon(z) = 0.$$

2. *If $q \neq 0$, then $E^2_q(k)$ is the free algebra with generators A, B and relations:*

$$B^3 = B^2$$

   The comultiplication Δ and the counity ε are given by:

$$\Delta(A) = A \otimes A, \quad \Delta(B) = B \otimes A + B^2 \otimes (B - A)$$

$$\varepsilon(A) = \varepsilon(B) = 1$$

Proof. We proceed as in Proposition 133. The (x^{ji}_{uv}) that are different from zero are:

$$x^{11}_{11} = 1, \quad x^{11}_{21} = x^{11}_{12} = q, \quad x^{11}_{22} = q^2.$$

Now the relations $\chi^{ij}_{kl} = 0$ are

$$c^1_1 c^1_1 + q c^2_1 c^1_1 + q c^1_1 c^2_1 + q^2 c^2_1 c^2_1 = c^1_1$$

$$c^1_1 c^1_2 + q c^2_1 c^1_2 + q c^1_1 c^2_2 + q^2 c^2_1 c^2_2 = q c^1_1$$

$$c^1_2 c^1_1 + q c^2_2 c^1_1 + q c^1_2 c^2_1 + q^2 c^2_2 c^2_1 = q c^1_1$$

$$c^1_2 c^1_2 + q c^2_2 c^1_2 + q c^1_2 c^2_2 + q^2 c^2_2 c^2_2 = q^2 c^1_1$$

$$0 = 0, \quad 0 = 0, \quad 0 = 0, \quad 0 = 0,$$

$$0 = c^2_1, \quad 0 = q c^2_1, \quad 0 = q c^2_1, \quad 0 = q^2 c^2_1,$$

$$0 = 0, \quad 0 = 0, \quad 0 = 0, \quad 0 = 0,$$

Hence $c^2_1 = 0$. If we denote $c^1_1 = x$, $c^2_2 = y$, $c^1_2 = z$ then we obtain the description of $E^2_q(k)$ as a bialgebra:

- as an algebra, $E_q^2(k)$ has generators x, y, z and relations

$$x^2 = x, \quad xz + qxy = qx, \quad zx + qyx = qx, \quad z^2 + qyz + qzy + q^2y^2 = q^2x.$$

- the comultiplication Δ and the counit ε are given by the equations

$$\Delta(x) = x \otimes x, \quad \Delta(y) = y \otimes y, \quad \Delta(z) = x \otimes z + z \otimes y \qquad (6.28)$$

and

$$\varepsilon(x) = \varepsilon(y) = 1, \quad \varepsilon(z) = 0. \qquad (6.29)$$

For $q = 0$, we obtain the relations for $E_0^2(k)$. If $q \neq 0$, then x is in the free algebra generated by y and z and

$$x = y^2 + q^{-1}zy + q^{-1}yz + q^{-2}z^2 = (y + q^{-1}z)^2.$$

Substituting x in the other three relations, the only remaining defining relation is

$$(y + q^{-1}z)^3 = (y + q^{-1}z)^2$$

The other two are linearly dependent on it. Writing $A = y$ and $B = y + q^{-1}z$, we obtain the desired description of $E_q^2(k)$.

Remarks 19. 1. Let C be the three dimensional subcoalgebra of $E_q^2(k)$ with k-basis $\{x, y, z\}$. Then

$$\sigma : C \otimes E_q^2(k) \to k$$

defined by

$$\sigma(x \otimes 1) = 1, \sigma(x \otimes x) = 1, \sigma(x \otimes y) = 0, \sigma(x \otimes z) = q,$$
$$\sigma(y \otimes 1) = 1, \sigma(y \otimes x) = 0, \sigma(y \otimes y) = 0, \sigma(y \otimes z) = 0,$$
$$\sigma(z \otimes 1) = 0, \sigma(z \otimes x) = q, \sigma(z \otimes y) = 0, \sigma(z \otimes z) = q^2.$$

is a Hopf function.

2. If $q \neq 0$, then B^2 is a grouplike element of $E_q^2(k)$, since

$$\begin{aligned}
\Delta(B^2) &= B^2 \otimes A^2 + B^2 \otimes (AB - A^2) + B^2 \otimes (BA - A^2) \\
&\quad + B^4 \otimes (B^2 - BA - AB + A^2) \\
&= B^2 \otimes A^2 + B^2 \otimes (AB - A^2) + B^2 \otimes (BA - A^2) \\
&\quad + B^2 \otimes (B^2 - BA - AB + A^2) \\
&= B^2 \otimes B^2
\end{aligned}$$

3. Let $n \geq 2$ be a positive integer. The two-sided ideal I of $E_0^2(k)$ generated by $y^n - y$, zy, $xy - x$ and $yx - x$ is a biideal (cf. [96]) and

$$B_{2n+1}(k) = E_0^2(k)/I$$

is a $2n + 1$-dimensional noncommutative noncocommutative bialgebra. The bialgebra $B_{2n+1}(k)$ can be described as follows:

- $B_{2n+1}(k)$ is algebra with generators x, y, z and relations

$$x^2 = x, \quad xz = zx = z^2 = zy = 0, \quad y^n = y, \quad xy = yx = x.$$

- The comultiplication Δ and the counity ε are given by

$$\Delta(x) = x \otimes x, \quad \Delta(y) = y \otimes y, \quad \Delta(z) = x \otimes z + z \otimes y$$

$$\varepsilon(x) = \varepsilon(y) = 1, \quad \varepsilon(z) = 0.$$

4. Observe that y does not appear in the relations of $E_0^2(k)$. As $\Delta(y - 1) = (y - 1) \otimes y + 1 \otimes (y - 1)$ and $\varepsilon(y - 1) = 1$, we get that the two-sided ideal generated by $y - 1$ is also a coideal. We can add the new relation $y = 1$ in the definition of $E_0^2(k)$ and we obtain a three dimensional noncocommutative bialgebra $\mathcal{T}(k)$. More explicitly:

- $\{1, x, z\}$ is a basis of $\mathcal{T}(k)$ as a vector space.
- The multiplication is given by

$$x^2 = x, \quad xz = zx = z^2 = 0.$$

- The comultiplication Δ and the counit ε are given by

$$\Delta(x) = x \otimes x, \quad \Delta(z) = x \otimes z + z \otimes 1, \quad \varepsilon(x) = 1, \quad \varepsilon(z) = 0.$$

In [103], Kaplansky gives two examples of three dimensional bialgebras over a field of characteristic zero, both of them are commutative and cocommutative. The only difference between our $\mathcal{T}(k)$ and one of Kaplansky's bialgebras is the relation $\Delta(z) = x \otimes z + z \otimes 1$ (in [103]: $\Delta(z) = 1 \otimes z + z \otimes 1$). This minor change of Δ (in our case k being a field of arbitrary characteristic) makes $\mathcal{T}(k)$ noncocommutative.

The proofs of Propositions 135 and 136 are left to the reader, since they are similar to the proofs of Propositions 133 and 134.

Proposition 135. *Let $q \in k$ and R_q the solution of the Hopf equation given by*

$$R_q = \begin{pmatrix} 0 & -q & 0 & -q^2 \\ 0 & 1 & 0 & q \\ 0 & 0 & 0 & 0 \\ 0 & 0 & 0 & 0 \end{pmatrix}$$

Let $B_q^2(k)$ be the bialgebra $B(R_q)$.

1. If $q = 0$, the bialgebra $B_0^2(k)$ has generators x, y, z and relations

$$yx = x, \quad yz = 0$$

The comultiplication Δ and the counity ε are given by

$$\Delta(x) = x \otimes x, \quad \Delta(y) = y \otimes y, \quad \Delta(z) = x \otimes z + z \otimes y$$

$$\varepsilon(x) = \varepsilon(y) = 1, \quad \varepsilon(z) = 0$$

2. *If $q \neq 0$, the bialgebra $B_q^2(k)$ has generators A, B and relations*

$$A^2 B = AB$$

The comultiplication Δ and the counit ε are given by:

$$\Delta(A) = A \otimes A, \quad \Delta(B) = q^{-1} AB \otimes B + (B - AB) \otimes A,$$

$$\varepsilon(A) = 1, \quad \varepsilon(B) = q.$$

Remark 20. $B_q^2(k)$ is not a Hopf algebra. We can localize it to obtain a Hopf algebra. As $\Delta(x) = x \otimes x$, $\Delta(y) = y \otimes y$ and $\varepsilon(x) = \varepsilon(y) = 1$ we should add new generators which make x and y invertible. But then $y = 1$ and $z = q(x - 1)$. It follows that the localization of the bialgebra $B_q^2(k)$, is the usual Hopf algebra $k[X, X^{-1}]$, with $\Delta(X) = X \otimes X$, $\varepsilon(X) = 1$, and antipode $S(X) = X^{-1}$.

Example 24. Let C be the three dimensional subcoalgebra of $B_0^2(k)$ with basis $\{x, y, z\}$. An easy but long and boring computation shows that $\sigma :$ $C \otimes B_0^2(k) \to k$ is a Hopf function if and only if there exists $a, b \in k$ such that

$$\sigma(x \otimes 1) = 1, \sigma(x \otimes x) = 0, \sigma(x \otimes y) = a, \sigma(x \otimes z) = b,$$
$$\sigma(y \otimes 1) = 1, \sigma(y \otimes x) = 0, \sigma(y \otimes y) = 0, \sigma(y \otimes z) = 0,$$
$$\sigma(z \otimes 1) = 0, \sigma(z \otimes x) = 0, \sigma(z \otimes y) = 0, \sigma(z \otimes z) = 0.$$

and $ab = 0$.

Proposition 136. *For $q \in k$, let R_q' be the solution of the Hopf equation given by*

$$R_q' = \begin{pmatrix} 1 & 0 & q & 0 \\ 0 & 1 & 0 & q \\ 0 & 0 & 0 & 0 \\ 0 & 0 & 0 & 0 \end{pmatrix}$$

Write $B(R_q') = D_q^2(k)$.
The bialgebra $D_0^2(k)$ has generators x, y, z and relations

$$x^2 = x = yx, \quad zx = xz = z^2 = yz = 0$$

The comultiplication Δ and the counit ε are given by:

$$\Delta(x) = x \otimes x, \quad \Delta(y) = y \otimes y, \quad \Delta(z) = x \otimes z + z \otimes y$$

$$\varepsilon(x) = \varepsilon(y) = 1, \quad \varepsilon(z) = 0.$$

For $q \neq 0$, $D_q^2(k)$ has generators A, B and relations

$$A^3 = A^2, \qquad BA = 0.$$

The comultiplication Δ and the counit ε are given by:

$$\Delta(A) = A \otimes A + q^{-1}(A^2 - A) \otimes B, \quad \Delta(B) = A^2 \otimes B + B \otimes A - q^{-1}B \otimes B,$$

$$\varepsilon(A) = 1, \quad \varepsilon(B) = 0.$$

Remarks 20. 1. R_q' is also a solution of the quantum Yang-Baxter equation. The bialgebra $A(R_q')$ from Theorem 60 has generators x, y, z, t and relations

$$zx = xz = zy = z^2 = zt = 0, \quad xy - yx = qyz, \quad xt - tx = qtz,$$

$$xy + qxt = qx^2, \quad y^2 + qyt = qxy, \quad ty + qt^2 = qxt.$$

The comultiplication Δ and the counit ε are such that the matrix

$$\begin{pmatrix} x & y \\ z & t \end{pmatrix}$$

is comultiplicative.

2. The bialgebra $D_0^2(k)$ is the quotient of $B_0^2(k)$ by the two-sided ideal (which is also a coideal) generated by

$$x^2 - x, \quad zx, \quad xz, \quad z^2.$$

$D_0^2(k)$ is also the quotient of $E_0^2(k)$ by the two-sided ideal generated by

$$yx - x, \quad yz.$$

3. Let $n \geq 2$ be a positive integer. The bialgebras $B_0^2(k)$, $D_0^2(k)$ and $E_0^2(k)$ constructed in the previous Propositions can be generalized to $B_0^n(k)$, $D_0^n(k)$ and $E_0^n(k)$. We will describe $B_0^n(k)$.

Let $\pi_1 : k^n \to k^n$ be the projection of k^n onto the first component, i.e. $\pi_1((x_1, x_2, \cdots, x_n)) = (x_1, 0, \cdots, 0)$, and let $\pi^1 = \mathrm{Id}_{k^n} - \pi_1$ be the projection of k^n onto the hyperplane $x_1 = 0$, that is $\pi^1((x_1, x_2, \cdots, x_n)) = (0, x_2, \cdots, x_n)$. Then $\pi_1 \otimes \pi^1$ is a solution of the Hopf equation and the bialgebra $B_0^n(k) = B(\pi_1 \otimes \pi^1)$ can be described as follows:

- $B_0^n(k)$ has generators $(c_j^i)_{i,j=1,\cdots,n}$ and relations

$$c_i^i = 0, \qquad c_k^j c_1^i = \delta_k^j \delta_1^l c_1^i$$

for all $i, j \geq 2$ and $k, l \geq 1$. As before, δ_v^u is the Kronecker symbol.

– The comultiplcation Δ and the counity ε are such that the matrix (c_i^j) is comultiplicative.

The proof is similar to the one of Proposition 134. Among the elements (x_{vu}^{ij}), which define $\pi_1 \otimes \pi^1$, the only nonzero ones

$$x_{1t}^{1t} = 1, \quad \forall t \geq 2.$$

If $i \neq 1$, all the relations $\chi_{kl}^{ij} = 0$ are $0 = 0$, with the exception of the relations $\chi_{j1}^{ij} = 0$ for all $j \geq 2$, which give us $0 = c_1^i$ for all $i \geq 2$. The relations $\chi_{kl}^{1j} = 0$ give us $c_k^j c_l^1 = \delta_k^j \delta_l^1 c_1^1$ for all $j \geq 2$ and $k, l \geq 1$.
Other new examples of bialgebras can be constructed starting from projections of k^n on different intersections of hyperplanes.

We end this section with one more example communicated to us by G. Mititica.

Example 25. Let n be a positive integer and k a field such that n is invertible in k. Let $R = (x_{ij}^{kl})$ given by

$$x_{ij}^{kl} = \begin{cases} 0, & \text{if } i + j \not\equiv k + l \pmod n \\ n^{-1}, & \text{if } i + j \equiv k + l \pmod n \end{cases} \tag{6.30}$$

It is easy to show that $R = (x_{ij}^{kl})$ is a solution of the Hopf equation.
For $n = 2$, the bialgebra algebra $B(R)$ from Theorem 66 is given as follows: $B(R)$ has generators x and y and relations

$$x^2 + y^2 = x, \quad xy + yx = y.$$

The comultiplication Δ and the counit ε are given by:

$$\Delta(x) = x \otimes x + y \otimes y, \quad \Delta(y) = x \otimes y + y \otimes x, \quad \varepsilon(x) = 1, \quad \varepsilon(y) = 0.$$

6.4 The pentagon equation versus the structure and the classification of finite dimensional Hopf algebras

In this section we will present a fundamental construction related to the pentagon equation that associate to any solution of the pentagon equation a finite dimensional Hopf algebra. This construction is originally due to Baaj and Skandalis (see [10]) in the case of unitary multiplicatives $R \in \mathcal{L}(K \otimes K)$, where K is a separable Hilbert space and to Davydov (see [65]) for vector spaces K over arbitrary fields. We will follow [137], leading us to the structure and the classification of finite dimensional Hopf algebras. A key role is played by the canonical element of the Heisenberg double of a Hopf algebra.
Let M be a finite dimensional vector space and

$$\varphi: \ M \otimes M^* \to \operatorname{End}(M), \quad \varphi(v \otimes v^*)(w) = \langle v^*, w \rangle v$$

the canonical isomorphism. The element $R = \varphi^{-1}(\operatorname{Id}_M)$ is called the *canonical element* of $M \otimes M^*$. Of course

$$R = \sum_{i=1}^{n} e_i \otimes e_i^*$$

where $\{e_i, e_i^* \mid i = 1, \cdots, n\}$ is a dual basis and R is independent of the choice of the dual basis.

Throughout this Section, A will be a finite dimensional algebra and $R \in A \otimes A$ will be an invertible solution of the pentagon equation

$$R^{12} R^{13} R^{23} = R^{23} R^{12} \tag{6.31}$$

For later use, we remark that if $R \in A \otimes A$ is a solution of the pentagon equation and $f: \ A \to B$ is an algebra map, then $(f \otimes f)(R)$ is a solution of the pentagon equation in $B \otimes B \otimes B$.

We can define the category $\underline{\operatorname{Pent}}$ of the *pentagon objects*: the objects are pairs (A, R), where A is a finite dimensional algebra and $R \in A \otimes A$ is an invertible solution of the pentagon equation. A morphism $f: (A, R) \to (B, T)$ between two pentagon objects (A, R) and (B, T) is an algebra map $f: \ A \to B$ such that $(f \otimes f)(R) = T$. $\underline{\operatorname{Pent}}$ is a monoidal category under the product $(A, R) \otimes (B, T) = (A \otimes B, R^{13} T^{24})$.

The following is the pentagon equation version of Proposition 126.

Proposition 137. *Let A be an algebra and $R \in A \otimes A$ an invertible solution of the pentagon equation. Consider the comultiplications $\Delta_l, \Delta_r : \ A \to A \otimes A$ given by*

$$\Delta_r(a) = R^{-1}(1_A \otimes a)R = \sum U^1 R^1 \otimes U^2 a R^2 \tag{6.32}$$

$$\Delta_l(a) = R(a \otimes 1_A)R^{-1} = \sum R^1 a U^1 \otimes R^2 U^2 \tag{6.33}$$

where $U = U^1 \otimes U^2 = R^{-1}$. Then $A_r = (A, \cdot, \Delta_r)$ and $A_l = (A, \cdot, \Delta_l)$ are bialgebras without counit.

Proof. It is obvious that Δ_r and Δ_l are algebra maps. For $a \in A$ we have

$$(\operatorname{Id} \otimes \Delta_r)\Delta_r(a) = (R^{23})^{-1}(R^{13})^{-1}(1_A \otimes 1_A \otimes a)R^{13} R^{23}$$

and

$$(\Delta_r \otimes \operatorname{Id})\Delta_r(a) = (R^{12})^{-1}(R^{23})^{-1}(1_A \otimes 1_A \otimes a)R^{23} R^{12}$$

so Δ_r is coassociative if and only if

$$R^{23} R^{12}(R^{23})^{-1}(R^{13})^{-1}(1_A \otimes 1_A \otimes a) = (1_A \otimes 1_A \otimes a)R^{23} R^{12}(R^{23})^{-1}(R^{13})^{-1}$$

using the pentagon equation (6.31), we find that this is equivalent to

$$R^{12}(1_A \otimes 1_A \otimes a) = (1_A \otimes 1_A \otimes a)R^{12}$$

and this equality holds for any $a \in A$. In a similar way we can prove that Δ_l is also coassociative.

It follows from Proposition 137 that we can put two different algebra structures (without unit) on A^*: the multiplications are the convolutions $*_l$ and $*_r$ which are the dual maps of Δ_l and Δ_r, i.e.

$$\langle \omega *_l \omega', a \rangle = \sum \langle \omega, R^1 a U^1 \rangle \langle \omega', R^2 U^2 \rangle$$

$$\langle \omega *_r \omega', a \rangle = \sum \langle \omega, U^1 R^1 \rangle \langle \omega', U^2 a R^2 \rangle$$

for all $\omega, \omega' \in A^*$, $a \in A$.

We have seen in Example 9 2) that the canonical element of the Drinfeld double is a solution of the QYBE. If the case of the pentagon equation, a similar result holds for the *Heisenberg double* $\mathcal{H}(L)$ of a Hopf algebra L (cf. Section 4.1).

Let L be a Hopf algebra. Recall that L^* is a left L-module algebra in the usual way

$$\langle h \cdot g^*, h' \rangle = \langle g^*, h'h \rangle$$

and the Heisenberg double is by definition the smash product

$$\mathcal{H}(L) = L \# L^*$$

The multiplication is given by

$$(h \# h^*)(g \# g^*) = h_{(2)} g \# h^* * (h_{(1)} \cdot g^*)$$

Recall also that the maps

$$i_L : L \to \mathcal{H}(L), \quad i_L(l) = l \# \varepsilon_L$$

and

$$i_{L^*} : L^* \to \mathcal{H}(L), \quad i_{L^*}(l^*) = 1_L \# l^*$$

are injective algebra maps. The Heisenberg double $\mathcal{H}(L)$ satisfies the following universal property: given a k-algebra A and algebra maps $u : L \to A$, $v : L^* \to A$ such that

$$u(l)v(l^*) = \sum v(l_{(1)} \cdot l^*)u(l_{(2)}) \tag{6.34}$$

there exists a unique algebra map $F : \mathcal{H}(L) \to A$ (given by $F(l \# l^*) = v(l^*)u(l)$, for all $l \in L$, $l^* \in L^*$) such that the following diagram commutes

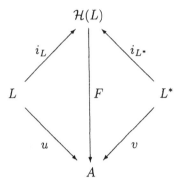

The Heisenberg double $\mathcal{H}(L)$ presented above, differs from $\mathcal{H}(L)'$ introduced in [140, Example 4.1.10], where $\mathcal{H}(L)' = L\#'L^*$, with multiplication given by

$$(h\#'h^*)(g\#'g^*) = \sum h\langle h_{(1)}^*, g_{(2)}\rangle g_{(1)}\#'h_{(2)}^*g^*$$

for all h, $g \in L$, h^*, $g^* \in L^*$. However, [140, Corollary 9.4.3] and Proposition 138 show that the two descriptions of the Heisenberg double $\mathcal{H}(L)$ and $\mathcal{H}(L)'$ are isomorphic as algebras, both of them being isomorphic to the matrix algebra $\mathcal{M}_n(k)$, where $n = \dim(L)$.

Proposition 138. *Let L be a finite dimensional Hopf algebra. Then there exists an algebra isomorphism*

$$\mathcal{H}(L) \cong \mathcal{M}_{\dim(L)}(k).$$

Proof. As L is finite dimensional, the functor

$$T : {}^L\mathcal{M}_L \to \mathcal{M}_{\mathcal{H}(L)}, \quad T(M) = M$$

where the right $\mathcal{H}(L)$-action on M is given by

$$m \bullet (l\#l^*) = \sum\langle l^*, m_{<-1>}\rangle m_{<0>} \cdot l$$

is an equivalence of categories (see Theorem 8). As the antipode of L is bijective ([172]) we have the following equivalences of categories

$$\mathcal{M}_{\mathcal{H}(L)} \cong {}^L\mathcal{M}_L \cong \mathcal{M}_k$$

i.e. $\mathcal{H}(L)$ is Morita equivalent to k. It follows from the Morita theory (see, for instance, [2], pag. 265) that there exists an algebra isomorphism $\mathcal{H}(L) \cong \mathcal{M}_n(k)$. Taking into account that $\dim(\mathcal{H}(L)) = \dim(L)^2$, we obtain that $n = \dim(L)$.

Remark 21. Let L be a finite dimensional Hopf algebra. Kashaev ([106]) proved that the Drinfeld double $D(L)$ can be realized as a subalgebra in the tensor product of two Heisenbergs $\mathcal{H}(L) \otimes \mathcal{H}(L^*)$. This can be proved immediately using Proposition 138: if $\dim(L) = n$ then $\dim(D(L)) = n^2$ and hence

$$D(L) \subset M_{n^2}(k) \cong M_n(k) \otimes M_n(k) \cong \mathcal{H}(L) \otimes \mathcal{H}(L) \cong \mathcal{H}(L) \otimes \mathcal{H}(L^*).$$

The following theorem is [181, Theorem 5.2] and [106, Theorem 1]. In [181], the Heisenberg double does not appear explicitly, and in [106] Heisengerg double is described in terms of structure constants, and not as a smash product.

Theorem 69. *Let L be a finite dimensional Hopf algebra and $\{e_i, e_i^* \mid i = 1, \cdots, n\}$ a dual basis of L. Then the canonical element*

$$\mathcal{R} = \sum_i (e_i \# \varepsilon) \otimes (1 \# e_i^*) \in \mathcal{H}(L) \otimes \mathcal{H}(L)$$

is an invertible solution of the pentagon equation in $\mathcal{H}(L) \otimes \mathcal{H}(L) \otimes \mathcal{H}(L)$. Consequently, if A is and algebra, and $f : \mathcal{H}(L) \to A$ an algebra map, then $(f \otimes f)(\mathcal{R})$ is an invertible solution of the pentagon equation in $A \otimes A \otimes A$.

Proof. Taking into account the multiplication rule of $\mathcal{H}(L)$ we find

$$\mathcal{R}^{23}\mathcal{R}^{12} = \sum_{i,j} (e_j \# \varepsilon) \otimes (e_{i_{(2)}} \# e_{i_{(1)}} \cdot e_j^*) \otimes (1 \# e_i^*)$$

and

$$\mathcal{R}^{12}\mathcal{R}^{13}\mathcal{R}^{23} = \sum_{a,b,c} (e_a e_b \# \varepsilon) \otimes (e_c \# e_a^*) \otimes (1 \# e_b^* * e_c^*)$$

so we have to prove the equality

$$\sum_{i,j} e_j \otimes e_{i_{(2)}} \otimes e_{i_{(1)}} \cdot e_j^* \otimes e_i^* = \sum_{a,b,c} e_a e_b \otimes e_c \otimes e_a^* \otimes e_b^* * e_c^* \qquad (6.35)$$

in $L \otimes L \otimes L^* \otimes L^*$. Fix indices $x, y, z, t \in \{1, \cdots, n\}$, and evaluate (6.35) at $e_x^* \otimes e_y^* \otimes e_z \otimes e_t$. (6.35) is then equivalent to

$$\langle e_y^*, e_{t_{(2)}} \rangle \langle e_x^*, e_z e_{t_{(1)}} \rangle = \sum_b \langle e_x^*, e_z e_b \rangle \langle e_b^* * e_y^*, e_t \rangle$$

Applying the definition of the convolution product

$$\langle e_b^* e_y^*, e_t \rangle = \langle e_b^*, e_{t_{(1)}} \rangle \langle e_y^*, e_{t_{(2)}} \rangle$$

and the dual basis formula

$$\sum_b e_b \langle e_b^*, e_{t_{(1)}} \rangle = e_{t_{(1)}}$$

we find that (6.35) holds, as needed. We will now prove that

$$U = \sum_i (S(e_i)\#\varepsilon) \otimes (1\#e_i^*)$$

is the inverse of R, where S is the antipode of L. As $\mathcal{H}(L)\otimes\mathcal{H}(L)$ is isomorphic to $n^2 \times n^2$-matrix algebra over k, it is enough to prove that $RU = 1\otimes 1$. Indeed,

$$RU = \sum_{i,j}(e_i S(e_j)\#\varepsilon) \otimes (1\#e_i^* e_j^*)$$

Hence, we have to prove the formula

$$\sum_{i,j} e_i S(e_j) \otimes e_i^* e_j^* = 1 \otimes \varepsilon$$

which holds, as for indices $x, y = 1, \cdots, n$ we have

$$\sum_{i,j}\langle e_x^*, e_i S(e_j)\rangle\langle e_i^* e_j^*, e_y\rangle = \sum_{i,j}\langle e_x^*, e_i S(e_j)\rangle\langle e_i^*, e_{y_{(1)}}\rangle\langle e_j^*, e_{y_{(2)}}\rangle$$

$$= \sum_{i,j}\langle e_x^*, e_i\langle e_i^*, e_{y_{(1)}}\rangle S(e_j\langle e_j^*, e_{y_{(2)}}\rangle)\rangle$$

$$= \sum\langle e_x^*, e_{y_{(1)}} S(e_{y_{(2)}})\rangle$$

$$= \langle e_x^*, 1\rangle\langle \varepsilon, e_y\rangle$$

From now let $R = \sum R^1 \otimes R^2 \in A \otimes A$ be an invertible solution of the pentagon equation, on a finite dimensional algebra A. The subspaces

$$A^{R,l} = \{a \in A \mid R(a\otimes 1_A) = a\otimes 1_A\} \text{ and } A^{R,r} = \{a \in A \mid (1_A\otimes a)R = 1_A\otimes a\}$$

are called the spaces of left, respectively right, R-invariants of A.

$$R_{(l)} = \{\sum\langle a^*, R^2\rangle R^1 \mid a^* \in A^*\} \text{ and } R_{(r)} = \{\sum\langle a^*, R^1\rangle R^2 \mid a^* \in A^*\}$$

are called the spaces of left, respectively right coefficients of R. We will denote them as follows

$$P = P(A, R) = R_{(l)} \; ; \; H = H(A, R) = R_{(r)}.$$

Assume now that $R = \sum_{i=1}^m a_i \otimes b_i$, where m is as small as possible. Then m is called the length of R and will be denoted $l(R) = m$. From the choice of m, the sets $\{a_i \mid i = 1, \cdots, m\}$, respectively $\{b_i \mid i = 1, \cdots, m\}$ are linear independent in A and hence bases of $R_{(l)}$, respectively $R_{(r)}$. In particular, $\dim(R_{(l)})=\dim(R_{(r)})=l(R)$. Two elements R and $S \in A\otimes A$ are called equivalent (we will write $R \sim S$) if there exists $u \in U(A)$ an invertible element of A such that $S = {}^u R := (u \otimes u)R(u \otimes u)^{-1}$. If $R \sim S$ then $l(R)=l(S)$. Indeed, let $R = \sum_{i=1}^m a_i \otimes b_i$ where $m = l(R)$. Then $S = \sum_{i=1}^m u a_i u^{-1} \otimes u b_i u^{-1}$ and hence $l(S) \leq l(R)$; in a similar way we obtain that $l(R) \leq l(S)$. In particular, if $\{a_i \mid i = 1, \cdots, m\}$ is a basis of $R_{(l)}$, then $\{ua_iu^{-1} \mid i = 1, \cdots, m\}$ is a basis of $S_{(l)} = {}^uR_{(l)}$.

Now consider $a_i^* \in P^*$ and $b_i^* \in H^*$ defined by

$$\langle a_i^*, a_j \rangle = \delta_{ij} = \langle b_i^*, b_j \rangle$$

i.e. $\{a_i, a_i^*\}$ and $\{b_i, b_i^*\}$ are dual bases of P and H. Extend $a_i^* : P \to k$ and $b_i^* : H \to k$ to respectively $\omega_i^* : A \to k$ and $\lambda_i^* : A \to k$. We then have

$$\sum_{i=1}^m \langle \omega_k, a_i \rangle b_i = b_k \quad \text{and} \quad \sum_{j=1}^m a_j \langle \lambda_k, b_j \rangle = a_k \qquad (6.36)$$

for all $k = 1, \cdots, m$. We will use two different notations for R:

$$R = \sum_{i=1}^m a_i \otimes b_i = \sum_{j=1}^m a_j \otimes b_j$$

when we are interested in the basis elements of P and H, and the generic notation

$$R = \sum R^1 \otimes R^2 = \sum r^1 \otimes r^2 = r \; ; \; U = \sum U^1 \otimes U^2 = R^{-1}$$

Theorem 70. *Let A be a finite dimensional algebra, $R = \sum R^1 \otimes R^2 \in A \otimes A$ an invertible solution of the pentagon equation and $P = P(A, R) = R_{(l)}$, $H = H(A, R) = R_{(r)}$ the subspaces of coefficients of R.*

1. *P and H are unitary subalgebras of A and Hopf algebras with comultiplication given by*

$$\Delta_P(x) = \Delta_r(x) = R^{-1}(1_A \otimes x)R \quad \text{and} \quad \Delta_H(y) = \Delta_l(y) = R(y \otimes 1_A)R^{-1} \qquad (6.37)$$

for all $x \in P$, $y \in H$. Furthermore, the k-linear map

$$f : P^* \to H, \quad f(p^*) = \sum \langle p^*, R^1 \rangle R^2 \qquad (6.38)$$

is an isomorphism of Hopf algebras.

2. *The k-linear map*

$$F : \mathcal{H}(P) \to A, \quad F(p \# p^*) = \sum \langle p^*, R^1 \rangle R^2 p$$

is an algebra map and

$$R = (F \otimes F)(\mathcal{R})$$

where $\mathcal{R} \in \mathcal{H}(P) \otimes \mathcal{H}(P)$ is the canonical element associate of the Heisenberg double.

3. *The multiplication on A defines isomorphisms*

$$A^{R,r} \otimes P \cong A \quad (\text{resp. } H \otimes A^{R,l} \cong A)$$

of right P-modules (resp. left H-modules). In particular, A is free as a right P-module and as a left H-module and

$$\dim(P) = \dim(H) = \frac{\dim(A)}{\dim(A^{R,l})} = \frac{\dim(A)}{\dim(A^{R,r})} = l(R)$$

4. *If $f : (A, R) \to (B, S)$ is an isomorphism in* <u>Pent</u>, *then the Hopf algebras $P(A, R)$ and $P(B, S)$ are isomorphic.*
Consequently, if $S \in A \otimes A$ is equivalent to R, then the Hopf algebras $P(A, R)$ and $P(A, S)$ are isomorphic.

Proof. 1. We will use the notation introduced above. First we will prove that P (resp. H) are unitary subalgebras in A and subcoalgebras of $A_r = (A, \Delta_r)$ (resp. $A_l = (A, \Delta_l)$). This will follow from the formulas:

$$a_p a_q = \sum_{j=1}^{m} \langle \lambda_p *_l \lambda_q, b_j \rangle a_j \in P \tag{6.39}$$

$$\Delta_r(a_p) = \sum_{i,j=1}^{m} \langle \lambda_p, b_i b_j \rangle a_i \otimes a_j \in P \otimes P \tag{6.40}$$

and

$$b_p b_q = \sum_{j=1}^{m} \langle \omega_p *_r \omega_q, a_j \rangle b_j \in H \tag{6.41}$$

$$\Delta_l(b_p) = \sum_{i,j=1}^{m} \langle \omega_p, a_i a_j \rangle b_i \otimes b_j \in H \otimes H \tag{6.42}$$

for all $p, q = 1, \cdots, m$. We prove (6.39) and (6.40), leaving (6.41) and (6.42) to the reader.

$$\sum_{j=1}^{m} a_j \langle \lambda_p *_l \lambda_q, b_j \rangle = \sum_{j=1}^{m} a_j \langle \lambda_p, R^1 b_j U^1 \rangle \langle \lambda_q, R^2 U^2 \rangle$$

$$= (\mathrm{Id} \otimes \lambda_p \otimes \lambda_q)(\sum_{j=1}^{m} a_j \otimes R^1 b_j U^1 \otimes R^2 U^2))$$

$$= (\mathrm{Id} \otimes \lambda_p \otimes \lambda_q)(R^{23} R^{12}(R^{23})^{-1})$$

$$(6.31) \quad = (\mathrm{Id} \otimes \lambda_p \otimes \lambda_q)(R^{12} R^{13})$$

$$= (\mathrm{Id} \otimes \lambda_p \otimes \lambda_q)(\sum_{j,k=1}^{m} a_j a_k \otimes b_j \otimes b_k)$$

$$= \sum_{j,k=1}^{m} a_j \langle \lambda_p, b_j \rangle a_k \langle \lambda_q, b_k \rangle = a_p a_q$$

i.e. P is a subalgebra of A. On the other hand

$$\Delta_r(a_p) = \Delta_r(\sum_{j=1}^{m} a_j \langle \lambda_p, b_j \rangle) \quad (6.36)$$

$$(6.32) \quad = \sum_{j=1}^{m} U^1 R^1 \otimes U^2 a_j R^2 \langle \lambda_p, b_j \rangle$$

$$= (\mathrm{Id} \otimes \mathrm{Id} \otimes \lambda_p)(\sum_{j=1}^{m} U^1 R^1 \otimes U^2 a_j R^2 \otimes b_j)$$

$$= (\mathrm{Id} \otimes \mathrm{Id} \otimes \lambda_p)((R^{12})^{-1} R^{23} R^{12})$$

$$(6.31) \quad = (\mathrm{Id} \otimes \mathrm{Id} \otimes \lambda_p)(R^{13} R^{23})$$

$$= (\mathrm{Id} \otimes \mathrm{Id} \otimes \lambda_p)(\sum_{i,j=1}^{m} a_i \otimes a_j \otimes b_i b_j)$$

$$= \sum_{i,j=1}^{m} a_i \otimes a_j \langle \lambda_p, b_i b_j \rangle$$

and P is subcoalgebra of (A, Δ_r). A similar computation yields (6.41) and (6.42), proving that H is a subalgebra of A and a subcoalgebra of (A, Δ_l). Moreover $R \in P \otimes H$ so, for any positive integer t, there exist scalars $\alpha_{ij} \in k$ such that

$$R^t = \sum_{i,j=1}^{m} \alpha_{ij} a_i \otimes b_j \quad (6.43)$$

We will prove now that $1_A \in P$ and $1_A \in H$. As A is finite dimensional, A can embeded into a matrix algebra $A \subset \mathcal{M}_n(k)$, where $n = \dim(A)$. We view

$$R \in A \otimes A \subset \mathcal{M}_n(k) \otimes \mathcal{M}_n(k) \cong \mathcal{M}_{n^2}(k)$$

R is invertible, and it follows from the Cayley-Hamilton Theorem that the identity matrix I_{n^2} can be written as a linear combination of powers of R. Using (6.43), we find

$$1_A \otimes 1_A = \sum_{i,j=1}^{m} \gamma_{i,j} a_i \otimes a_j$$

for some $\gamma_{i,j} \in k$. Hence in $\mathcal{M}_{n^2}(k)$, the identity matrix I_{n^2} can be representated as a linear combinations of powers of R. Hence, using (6.43), we obtain in $A \otimes A$ a linear combination

$$1_A \otimes 1_A = \sum_{i,j=1}^{m} \gamma_{i,j} a_i \otimes b_j$$

for some $\gamma_{i,j} \in k$. Let $p : A \to k$ be the projection of A onto the one dimensional subspace spanned by 1_A. Then

$$1_A = \sum_{i,j=1}^{m} a_i \langle p, \gamma_{i,j} b_j \rangle = \sum_{i,j=1}^{m} \langle p, \gamma_{i,j} a_i \rangle b_j \in P \cap H$$

i.e. P and H are unitary subalgebras of A and hence we can view $U = R^{-1} \in P \otimes H$.

The counit and the antipode of P and H are defined by the formulas:

$$\varepsilon_P : P \to k, \quad \varepsilon_P(a_k) = \langle b_k^*, 1_A \rangle, \quad S_P : P \to P, \quad S_P(a_k) = \sum U^1 \langle b_k^*, U^2 \rangle \tag{6.44}$$

and

$$\varepsilon_H : H \to k, \quad \varepsilon_H(b_k) = \langle a_k^*, 1_A \rangle, \quad S_H : H \to H, \quad S_H(b_k) = \sum \langle a_k^*, U^1 \rangle U^2 \tag{6.45}$$

for all $k = 1, \cdots, m$. We will prove that P is a Hopf algebra, the fact that H is a Hopf algebra is proved in a similar way. First, we remark that, as H is a subalgebra of A, (6.40) can be rewritten as

$$\Delta_r(a_p) = \sum_{i,j=1}^{m} \langle b_p^*, b_i b_j \rangle a_i \otimes a_j \tag{6.46}$$

Now, for $p = 1, \cdots, m$ we have

$$(\mathrm{Id} \otimes \varepsilon_P)\Delta_r(a_p) = \sum_{i,j=1}^{m} a_i \langle b_p^*, b_i b_j \rangle \langle b_j^*, 1_A \rangle \tag{6.46}$$

$$= \sum_{i,j=1}^{m} a_i \langle b_p^*, b_i b_j \langle b_j^*, 1_A \rangle \rangle = \sum_{i=1}^{m} a_i \langle b_p^*, b_i \rangle = a_p$$

i.e. $(I_P \otimes \varepsilon_P)\Delta_r = \mathrm{Id}$. A similar computation shows that $(\varepsilon_P \otimes I_P)\Delta_r = \mathrm{Id}$, and ε_P is a counit. Using (6.46), we find that S_P is a right convolution inverse of I_P:

$$(\mathrm{Id} \otimes S_P)\Delta_r(a_p) = \sum_{i,j=1}^{n} a_i \langle b_p^*, b_i b_j \rangle U^1 \langle b_j^*, U^2 \rangle$$

$$= \sum_{i,j=1}^{n} a_i U^1 \langle b_p^*, b_i b_j \langle b_j^*, U^2 \rangle \rangle$$

$$= \sum_{i=1}^{n} a_i U^1 \langle b_p^*, b_i U^2 \rangle = (\mathrm{Id} \otimes b_p^*)(RR^{-1})$$

$$= 1_A \langle b_p^*, 1_A \rangle = \varepsilon_P(a_p) 1_A$$

From the fact that P is finite dimensional, it follows that S_P is an antipode of P.

Let us prove now that $f : P^* \to H$ is an isomorphism of Hopf algebras.

$$f(a_j^*) = \sum_{i=1}^{m} \langle a_j^*, a_i \rangle b_i = b_j$$

so f is an isomorphism of vector spaces. Let us prove that f is an algebra map:

$$f(1_{P^*}) = f(\varepsilon_P) = \sum_{i=1}^{m} \langle \varepsilon_P, a_i \rangle b_i = \sum_{i=1}^{m} \langle b_i^*, 1_A \rangle b_i = 1_A$$

(6.41) can be rewritten as

$$b_p b_q = \sum_{j=1}^{m} \langle a_p^* *_r a_q^*, a_j \rangle b_j$$

which means that

$$f(a_p^*) f(a_q^*) = f(a_p^* *_r a_q^*)$$

for all $p, q = 1, \cdots, m$, i.e. f is an algebra isomorphism. Let us prove now that f is also a coalgebra map. We recall the definition of the comultiplication Δ_{P^*}:

$$\Delta_{P^*}(a_p^*) = \sum X^1 \otimes X^2 \in P^* \otimes P^*$$

if and only if

$$\langle a_p^*, xy \rangle = \sum \langle X^1, x \rangle \langle X^2, y \rangle$$

for all $x, y \in P$. It follows that

$$(f \otimes f)\Delta_{P^*}(a_p^*) = \sum f(X^1) \otimes f(X^2) = \sum \langle X^1, R^1 \rangle R^2 \otimes \langle X^2, r^1 \rangle r^2$$

$$= \sum \langle a_p^*, R^1 r^1 \rangle R^2 \otimes r^2 = \sum_{i,j=1}^{m} \langle a_p^*, a_i a_j \rangle b_i \otimes b_j$$

$$(6.42) \qquad = \Delta_H(b_p) = (\Delta_H \circ f)(a_p^*)$$

i.e. f is also a coalgebra map. Hence, we have proved that f is an isomorphism of bialgebras, and as P and H are finite dimensional it is also a isomorphism of Hopf algebras (see [172]).

2. We remark that

$$F(a_i \# a_j^*) = \sum_{t=1}^{m} \langle a_j^*, a_t \rangle b_t a_i = b_j a_i$$

for all $i, j = 1, \cdots, m$. The fact that F is an algebra map can be proved directly using this formula; another way to proceed is to use the universal property of the Heisenberg double $\mathcal{H}(P)$ for the diagram

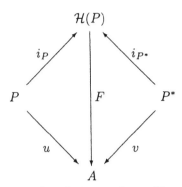

Here $u : P \to A$ is the usual inclusion and $v : P^* \to A$ is the composition
$v = f \circ j$, where $f : P^* \to H$ is the isomorphism from part 1) and $j : H \to A$
is the usual inclusion. We only have to prove that the compatibility condition
(6.34) holds, i.e.

$$hv(g^*) = \sum v(h_{(1)} \cdot g^*)h_{(2)}$$

for any $h \in P$ and $g^* \in P^*$, which turns out to be

$$\sum h\langle g^*, R^1\rangle R^2 = \sum \langle g^*, R^1 h_{(1)}\rangle R^2 h_{(2)}$$

or, equivalently

$$\sum R^1 \otimes hR^2 = \sum R^1 h_{(1)} \otimes R^2 h_{(2)}.$$

This equation holds, as $\Delta_P(h) = R^{-1}(1_H \otimes h)R$, for any $h \in P$.
Now let $\mathcal{R} = \sum_{i=1}^{m}(a_i \# \varepsilon_P) \otimes (1_A \# a_i^*)$ be the canonical element of $\mathcal{H}(P) \otimes \mathcal{H}(P)$. Then

$$(F \otimes F)(\mathcal{R}) = \sum_{i,t=1}^{m} \langle \varepsilon_P, a_t\rangle b_t a_i \otimes b_i = \sum_{i,t=1}^{m} \langle b_t^*, 1_A\rangle b_t a_i \otimes b_i = \sum_{i=1}^{m} a_i \otimes b_i = R.$$

3. Consider the map

$$\rho = \rho_P : A \to P \otimes A, \quad \rho(a) = (1_A \otimes a)R = \sum R^1 \otimes aR^2 = \sum_{i=1}^{m} a_i \otimes ab_i$$

for all $a \in A$. We will show that $(A, \cdot, \rho_P) \in {}^P\mathcal{M}_P$ is a right-left P-Hopf
module, where the structure of right P-module is just the multiplication $\cdot$ of
A. Indeed, for $a \in A$ we have

$$(\mathrm{Id} \otimes \rho)\rho(a) = \sum R^1 \otimes \rho(aR^2)$$

$$= \sum R^1 \otimes r^1 \otimes aR^2 r^2 = (1_A \otimes 1_A \otimes a)R^{13}R^{23}$$

$$(6.31) \quad = (1_A \otimes 1_A \otimes a)(R^{12})^{-1}R^{23}R^{12} = \sum U^1 r^1 \otimes U^2 R^1 r^2 \otimes aR^2$$

$$= \sum R^{-1}(1_A \otimes R^1)R \otimes aR^2 = \sum \Delta_P(R^1) \otimes aR^2$$

$$= (\Delta_P \otimes \mathrm{Id})\rho(a)$$

and

$$\sum_{i=1}^{m} \langle \varepsilon_P, a_i \rangle ab_i = \sum_{i=1}^{m} \langle b_i^*, 1_A \rangle ab_i = a$$

so (A, ρ) is a left P-comodule. We still have to prove the compatibility relation

$$\rho(a)\Delta_P(a_i) = (1_A \otimes a)RR^{-1}(1_A \otimes a_i)R = (1_A \otimes aa_i)R = \rho(aa_i)$$

for all $i = 1, \cdots, m$. Hence, $(A, \cdot, \rho_P) \in {}^P\mathcal{M}_P$ and the coinvariants

$$A^{\mathrm{co}(P)} = \{a \in A \mid \rho(a) = 1 \otimes a\} = A^{R,r}$$

the right R-invariants of A. From the right-left version of the Fundamental Theorem of Hopf modules it follows that the multiplication on A,

$$\mu : \ A^{R,r} \otimes P \to A, \quad \mu(a \otimes x) = ax$$

defines an isomorphism of P-Hopf modules, and, in particular, of right P-modules. We recall that $A^{R,r} \otimes P$ is a right P-module via $(a \otimes x) \cdot y = a \otimes xy$, for all $a \in A^{R,r}$, x, $y \in P$. It follows that A is free as a right P-module and

$$\dim(A) = \dim(P)\dim(A^{R,r}).$$

In a similar way we can show that $(A, \cdot, \rho_H) \in {}_H\mathcal{M}^H$, where $\cdot$ is the multiplication on A and

$$\rho_H : \ A \to A \otimes H, \quad \rho_H(a) = R(a \otimes 1_A) = \sum R^1 a \otimes R^2$$

for all $a \in A$. Moreover, $A^{\mathrm{co}(H)} = A^{R,l}$. If we apply once again the fundamental Theorem of Hopf modules (this time the left-right version) we obtain the other part of the statement.

4. Let $f : A \to B$ is an algebra isomorphism such that $S = (f \otimes f)(R) = \sum_{i=1}^{n} f(a_i) \otimes f(b_i)$. Then $S^{-1} = \sum f(U^1) \otimes f(U^2)$. It follows that $\{f(a_i) \mid i = 1, \cdots, n\}$ is a basis of $P(B, S)$ and hence the restriction of f to $P(A, R)$ gives an algebra isomorphism between $P(A, R)$ and $P(B, S)$ that is also a coalgebra map since

$$(f \otimes f)\Delta_{P(A,R)}(a_i) = \sum f(U^1)f(R^1) \otimes f(U^2)f(a_i)f(R^2)$$
$$= S^{-1}(1 \otimes f(a_i))S = \Delta_{P(B,S)}(f(a_i))$$

for all $i = 1, \cdots, n$. The last statement is obtain taking $B = A$ and $f : A \to A$, $f(x) = uxu^{-1}$ for all $x \in A$.

Using Theorems 69 and 70, we obtain the following Corollary, which is a pure algebraic version of [10, Theorem 4.7]: the role of the operator V_G on a local compact group G is played by the canonical element of the Heisenberg double.

Corollary 38. *Let A be a finite dimensional algebra and $R \in A \otimes A$ an invertible element. Then R is a solution of the pentagon equation if and only if there exists a finite dimensional Hopf algebra L and an algebra map $F : \mathcal{H}(L) \to A$ such that $R = (F \otimes F)(\mathcal{R})$, where $\mathcal{R}$ is the canonical element associated to the Heisenberg double $\mathcal{H}(L)$.*

Remarks 21. 1. Part 3. of Theorem 70 is a Lagrange type theorem, useful to evaluate the dimension of the Hopf algebra $P(A, R)$ coming from a solution of the pentagon equation. Let $(A, R) \in \underline{\text{Pent}}$. It follows from Corollary 38 and Proposition 138 that there exists an algebra map $F : \mathcal{M}_n(k) \to A$, where $n = l(R)$. As $\mathcal{M}_n(k)$ is a simple algebra, F is injective. Hence $a_{ij} = F(e_i^j) \neq 0 \in A$, $i, j = 1, \cdots, n$; then $a_{ij}a_{kl} = \delta_{jk}a_{il}$ and $1_A = \sum_{i=1}^{n} a_{ii}$. It follows from the Reconstruction Theorem of the matrix algebra ([113, Theorem 17.5]) that there exists an algebra isomorphism

$$A \cong \mathcal{M}_n(B), \text{ where } B = \{x \in A \mid xa_{ij} = a_{ij}x, \; \forall i, j = 1, \cdots, n\}.$$

Hence A is a matrix algebra if R is non-trivial ($l(R) > 1$ or, equivalently, $R \neq 1_A \otimes 1_A$). Furthermore, $\dim(A) = n^2 \dim(B)$ and hence, $l(R)^2 | \dim(A)$.
2. We can compute the space of integrals on P, (resp. on H); we have to use the space $A^{R,r}$ of right R-coinvariants of A (resp. $A^{R,l}$). Let $a \in A^{R,r}$ and $\chi : A \to H$ be an arbitrary right H-linear map. Then

$$\varphi : P \to k, \quad \varphi(a_i) = \langle b_i^*, \chi(a) \rangle$$

is a right integral on P. Indeed, $a \in A^{R,r}$, hence $a \otimes 1_A = \sum_{i=1}^{n} ab_i \otimes a_i$. As χ is right H-linear we obtain

$$\chi(a) \otimes 1_A = \sum_{i=1}^{n} \chi(a)b_i \otimes a_i \tag{6.47}$$

Now, for $a_p \in P$ we have

$$\sum \varphi((a_p)_{(1)}) \otimes (a_p)_{(2)} = \sum_{i,j} \langle b_p^*, b_i b_j \rangle \langle b_i^*, \chi(a) \rangle \otimes a_j$$

$$= \sum_j \langle b_p^*, \chi(a)b_j \rangle \otimes a_j$$

$$\overset{(6.47)}{=} \langle b_p^*, \chi(a) \rangle \otimes 1_A = \varphi(a_p) \otimes 1_A$$

which shows that φ is right P-colinear i.e. a right integral on P. Similarly, if $b \in A^{R,l}$ and $\psi : A \to P$ is an arbitrary left P-linear map,

$$\gamma : H \to k, \quad \gamma(b_i) = \langle a_i^*, \chi(b) \rangle$$

is a left integral on H.

Theorem 71. *Let L be a finite dimensional Hopf algebra. Then there exists an isomorphism of Hopf algebras*

$$L \cong P(\mathcal{H}(L), \mathcal{R})$$

where $\mathcal{R}$ is the canonical element of the Heisenberg double $\mathcal{H}(L)$.

Proof. Let $\{e_i \mid i = 1, \cdots, m\}$ be a basis of L, $\{e_i^* \mid i = 1, \cdots, m\}$ the dual basis of L^*, and

$$\mathcal{R} = \sum_{i=1}^{m} (e_i \# \varepsilon_L) \otimes (1_L \# e_i^*) \in \mathcal{H}(L) \otimes \mathcal{H}(L)$$

the canonical element. We have to prove that the Hopf algebra $P(\mathcal{H}(L), \mathcal{R})$ extracted from part 1. of Theorem 70 is isomorphic to L, with the initial structure of Hopf algebra. Of course,

$$i_L : L \to \mathcal{H}(L), \quad i_L(l) = l \# \varepsilon_L$$

is an injective algebra map. We identify

$$L \cong \text{Im}(i_L) = L \# \varepsilon_L$$

From the construction, $P(\mathcal{H}(L), \mathcal{R})$ is the subalgebra of $\mathcal{H}(L)$ having $\{e_i \# \varepsilon_L \mid i = 1, \cdots, m\}$ as a basis; i.e. there exists an algebra isomorphism $L \cong \text{Im}(i_L) = P(\mathcal{H}(L), \mathcal{R})$. It remains to prove that the coalgebra structure (resp. the antipode) of $P(\mathcal{H}(L), \mathcal{R})$ extracted from Theorem 70 is exactly the original coalgebra structure (resp. the antipode) of L. As the counit and the antipode of a Hopf algebra are uniquely determined by the multiplication and the comultiplication, the only thing left to show is the fact that, via the above identification, $\Delta_P = \Delta_L$. This means that

$$\Delta_L(e_i \# \varepsilon_L) = \mathcal{R}^{-1}(1_{\mathcal{H}(L)} \otimes e_i \# \varepsilon_L)\mathcal{R}$$

or, equivalently,

$$\mathcal{R}\Delta_L(e_i \# \varepsilon_L) = \Big((1_L \# \varepsilon_L) \otimes (e_i \# \varepsilon_L) \Big)\mathcal{R}$$

Now we compute

$$\Big((1_L \# \varepsilon_L) \otimes (e_i \# \varepsilon_L) \Big)\mathcal{R} = \Big((1_L \# \varepsilon_L) \otimes (e_i \# \varepsilon_L) \Big) \Big(\sum_{j=1}^{m} (e_j \# \varepsilon_L) \otimes (1_L \# e_j^*) \Big)$$

$$= \sum_{j=1}^{m} (e_j \# \varepsilon_L) \otimes (e_{i_{(2)}} \# e_{i_{(1)}} \cdot e_j^*)$$

On the other hand

$$\mathcal{R}\Delta_L(e_i \# \varepsilon_L) = \sum_{j=1}^{m} \Big((e_j \# \varepsilon_L) \otimes (1_L \# e_j^*) \Big) \Big((e_{i_{(1)}} \# \varepsilon_L) \otimes (e_{i_{(2)}} \# \varepsilon_L) \Big)$$

$$= \sum_{j=1}^{m} (e_j e_{i_{(1)}} \# \varepsilon_L) \otimes (e_{i_{(2)}} \# e_j^*)$$

Hence, we have to show

$$\sum_{j=1}^{m} e_j e_{i_{(1)}} \otimes e_{i_{(2)}} \otimes e_j^* = \sum_{j=1}^{m} e_j \otimes e_{i_{(2)}} \otimes e_{i_{(1)}} \cdot e_j^* \qquad (6.48)$$

For indices $a, b, k \in \{1, \cdots, m\}$, evaluate (6.48) at $e_a^* \otimes e_b^* \otimes e_k$. (6.48) is then equivalent to

$$\langle e_a^*, e_k e_{i_{(1)}} \rangle \langle e_b^*, e_{i_{(2)}} \rangle = \sum_{j=1}^{m} \langle e_a^*, e_j \rangle \langle e_b^*, e_{i_{(2)}} \rangle \langle e_{i_{(1)}} \cdot e_j^*, e_k \rangle$$

and this is easily verified:

$$\sum_{j=1}^{m} \langle e_a^*, e_j \rangle \langle e_b^*, e_{i_{(2)}} \rangle \langle e_{i_{(1)}} \cdot e_j^*, e_k \rangle = \sum_{j=1}^{m} \langle e_a^*, e_j \rangle \langle e_b^*, e_{i_{(2)}} \rangle \langle e_j^*, e_k e_{i_{(1)}} \rangle$$

$$= \sum_{j=1}^{m} \langle e_a^*, e_j \langle e_j^*, e_k e_{i_{(1)}} \rangle \rangle \langle e_b^*, e_{i_{(2)}} \rangle$$

$$= \langle e_a^*, e_k e_{i_{(1)}} \rangle \langle e_b^*, e_{i_{(2)}} \rangle$$

It follows that $\Delta_L = \Delta_P$ and $L \cong P(\mathcal{H}(L), \mathcal{R})$ as Hopf algebras.

Let L be a finite dimensional Hopf algebra. Proposition 138 proves that there exists an algebra isomorphism $\mathcal{H}(L) \cong \mathcal{M}_n(k)$, where $n = \dim(L)$. Via this isomorphism the canonical element $\mathcal{R} \in \mathcal{H}(L) \otimes \mathcal{H}(L)$ is viewed as an element of $\mathcal{M}_n(k) \otimes \mathcal{M}_n(k)$, or as a matrix of $\mathcal{M}_{n^2}(k)$.

We will now give the data which show us how any finite dimensional Hopf algebra is constructed. Let $R = \sum_{i=1}^{m} A_i \otimes B_i \in \mathcal{M}_n(k) \otimes \mathcal{M}_n(k)$ be an invertible solution of the pentagon equation such that the sets of matrices $\{A_i \mid i = 1, \cdots m\}$ and $\{B_i \mid i = 1, \cdots m\}$ are linearly independent over k. Let $\{B_i^* \mid i = 1, \cdots m\}$ be the dual basis of $\{B_i \mid i = 1, \cdots m\}$ and write $U = R^{-1} = \sum U^1 \otimes U^1$.

The Hopf algebra $P(\mathcal{M}_n(k), R)$ is described as follows:

– as an algebra $P(\mathcal{M}_n(k), R)$ is the subalgebra of the $n \times n$-matrix algebra $\mathcal{M}_n(k)$ with $\{A_i \mid i = 1, \cdots m\}$ as a k-basis;
– the coalgebra structure and the antipode of $P(\mathcal{M}_n(k), R)$ are given by the following formulas:

$$\Delta : P(\mathcal{M}_n(k), R) \to P(\mathcal{M}_n(k), R) \otimes P(\mathcal{M}_n(k), R),$$
$$\Delta(A_i) = R^{-1}(I_n \otimes A_i)R \tag{6.49}$$

$$\varepsilon : P(\mathcal{M}_n(k), R) \to k, \quad \varepsilon(A_i) = \langle B_i^*, I_n \rangle \tag{6.50}$$

$$S : P(\mathcal{M}_n(k), R) \to P(\mathcal{M}_n(k), R), \quad S(A_i) = \sum \langle B_i^*, U^2 \rangle U^1 \tag{6.51}$$

for all $i = 1, \cdots, m$.

Theorems 70 and 71 imply the following Structure Theorem for finite dimensional Hopf algebras.

Theorem 72. *L is a finite dimensional Hopf algebra if and only if there exist a positive integer n and an invertible solution of the pentagon equation $R \in \mathcal{M}_n(k) \otimes \mathcal{M}_n(k) \cong \mathcal{M}_{n^2}(k)$ such that $L \cong P(\mathcal{M}_n(k), R)$. Furthermore,*

$$\dim(L) = l(R) = \frac{n^2}{\dim(\mathcal{M}_n(k)^{R,r})}$$

where $\mathcal{M}_n(k)^{R,r}$ is the subspace of right R-invariants of $\mathcal{M}_n(k)$.

Remark 22. Let L be a Hopf algebra with a comultiplication Δ and $R \in L \otimes L$ an invertible element. On the algebra L, Drinfeld ([75]) introduced a new comultiplication Δ_R given by

$$\Delta_R(l) = R^{-1}\Delta(l)R$$

for all $l \in L$. Let $L_R := L$ as an algebra and having Δ_R as a comultiplication. If L_R is a structure of Hopf algebra it is called a *twist* of L. It was proved in [75] that if R is a *Harrison cocycle*[1], i.e.

$$(\Delta \otimes Id)(R)(R \otimes 1) = (Id \otimes \Delta)(R)(1 \otimes R)$$
$$(\varepsilon \otimes Id)(R) = (Id \otimes \varepsilon)(R) = 1 \tag{6.52}$$

then L_R is a Hopf algebra, i.e. a twist of L. The twist construction plays a crucial role in the theory of finite dimensional triangular semisimple Hopf algebras classification ([82]).

Let $\mathcal{M}_n(k)$ be the matrix algebra having the trivial bialgebra structure (without counit)

$$\Delta : \mathcal{M}_n(k) \to \mathcal{M}_n(k) \otimes \mathcal{M}_n(k), \quad \Delta(x) = I_n \otimes x$$

for all $x \in \mathcal{M}_n(k)$. Any subalgebra of $\mathcal{M}_n(k)$ is a subbialgebra. Theorem 72 and the comultiplication (6.37) show that any finite dimensional Hopf algebra L, viewed as a subalgebra of the matrix algebra, is obtained as a twist in the sense of Drinfeld: the trivial bialgebra structure of $\mathcal{M}_n(k)$ is twisted by an invertible element R. An important difference with the previous situation is that R is not a Harrison cocycle in the sense of (6.52): R is not a solution of the equation $R^{23}R^{12} = R^{13}R^{23}$) but a solution of the pentagon equation.

---

[1] The fact that (6.52) is a Harrison cocycle condition is shown in [37]: in the literature (see e.g. [82]), (6.52) called the twist equation.

Let n be a positive integer. We have proved that an n-dimensional Hopf algebra L is isomorphic to a $P(n, R)$ for $R \in M_n(k) \otimes M_n(k)$ an invertible solution of the pentagon equation such that $l(R) = n$. We are now going to prove the Classification Theorem for finite dimensional Hopf algebras. Let $\underline{\text{Pent}}_n$ be the set

$$\underline{\text{Pent}}_n = \{R \in M_n(k) \otimes M_n(k) \mid (M_n(k), R) \in \underline{\text{Pent}} \text{ and } l(R) = n\}.$$

Theorem 73. *Let n be a positive integer. Then there exists a one to one correspondence between the set of types of n-dimensional Hopf algebras and the set of the orbits of the action*

$$GL_n(k) \times \underline{\text{Pent}}_n \to \underline{\text{Pent}}_n, \quad (u, R) \to (u \otimes u)R(u \otimes u)^{-1}. \tag{6.53}$$

Proof. In part 5. of the Theorem 70 we proved that there exists a Hopf algebra isomorphism $P(M_n(k), R) \cong P(M_n(k), {}^uR)$ for any $u \in GL_n(k)$, which means that all the Hopf algebras associated to the elements of an orbit of the action (6.53) are isomorphic. We will now prove the converse. First we show that two finite dimensional Hopf algebras L_1 and L_2 are isomorphic if and only if $(\mathcal{H}(L_1), \mathcal{R}_{L_1})$ and $(\mathcal{H}(L_2), \mathcal{R}_{L_2})$ are isomorphic as objects in $\underline{\text{Pent}}$. Indeed, let $f : L_1 \to L_2$ be a Hopf algebra isomorphism. Then, $f^* : L_2^* \to L_1^*$, $f^*(l^*) = l^* \circ f$ is an isomorphism of Hopf algebras and

$$\tilde{f} : \mathcal{H}(L_1) \to \mathcal{H}(L_2), \quad \tilde{f}(h\#h^*) := f(h)\#(f^*)^{-1}(h^*) = f(h)\#h^* \circ f^{-1}$$

for all $h \in L_1$, $h^* \in L_1^*$ is an algebra isomorphism. Indeed, let $h, g \in L_1$ and $h^*, g^* \in L_1^*$; using the fact that f is an algebra map, we have

$$\tilde{f}\Big((h\#h^*)(g\#g^*)\Big) = \sum f(h_{(2)})f(g)\#\Big(h^*(h_{(1)} \cdot g^*)\Big) \circ f^{-1}$$

and as f is a coalgebra map

$$\tilde{f}(h\#h^*)\tilde{f}(g\#g^*) = \sum f(h_{(2)})f(g)\#\Big(h^* \circ f^{-1}\Big)\Big(f(h_{(1)}) \cdot (g^* \circ f^{-1})\Big)$$

It follows that $\tilde{f}$ is an algebra map, since for any $l \in L_2$ we have

$$\langle \Big(h^* \circ f^{-1}\Big)\Big(f(h_{(1)}) \cdot (g^* \circ f^{-1})\Big), l \rangle$$

$$= \sum \langle h^* \circ f^{-1}, l_{(1)} \rangle \langle g^* \circ f^{-1}, l_{(2)} f(h_{(1)}) \rangle$$

$$= \sum \langle h^*, f^{-1}(l_{(1)}) \rangle \langle g^*, f^{-1}(l_{(2)}) h_{(1)} \rangle$$

$$= \langle \Big(h^*(h_{(1)} \cdot g^*)\Big), f^{-1}(l) \rangle$$

$$= \langle \Big(h^*(h_{(1)} \cdot g^*)\Big) \circ f^{-1}, l \rangle$$

On the other hand, if $\{e_i, e_i^*\}$ is a dual basis of L_1, then $\{f(e_i), e_i^* \circ f^{-1}\}$ is a dual basis of L_2 and hence $(\tilde{f} \otimes \tilde{f})(\mathcal{R}_{L_1}) = \mathcal{R}_{L_2}$, and this proves that $\tilde{f}$ is an isomorphism in Pent. Let $n_i = \dim(L_i)$, $i = 1, 2$. Using Proposition 138 we obtain that $(\mathcal{H}(L_1), \mathcal{R}_{L_1}) \cong (\mathcal{H}(L_2), \mathcal{R}_{L_2})$ if and only if $(\mathcal{M}_{n_1}(k), R_1) \cong (\mathcal{M}_{n_2}(k), R_2)$ in Pent, where R_i is the image of $\mathcal{R}_{L_i}$ under the algebra isomorphism $\mathcal{H}(L_i) \cong \mathcal{M}_{n_i}(k)$. Now, the two matrix algebras $\mathcal{M}_{n_1}(k)$ and $\mathcal{M}_{n_2}(k)$ are isomorphic if and only if $n_1 = n_2$ and the Skolem-Noether theorem tells us that any automorphism g of the matrix algebra $\mathcal{M}_{n_1}(k)$ is an inner one: there exists $u \in GL_{n_1}(k)$ such that $g(x) = g_u(x) = uxu^{-1}$. Hence we obtain that $(\mathcal{M}_{n_1}(k), R_1) \cong (\mathcal{M}_{n_2}(k), R_2)$ in Pent if and only if $n_1 = n_2$ and there exists $u \in GL_{n_1}(k)$ such that $R_2 = (g_u \otimes g_u)(R_1) = (u \otimes u)R_1(u \otimes u)^{-1}$, i.e. R_2 is equivalent to R_1, as needed.

We conclude with a few examples, evidencing our general method to determine invertible solutions of the pentagon equation $R \in \mathcal{M}_n(k) \otimes \mathcal{M}_n(k)$. Let $A = (a_{ij})$, $B = (b_{ij}) \in \mathcal{M}_n(k)$. We recall that, via the canonical isomorphism $\mathcal{M}_n(k) \otimes \mathcal{M}_n(k) \cong \mathcal{M}_{n^2}(k)$, $A \otimes B$ viewed as a matrix of $\mathcal{M}_{n^2}(k)$ is given by:

$$A \otimes B = \begin{pmatrix} a_{11}B & a_{12}B & \cdots & a_{1n}B \\ a_{21}B & a_{22}B & \cdots & a_{2n}B \\ \cdot & \cdot & \cdots & \cdot \\ a_{n1}B & a_{n2}B & \cdots & a_{nn}B \end{pmatrix} \tag{6.54}$$

Let $(e_i^j)_{i,j=1,n}$ be the canonical basis of $\mathcal{M}_n(k)$. An element $R \in \mathcal{M}_n(k) \otimes \mathcal{M}_n(k)$ can be written as follows

$$R = \sum_{i,j=1}^{n} e_i^j \otimes A_{ij} \tag{6.55}$$

for some matrices $A_{ij} \in \mathcal{M}_n(k)$.

Using the formula (6.54), R viewed as a matrix in $\mathcal{M}_{n^2}(k)$ is given by the Kronecker product:

$$R = \begin{pmatrix} A_{11} & A_{12} & \cdots & A_{1n} \\ A_{21} & A_{22} & \cdots & A_{2n} \\ \cdot & \cdot & \cdots & \cdot \\ A_{n1} & A_{n2} & \cdots & A_{nn} \end{pmatrix} \tag{6.56}$$

and we can quickly check if R is invertible ($\det(R) \neq 0$). A large class of invertible R is given by choosing (A_{ij}) such that R is upper triangular (i.e. $A_{ij} = 0$ for all $i > j$) and A_{ii} is invertible in $\mathcal{M}_n(k)$ for all $i = 1, \cdots n$. The next Proposition clarify the condition for $R = (A_{ij})_{i,j=1,n} \in \mathcal{M}_{n^2}(k) \cong \mathcal{M}_n(k) \otimes \mathcal{M}_n(k)$ to be a solution of the pentagon equation.

Proposition 139. *Let n be a positive integer and $R = (A_{ij})_{i,j=1,n} \in \mathcal{M}_{n^2}(k) \cong \mathcal{M}_n(k) \otimes \mathcal{M}_n(k)$, $A_{ij} \in \mathcal{M}_n(k)$, be an invertible matrix. Then R is a solution of the pentagon equation if and only if*

$$\sum_{j=1}^{n} A_{ij} \otimes A_{jp} = R(A_{ip} \otimes I_n)R^{-1} \tag{6.57}$$

for all $i, p = 1, \cdots, n$.

Proof. Taking into account the multiplication rule for elementary matrices, we find

$$R^{12}R^{13}R^{23} = \sum_{i,j,p,r,s=1}^{n} e_i^p \otimes A_{ij}e_r^s \otimes A_{jp}A_{rs}$$

$$R^{23}R^{12} = \sum_{a,b,i,p=1}^{n} e_i^p \otimes e_a^b A_{ip} \otimes A_{ab}.$$

Hence, R is a solution of the pentagon equation if and only if

$$\sum_{a,b=1}^{n} e_a^b A_{ip} \otimes A_{ab} = \sum_{i,r,s=1}^{n} A_{ij}e_r^s \otimes A_{jp}A_{rs} \tag{6.58}$$

or, equivalently,

$$R(A_{ip} \otimes I_n) = (\sum_{j=1}^{n} A_{ij} \otimes A_{jp})R \tag{6.59}$$

for all $i, p = 1, \cdots n$. Viewing R as a matrix in $\mathcal{M}_{n^2}(k)$ (see (6.56)) and using (6.54), we find that the pentagon equation (6.59) can be rewritten in $\mathcal{M}_{n^2}(k)$. As R is invertible, (6.59) is equivalent to

$$\sum_{j=1}^{n} A_{ij} \otimes A_{jp} = R(A_{ip} \otimes I_n)R^{-1} = \Delta_l(A_{ip})$$

which means that the matrix (A_{ij}) is comultiplicative with respect to Δ_l.

Example 26. In this example we will show that the two constructions of bialgebras that follow from the solutions of the pentagon equation using Theorems 67 and 70 give very different objects.
Let $R \in \mathcal{M}_4(k) \cong \mathcal{M}_2(k) \otimes \mathcal{M}_2(k)$ be given by

$$R = \begin{pmatrix} 1 & 0 & 0 & 0 \\ 0 & 1 & 0 & 0 \\ 0 & 1 & 1 & 0 \\ 0 & 0 & 0 & 1 \end{pmatrix}$$

Viewing R as an element of $\mathcal{M}_2(k) \otimes \mathcal{M}_2(k)$, we have

$$R = I_2 \otimes I_2 + e_2^1 \otimes e_1^2$$

where $\{e_j^i\}$ is the canonical basis of the matrix algebra $\mathcal{M}_2(k)$ (see Section 5.1). As $\text{char}(k) = 2$, $R^{-1} = R$.

It follows from Proposition 133 that R is is an invertible solution of the pentagon equation if and only if $\text{char}(k) = 2$, and in this case $P(R) = B(\tau \circ R \circ \tau)$ is a five dimensional noncommutative noncocommutative bialgebra.

It is easy to see that the Hopf algebra $P(\mathcal{M}_2(k), R)$ obtained applying Theorem 70 to R, is the groupring kC_2 where $C_2 = \{1, g\}$ is the group with two elements. Indeed, from the construction it follows that $P(\mathcal{M}_2(k), R)$ is the subalgebra of $\mathcal{M}_2(k)$ with $\{I_2, e_{21}\}$ as a basis. If we denote $g = I_2 - e_{21}$, then $g^2 = I_2$ and $P(\mathcal{M}_2(k), R)$ and kC_2 are isomorphiic as Hopf algebras (using the fact that $\text{char}(k) = 2$ in the formula for Δ_P).

Example 27. Let n be a positive integer and

$$A = e_2^1 + e_3^2 + \cdots + e_n^{n-1} + e_1^n \in \mathcal{M}_n(k)$$

Let $R \in \mathcal{M}_n(k) \otimes \mathcal{M}_n(k)$ be given by

$$R = e_1^1 \otimes I_n + e_2^2 \otimes A + \cdots + e_n^n \otimes A^{n-1} \tag{6.60}$$

A routine computation shows that R is an invertible solution of the pentagon equation with inverse

$$R^{-1} = e_1^1 \otimes I_n + e_2^2 \otimes A^{n-1} + \cdots + e_n^n \otimes A. \tag{6.61}$$

Then $P(\mathcal{M}_n(k), R) \cong (kG)^*$, the Hopf algebra of functions on a cyclic group with n elements G.

Proof. We will prove that $H = H(\mathcal{M}_n(k), R) \cong kG$, the group algebra of G and then we use the duality between $P(\mathcal{M}_n(k), R)$ and $H(\mathcal{M}_n(k), R)$ given by Theorem 70. We remark that R is already written in the form $R = \sum_{i=1}^n A_i \otimes B_i$, with (A_i) and (B_i) are linear independent. Then $H(\mathcal{M}_n(k), R)$ is the commutative subalgebra of $\mathcal{M}_n(k)$ with basis $\{I_n, A, A^2, \cdots, A^{n-1}\}$. We also note that $A^n = I_n$. Using (6.61), (6.37), and (6.45), we obtain that the comultiplication, the counit and the antipode of H are given by

$$\Delta_H(A) = A \otimes A, \quad \varepsilon_H(A) = I_n, \quad S_H(A) = A^{n-1} = A^{-1}$$

i.e. $H \cong kG$.

Example 28. Let $R \in \mathcal{M}_{16}(k) \cong \mathcal{M}_4(k) \otimes \mathcal{M}_4(k)$ be the upper triangular matrix given by

$$R = \begin{pmatrix} I_4 & 0 & B & 0 \\ 0 & A & 0 & C \\ 0 & 0 & A & 0 \\ 0 & 0 & 0 & I_4 \end{pmatrix}$$

where

$$A = \begin{pmatrix} 0 & 1 & 0 & 0 \\ 1 & 0 & 0 & 0 \\ 0 & 0 & 0 & 1 \\ 0 & 0 & 1 & 0 \end{pmatrix}, \quad B = \begin{pmatrix} 0 & 0 & 0 & 0 \\ 0 & 0 & 0 & 0 \\ 1 & 0 & 0 & 0 \\ 0 & -1 & 0 & 0 \end{pmatrix}, \quad C = \begin{pmatrix} 0 & 0 & 0 & 0 \\ 0 & 0 & 0 & 0 \\ 0 & -1 & 0 & 0 \\ 1 & 0 & 0 & 0 \end{pmatrix}$$

If we view $R \in \mathcal{M}_4(k) \otimes \mathcal{M}_4(k)$, then R is given by

$$\begin{aligned} R = (e_1^1 + e_4^4) \otimes I_4 + (e_2^2 + e_3^3) \otimes (e_1^2 + e_2^1 + e_3^4 + e_4^3) \\ + e_1^3 \otimes (e_3^1 - e_4^2) + e_2^4 \otimes (e_4^1 - e_3^2) \end{aligned} \tag{6.62}$$

R is an invertible solution of the pentagon equation, $H(\mathcal{M}_4(k), R) \cong H_4$, and hence $P(\mathcal{M}_4(k), R) \cong H_4^* \cong H_4$, where H_4, Sweedler's four dimensional Hopf algebra.

Proof. Similar to the proof in Example 27. The inverse of R is

$$R^{-1} = (e_1^1 + e_4^4) \otimes I_4 + (e_2^2 + e_3^3) \otimes (e_1^2 + e_2^1 + e_3^4 + e_4^3) + e_1^3 \otimes (e_4^1 - e_3^2) + e_2^4 \otimes (e_3^1 - e_4^2)$$

and therefore the Hopf algebra $H = H(\mathcal{M}_4(k), R)$ is the four dimensional subalgebra of $\mathcal{M}_4(k)$ with k-basis

$$\{I_4, e_1^2 + e_2^1 + e_3^4 + e_4^3, e_4^1 - e_3^2, e_3^1 - e_4^2\}$$

Now, writing $x = e_3^1 - e_4^2$ and $g = e_1^2 + e_2^1 + e_3^4 + e_4^3$ we find that

$$x^2 = 0, \quad g^2 = I_4, \quad gx = -xg = e_4^1 - e_3^2$$

On the other hand the formula of the comultiplication of H given by (6.37), namely $\Delta(A) = R(A \otimes I_4)R^{-1}$, for all $A \in H$ gives, using the expresion of R^{-1},

$$\Delta(g) = g \otimes g, \quad \Delta(x) = x \otimes g + I_4 \otimes x$$

i.e. $H \cong H_4$, the Sweedler four dimensional Hopf algebra (see Example 9 4)).

Theorem 73 opens a new road for describing the types of isomorphisms for Hopf algebras of a certain dimension. The first step, and the most impor- tant, is however the development of a new Jordan type theory (we called it *restricted Jordan theory*). From the point of view of actions, the classical Jordan theory gives the most elementary description of the representatives of the orbits of the action

$$GL_n(k) \times \mathcal{M}_n(k) \to \mathcal{M}_n(k), \quad (U, A) \to UAU^{-1}.$$

The restricted Jordan theory refers to the following open problem:
Problem: *Describe the orbits of the action*

$$GL_n(k) \times (\mathcal{M}_n(k) \otimes \mathcal{M}_n(k)) \to \mathcal{M}_n(k) \otimes \mathcal{M}_n(k), \ (U, R) \to (U \otimes U)R(U \otimes U)^{-1}.$$

We recall that the canonical Jordan form J_A of a matrix A is the matrix equivalent to A which has the greatest number of zeros. For practical reasons, in the restricted Jordan theory we are in fact interested in finding the elements of each orbit that have the greatest number of zeros. Of these, we retain only those which are invertible solutions of the pentagon equation. The set of types of n-dimensional Hopf algebras will be those Hopf algebras associated (using Theorem 72) to the solutions of length n (or, equivalently, the space of coinvariant elements is n-dimensional); all other Hopf algebras will have a dimension that is a divisor of n^2. We mention that, as a general rule, the set of types of n-dimensional Hopf algebras is infinite (this was proved recently in [7], [12], [90]). If however we limit ourselves to classifying certain special types of Hopf algebras, then this set can be finite. For instance, the set of types of n-dimensional semisimple and cosemisimple Hopf algebras is finite ([168]).

The restricted Jordan Theory, is also involved in the theory of classification of separable algebras (see the last chapter). This has to do with those orbits whose representatives $R = \sum R^1 \otimes R^2 \in \mathcal{M}_n(k) \otimes \mathcal{M}_n(k)$ are solutions of the separability equation $R^{12}R^{23} = R^{23}R^{13} = R^{13}R^{12}$ and $\sum R^1 R^2 = I_n$.

A new interesting direction related to the study of the pentagon equation is a general representation theory, whose objects are defined bellow. Let (M, R) be a pair, where M is a vector space and $R \in \text{End}(M \otimes M)$ is a solution of the pentagon equation. A representation of (M, R), or an (M, R)-module is a pair (V, ψ_V), where V is a vector space and $\psi_V : M \otimes V \to V \otimes M$ is a k-linear map such that

$$\psi_V^{12} \tau_{M,V}^{23} (\tau_{M,M} R)^{12} = (\tau_{M,M} R)^{23} \psi_V^{12} \psi_V^{23}$$

as maps $M \otimes M \otimes V \to V \otimes M \otimes M$. A morphism of two (M, R)-modules (V, ψ_V), (W, ψ_W) is a k-linear map $f : V \to W$ such that $(f \otimes \text{Id}_M)\psi_V = \psi_W(\text{Id}_M \otimes f)$. $_{(M,R)}\mathcal{M}$ will be the monoidal category of (M, R)-modules, where the monoidal structure is given by

$$(V, \psi_V) \otimes (W, \psi_W) = (V \otimes W, \psi_W^{23} \psi_V^{12}).$$

We will see that the category of representations of a Hopf algebra, or more generally of an algebra, is a subcategory of $_{(M,R)}\mathcal{M}$.

Examples 12. 1. Let L be a Hopf algebra and $R_L : L \otimes L \to L \otimes L$ the canonical solution of the pentagon equation, $R_L(g \otimes h) = g_{(1)} \otimes g_{(1)}h$, for all g, $h \in L$. Then any left L-module $(V, \cdot)$ has a natural structure of (L, R_L)-module with the map

$$\psi_V : L \otimes V \to V \otimes L, \quad \psi_V(g \otimes v) = g_{(2)} \cdot v \otimes g_{(1)}$$

Indeed, for g, $h \in L$ and $v \in V$, we have that

$$\psi_V^{12} \tau_{L,V}^{23} (\tau_{L,L} R_L)^{12} (g \otimes h \otimes v) = (\tau_{L,L} R_L)^{23} \psi_V^{12} \psi_V^{23} (g \otimes h \otimes v)$$
$$= g_{(3)} h_{(2)} \cdot v \otimes g_{(2)} h_{(1)} \otimes g_{(1)}$$

2. Now let A be a k-algebra and

$$R = R_A : \ A \otimes A \to A \otimes A, \quad R(a \otimes b) = a \otimes ba$$

the solution of the pentagon equation constructed in Proposition 127. Let V be a vector space and $\cdot : \ V \otimes A \to V$ a k-linear map. We define

$$\psi_V = \psi_{(V,\cdot)} : A \otimes V \to V \otimes A, \quad \psi_V(a \otimes v) = v \cdot a \otimes a$$

Then $(V, \psi_{(V,\cdot)})$ is an (A, R_A)-module if and only if $(V, \cdot)$ is a right A-module in the usual sense. Indeed, the statement follows from the formulas

$$(\tau_{A,A} R_A)^{23} \psi_V^{12} \psi_V^{23}(a \otimes b \otimes v) = (v \cdot b) \cdot a \otimes ba \otimes a$$

and

$$\psi_V^{12} \tau_{A,V}^{23}(\tau_{A,A} R_A)^{12}(a \otimes b \otimes v) = v \cdot (ba) \otimes ba \otimes a$$

for all a, $b \in A$, $v \in V$. Hence the category $_{(M,R)}\mathcal{M}$ generalizes the usual category of modules over an algebra A.

7 Long dimodules and the Long equation

In this Chapter, we will show that the nonlinear equation $R^{12}R^{23} = R^{23}R^{12}$ (called Long equation) can be associated to the category of Long dimodules $_H\mathcal{L}^H$ over a bialgebra H. The Long equation is obtained from the quantum Yang-Baxter equation by deleting the middle term from both sides. Our theory is similar to the one developed in Chapters 5 and 6, where we have discussed how Yetter-Drinfeld modules and Hopf modules are connected to respectively the quantum Yang-Baxter equation and the pentagon equation. A different approach to solving the Long equation, in a general monoidal category, is given in [16].

7.1 The Long equation

Definition 15. *Let M be a vector space over a field k. $R \in \text{End}(M \otimes M)$ is called a solution of the Long equation if*

$$R^{12}R^{23} = R^{23}R^{12} \tag{7.1}$$

in $\text{End}(M \otimes M \otimes M)$.

For later use, we rewrite the Long equation in matrix form. The proof is left to the reader.

Proposition 140. *Let $\{m_1, \cdots, m_n\}$ be a basis of M, and let R and $S \in \text{End}(M \otimes M)$ be given by their matrices (x_{uv}^{ij}) and (y_{uv}^{ij}), i.e.*

$$R(m_u \otimes m_v) = x_{uv}^{ij} m_i \otimes m_j \quad \text{and} \quad S(m_u \otimes m_v) = y_{uv}^{ij} m_i \otimes m_j$$

Then

$$R^{23}S^{12} = S^{12}R^{23}$$

if and only if

$$x_{vk}^{ij} y_{ql}^{pv} = x_{lk}^{\alpha j} y_{q\alpha}^{pi} \tag{7.2}$$

In particular, R is a solution for the Long equation if and only if

$$x_{vk}^{ij} x_{ql}^{pv} = x_{lk}^{\alpha j} x_{q\alpha}^{pi} \tag{7.3}$$

In Proposition 141, we give some other nonlinear equations that are equivalent to the Long equation.

Proposition 141. *Let M be a vector space and $R \in \mathrm{End}(M \otimes M)$. The following statements are equivalent*

1. *R is a solution of the Long equation;*
2. *$T = R\tau$ is a solution of the equation*

$$T^{12}T^{13} = T^{23}T^{13}\tau^{(123)}$$

3. *$U = \tau R$ is a solution of the equation*

$$U^{13}U^{23} = \tau^{(123)}U^{13}U^{12}$$

4. *$W = \tau R\tau$ is a solution of the equation*

$$\tau^{(123)}W^{23}W^{13} = W^{12}W^{13}\tau^{(123)}$$

Proof. 1. $\Leftrightarrow$ 2. As $R = T\tau$, R is a solution of the Long equation if and only if

$$T^{12}\tau^{12}T^{23}\tau^{23} = T^{23}\tau^{23}T^{12}\tau^{12} \tag{7.4}$$

Now

$$\tau^{12}T^{23}\tau^{23} = T^{13}\tau^{13}\tau^{12}, \quad \text{and} \quad \tau^{23}T^{12}\tau^{12} = T^{13}\tau^{12}\tau^{13}$$

and (7.4) is equivalent to

$$T^{12}T^{13}\tau^{13}\tau^{12} = T^{23}T^{13}\tau^{12}\tau^{13}$$

The equivalence of 1. and 2. follows since $\tau^{12}\tau^{13}\tau^{12}\tau^{13} = \tau^{(123)}$.
1. $\Leftrightarrow$ 3. $R = \tau U$, so R is a solution of the Long equation if and only if

$$\tau^{12}U^{12}\tau^{23}U^{23} = \tau^{23}U^{23}\tau^{12}U^{12} \tag{7.5}$$

Using the fact that

$$\tau^{12}U^{12}\tau^{23} = \tau^{23}\tau^{13}U^{13}, \quad \text{and} \quad \tau^{23}U^{23}\tau^{12} = \tau^{23}\tau^{12}U^{13}.$$

we find that (7.5) is equivalent to

$$U^{13}U^{23} = \tau^{13}\tau^{12}U^{13}U^{12}$$

and we are done since $\tau^{13}\tau^{12} = \tau^{(123)}$.
1. $\Leftrightarrow$ 4. $R = \tau W\tau$ is a solution of the Long equation if and only if

$$\tau^{12}W^{12}\tau^{12}\tau^{23}W^{23}\tau^{23} = \tau^{23}W^{23}\tau^{23}\tau^{12}W^{12}\tau^{12} \tag{7.6}$$

Using the formulas

$$\tau^{12}\tau^{23} = \tau^{13}\tau^{12}, \quad \tau^{23}\tau^{12} = \tau^{13}\tau^{23},$$

$$\tau^{12}W^{12}\tau^{13} = \tau^{12}\tau^{13}W^{23}, \quad \tau^{12}W^{23}\tau^{23} = W^{13}\tau^{12}\tau^{23}$$

$$\tau^{23}W^{23}\tau^{13} = \tau^{23}\tau^{13}W^{12}, \quad \tau^{23}W^{12}\tau^{12} = W^{13}\tau^{23}\tau^{12}$$

we find that (7.6) is equivalent to

$$\tau^{12}\tau^{13}W^{23}W^{13}\tau^{12}\tau^{23} = \tau^{23}\tau^{13}W^{12}W^{13}\tau^{23}\tau^{12}$$

proving the equivalence of 1. and 4. since

$$\tau^{13}\tau^{23}\tau^{12}\tau^{13} = \tau^{23}\tau^{12}\tau^{23}\tau^{12} = \tau^{(123)}$$

Examples 13. 1. If $R \in \mathrm{End}(M \otimes M)$ is bijective, then R is a solution of the Long equation if and only if R^{-1} is also a solution of the Long equation.

2. Let $(m_i)_{i \in I}$ be a basis of M and $(a_{ij})_{i,j \in I}$ be a family of scalars of k. Then, $R : M \otimes M \to M \otimes M$, $R(m_i \otimes m_j) = a_{ij}m_i \otimes m_j$, for all $i, j \in I$, is a solution of the Long equation. In particular, the identity map $Id_{M \otimes M}$ is a solution of the Long equation.

3. Let M be a finite dimensional vector space and u an automorphism of M. If R is a solution of the Long equation then ${}^u R = (u \otimes u)R(u \otimes u)^{-1}$ is also a solution of the Long equation.

4. Let $R \in \mathcal{M}_4(k)$ given by

$$R = \begin{pmatrix} a & 0 & 0 & 0 \\ 0 & b & c & 0 \\ 0 & d & e & 0 \\ 0 & 0 & 0 & f \end{pmatrix}$$

A direct computation shows that R is a solution of the Long equation if and only if $c = d = 0$. In particular, if $q \in k$, $q \neq 0$, $q \neq 1$, the two dimensional Yang-Baxter operator R_q is a solution of the QYBE and is not a solution for the Long equation.

5. Let G be a group and M be a left kG-module. Assume also that M is G-graded, i.e.

$$M = \oplus_{\sigma \in G}M_\sigma$$

where the M_σ are subspaces of M. If the M_σ are kG-submodules of M, then the map

$$R : M \otimes M \to M \otimes M, \quad R(n \otimes m) = \sum_\sigma \sigma \cdot n \otimes m_\sigma, \ \forall n, m \in M \quad (7.7)$$

is a solution of the Long equation. If G is non-abelian, then R is not a solution of the QYBE.

It suffices to show that (7.1) holds for homogenous elements. Let $m_\sigma \in M_\sigma$, $m_\tau \in M_\tau$ and $m_\theta \in M_\theta$. Then

$$R^{23}R^{12}(m_\sigma \otimes m_\tau \otimes m_\theta) = R^{23}(\tau \cdot m_\sigma \otimes m_\tau \otimes m_\theta)$$
$$= \tau \cdot m_\sigma \otimes \theta \cdot m_\tau \otimes m_\theta$$

and

$$R^{12}R^{23}(m_\sigma \otimes m_\tau \otimes m_\theta) = R^{12}(m_\sigma \otimes \theta \cdot m_\tau \otimes m_\theta)$$
$$= \tau \cdot m_\sigma \otimes \theta \cdot m_\tau \otimes m_\theta$$

and it follows that R is a solution of the Long equation. On the other hand

$$R^{12}R^{13}R^{23}(m_\sigma \otimes m_\tau \otimes m_\theta) = \tau\theta \cdot m_\sigma \otimes \theta \cdot m_\tau \otimes m_\theta$$

and

$$R^{23}R^{13}R^{12}(m_\sigma \otimes m_\tau \otimes m_\theta) = \theta\tau \cdot m_\sigma \otimes \theta \cdot m_\tau \otimes m_\theta$$

and we see that R is not a solution of the QYBE if G is not abelian.

7.2 The FRT Theorem for the Long equation

We reconsider the Long dimodules introduced in Section 4.5, in the situation where the underlying algebra A and coalgebra C are equal to a given bialgebra H. For completeness sake, we recall the Definition. In the case where H is commutative and cocommutative, it is due to Long [118].

Definition 16. *Let H be a bialgebra. A (left-right) Long H-dimodule is a threetuple $(M, \cdot, \rho^r)$, where $(M, \cdot)$ is a left H-module and (M, ρ^r) is a right H-comodule such that*

$$\rho(h \cdot m) = h \cdot m_{[0]} \otimes m_{[1]} \tag{7.8}$$

for all $h \in H$ and $m \in M$.

The category of H-dimodules and H-linear H-colinear maps will be denoted by $_H\mathcal{L}^H$.

Examples 14. 1. Let G be a group. A left-right kG-dimodule M is a kG-module M, together with a family $\{M_\sigma \mid \sigma \in G\}$ of kG-submodules of M such that $M = \oplus_{\sigma \in G} M_\sigma$ (cf. Example 13 5)).
Indeed, we know that M is a right kG-comodule if $M = \oplus_{\sigma \in G} M_\sigma$, with every M_σ a subspace of M. The compatibility relation (7.8) means exactly that the M_σ are kG-submodules of M.
Let us now suppose that M is a decomposable representation on G with the Long length smaller or equal to the cardinal of G. Let X be a subset of G

and $\{M_x \mid x \in X\}$ a family of idecomposable $k[G]$-submodules of M such that $M = \oplus_{x \in X} M_x$. Then M is a right comodule over $k[X]$, and as $X \subseteq G$, M can be viewed as a right $k[G]$-comodule. Hence, $M \in {}_{k[G]}\mathcal{L}^{k[G]}$. We obtain that the category of representations on G with the Long length smaller or equal to the cardinal of G can be viewed as a subcategory of ${}_{k[G]}\mathcal{L}^{k[G]}$.

2. For any left H-module $(N, \cdot)$, $N \otimes H \in {}_H\mathcal{L}^H$ with structure:

$$h \cdot (n \otimes l) = h \cdot n \otimes l, \qquad \rho(n \otimes l) = n \otimes l_{(1)} \otimes l_{(2)}$$

for all $h, l \in H$, $n \in N$. Thus we have a functor

$$G = \bullet \otimes H : {}_H\mathcal{M} \to {}_H\mathcal{L}^H$$

which is a right adjoint of the forgetful functor

$$F : {}_H\mathcal{L}^H \to {}_H\mathcal{M}$$

3. For any (M, ρ) be a right H-comodule (M, ρ^r), $H \otimes M \in {}_H\mathcal{L}^H$ with structure

$$h \cdot (l \otimes m) = hl \otimes m, \qquad \rho_{H \otimes M}(l \otimes m) = l \otimes m_{[0]} \otimes m_{[1]}$$

for all $h, l \in H$, $m \in M$. We obtain a functor

$$F = H \otimes \bullet : \mathcal{M}^H \to {}_H\mathcal{L}^H$$

which is a left adjoint of the forgetful functor

$$G : {}_H\mathcal{L}^H \to \mathcal{M}^H$$

4. Let $(N, \cdot)$ be a left H-module. Then, with the trivial structure of right H-comodule, $\rho^r : N \to N \otimes H$, $\rho(n) = n \otimes 1$, for all $n \in N$, $(N, \cdot, \rho) \in {}_H\mathcal{L}^H$.

5. Let (M, ρ) be a right H-comodule. Then, with the trivial structure of left H-module, $h \cdot m = \varepsilon(h)m$, for all $h \in H$, $m \in M$, $(M, \cdot, \rho) \in {}_H\mathcal{L}^H$.

Remarks 22. 1. Let M, $N \in {}_H\mathcal{L}^H$. $M \otimes N$ is also an object in ${}_H\mathcal{L}^H$, the structure maps are given by

$$h \cdot (m \otimes n) = h_{(1)} \cdot m \otimes h_{(2)} \cdot n, \qquad \rho(m \otimes n) = m_{[0]} \otimes n_{[0]} \otimes m_{[1]}n_{[1]}$$

for all $h \in H$, $m \in M$, $n \in N$. $k \in {}_H\mathcal{L}^H$ with the trivial structure

$$h \cdot a = \varepsilon(h)a, \qquad \rho(a) = a \otimes 1$$

for all $h \in H$, $a \in k$. It is easy to show that $({}_H\mathcal{L}^H, \otimes, k)$ is a monoidal category.

2. We have a natural functor

$$_H\mathcal{L}^H \to {}_{H \otimes H^*}\mathcal{M}$$

Any Long dimodule M is a left $H \otimes H^*$-module; the left $H \otimes H^*$-action is given by

$$(h \otimes h^*) \cdot m := < h^*, m_{[1]} > h m_{[0]}$$

for all $h \in H$, $h^* \in H^*$ and $m \in M$. If H is finite dimensional, then the categories $_H\mathcal{L}^H$ and $_{H \otimes H^*}\mathcal{M}$ are isomorphic.

The next Proposition is a generalization of Example 13 5).

Proposition 142. *Let H be a bialgebra and $(M, \cdot, \rho)$ be a Long H-dimodule. Then the natural map*

$$R_{(M, \cdot, \rho)}(m \otimes n) = n_{[1]} \cdot m \otimes n_{[0]}$$

is a solution of the Long equation.

Proof. Let $R = R_{(M, \cdot, \rho)}$. For l, m, $n \in M$ we have

$$R^{12}R^{23}(l \otimes m \otimes n) = R^{12}\left(l \otimes n_{[1]} \cdot m \otimes n_{[0]}\right)$$
$$= (n_{[1]} \cdot m)_{[1]} \cdot l \otimes (n_{[1]} \cdot m)_{[0]} \otimes n_{[0]}$$
$$= m_{[1]} \cdot l \otimes n_{[1]} \cdot m_{[0]} \otimes n_{[0]}$$
$$= R^{23}\left(m_{[1]} \cdot l \otimes m_{[0]} \otimes n\right)$$
$$= R^{23}R^{12}(l \otimes m \otimes n)$$

and it follows that R is a solution of the Long equation.

Lemma 30. *Let H be a bialgebra, $(M, \cdot)$ a left H-module and (M, ρ) a right H-comodule. Then the set*

$$\{h \in H \mid \rho(h \cdot m) = h \cdot m_{[0]} \otimes m_{[1]}, \forall m \in M\}$$

is a subalgebra of H.

Proof. Straightforward.

As a direct consequence of Lemma 30, we find that a vector space M on which H acts and coacts is a Long dimodule if and only if the compatibility condition (7.8) is satisfied for m running through a basis of M, and h running through a set of algebra generators of H.

Now we will present the FRT Theorem for the Long equation: in the finite dimensional case, every solution R of the Long equation is of the form $R = R_{(M, \cdot, \rho)}$, where $(M, \cdot, \rho)$ is a Long dimodule over a bialgebra $D(R)$. As one might expect, the proof is similar to the corresponding proofs of the FRT Theorems for the quantum Yang-Baxter equation and the pentagon equation.

Theorem 74. *Let M be a finite dimensional vector space and $R \in \text{End}(M \otimes M)$ a solution of the Long equation. Then*

1. *There exists a bialgebra $D(R)$ acting and coacting on M (with structure maps $\cdot$ and ρ) such that $(M, \cdot, \rho) \in {}_{D(R)}\mathcal{L}^{D(R)}$, and*

$$R = R_{(M, \cdot, \rho)}$$

2. *The bialgebra $D(R)$ is universal with respect to this property: if H is a bialgebra acting and coacting on M (with structure maps $\cdot'$ and ρ') such that $(M, \cdot', \rho') \in {}_H\mathcal{L}^H$ and*

$$R = R_{(M, \cdot', \rho')}$$

then there exists a unique bialgebra map $f : D(R) \to H$ such that $\rho' = (I \otimes f)\rho$ and $a \cdot m = f(a) \cdot' m$, for all $a \in D(R)$, $m \in M$.

Proof. 1. As usual, let $\{m_1, \cdots, m_n\}$ be a basis for M and (x_{uv}^{ij}) the matrix of R, i.e.

$$R(m_u \otimes m_v) = x_{uv}^{ij} m_i \otimes m_j \tag{7.9}$$

Let $(C, \Delta, \varepsilon) = \mathcal{M}^n(k)$ be the comatrix coalgebra of order n. The tensor algebra $T(C)$ has a unique bialgebra structure, with comultiplication and counit extending Δ and ε. Following the arguments in the proof of Theorem 66, we find a left $T(C)$- action and a right $T(C)$-coaction on M such that $R = R_{(M, \cdot, \rho)}$. The action and coaction are given by the formulas

$$\rho(m_l) = m_v \otimes c_l^v \tag{7.10}$$
$$c_u^j \cdot m_l = x_{lu}^{ij} m_i \tag{7.11}$$

We now define the *obstructions* o_{kl}^{ij}, measuring how far M is from being a Long dimodule over $T(C)$. Keeping in mind the fact that $T(C)$ is generated as an algebra by the c_j^i, and using Lemma 30, we restrict ourselves to computing $h \cdot m_{[0]} \otimes m_{[1]} - \rho(h \cdot m)$ for $h = c_k^j$, and $m = m_l$:

$$
\begin{aligned}
h \cdot m_{[0]} \otimes m_{[1]} &= c_k^j \cdot m_{l[0]} \otimes m_{l[1]} \\
&= c_k^j \cdot m_v \otimes c_l^v = x_{vk}^{ij} m_i \otimes c_l^v = m_i \otimes x_{vk}^{ij} c_l^v
\end{aligned}
$$

and

$$\rho(h \cdot m) = \rho(c_k^j \cdot m_l) = x_{lk}^{\alpha j} m_{\alpha[0]} \otimes m_{\alpha[1]} = m_i \otimes x_{lk}^{\alpha j} c_\alpha^i$$

Writing

$$o_{kl}^{ij} = x_{vk}^{ij} c_l^v - x_{lk}^{\alpha j} c_\alpha^i \tag{7.12}$$

we find

$$h \cdot m_{[0]} \otimes m_{[1]} - \rho(h \cdot m) = m_i \otimes o_{kl}^{ij} \tag{7.13}$$

Let I be the two-sided ideal of $T(C)$ generated by all o_{kl}^{ij}. We claim that I is a bi-ideal of $T(C)$ and $I \cdot M = 0$.

The fact that I is also a coideal will result from the following formula:

$$\Delta(o_{kl}^{ij}) = o_{ku}^{ij} \otimes c_l^u + c_u^i \otimes o_{kl}^{uj} \tag{7.14}$$

First we observe that (7.12) can be rewritten as

$$o_{kl}^{ij} = x_{vk}^{ij} c_l^v - x_{lk}^{vj} c_v^i$$

so

$$\begin{aligned}
\Delta(o_{kl}^{ij}) &= x_{vk}^{ij} c_u^v \otimes c_l^u - x_{lk}^{vj} c_u^i \otimes c_v^u \\
&= x_{vk}^{ij} c_u^v \otimes c_l^u - c_u^i \otimes x_{lk}^{vj} c_v^u \\
&= \left(o_{ku}^{ij} + x_{uk}^{vj} c_v^i\right) \otimes c_{ul} - c_u^i \otimes \left(-o_{kl}^{uj} + x_{vk}^{uj} c_l^v\right) \\
&= o_{ku}^{ij} \otimes c_l^u + c_u^i \otimes o_{kl}^{uj}
\end{aligned}$$

proving (7.14) holds. Now

$$\varepsilon\left(o_{kl}^{ij}\right) = x_{lk}^{ij} - x_{lk}^{ij} = 0$$

so we have proved that I is a coideal of $T(C)$.
$I \cdot M = 0$ will follow from the fact that R is a solution of the Long equation.
For any $n \in M$, we compute

$$\begin{aligned}
\left(R^{23}R^{12}\right)(n \otimes m_k \otimes m_j) &= R^{23}(c_k^\alpha \cdot n \otimes m_\alpha \otimes m_j) \\
&= c_k^\alpha \cdot n \otimes c_j^s \cdot m_\alpha \otimes m_s \\
&= c_k^\alpha \cdot n \otimes x_{\alpha j}^{rs} m_r \otimes m_s \\
&= \left(x_{\alpha j}^{rs} c_k^\alpha\right) \cdot n \otimes m_r \otimes m_s
\end{aligned}$$

and

$$\begin{aligned}
\left(R^{12}R^{23}\right)(n \otimes m_k \otimes m_j) &= R^{12}(n \otimes c_j^s \cdot m_k \otimes m_s) \\
&= R^{12}(n \otimes x_{kj}^{\alpha s} m_\alpha \otimes m_s) \\
&= x_{kj}^{\alpha s} c_\alpha^r \cdot n \otimes m_r \otimes m_s
\end{aligned}$$

so it follows that

$$\begin{aligned}
\left(R^{23}R^{12} - R^{12}R^{23}\right)(n \otimes m_k \otimes m_j) &= \left(x_{\alpha j}^{rs} c_k^\alpha - x_{kj}^{\alpha s} c_\alpha^r\right) \cdot n \otimes m_r \otimes m_s \\
&= o_{jk}^{rs} \cdot n \otimes m_r \otimes m_s \tag{7.15}
\end{aligned}$$

Now R is a solution of the Long equation, hence $o_{jk}^{rs} \cdot n = 0$, for all $n \in M$,
so can and $I \cdot M = 0$. Now we define

$$D(R) = T(C)/I$$

$D(R)$ coacts on M via the canonical projection $T(C) \to D(R)$, and $D(R)$
acts on M since $I \cdot M = 0$. The c_j^i generate $D(R)$ and $o_{kl}^{ij} = 0$ in $D(R)$, so

we find from (7.13) that $(M, \cdot, \rho) \in {}_{D(R)}\mathcal{L}^{D(R)}$ and $R = R_{(M, \cdot, \rho)}$.

2. Let H be a bialgebra and suppose that $(M, \cdot', \rho') \in {}_H\mathcal{L}^H$ is such that $R = R_{(M, \cdot', \rho')}$. Let $(c'_{ij})_{i,j=1,\cdots,n}$ be a family of elements in H such that

$$\rho'(m_l) = m_v \otimes c_l^{v'}$$

Then

$$R(m_v \otimes m_u) = c_u^{j'} \cdot' m_v \otimes m_j$$

and

$$c_u^{j'} \cdot' m_v = x_{vu}^{ij} m_i = c_j^j \cdot m_v$$

Let

$$o'^{ij}_{kl} = x_{vk}^{ij} c_l^{v'} - x_{lk}^{\alpha j} c_\alpha^{i'}$$

From the universal property of the tensor algebra $T(C)$, it follows that there exists a unique algebra map $f_1 : T(C) \to H$ such that $f_1(c_j^i) = c_j^{i'}$, for all $i, j = 1, \cdots, n$. Now $(M, \cdot', \rho') \in {}_H\mathcal{L}^H$, so $0 = o'^{ij}_{kl} = f_1(o^{ij}_{kl})$, for all $i, j, k, l = 1, \cdots, n$. So the map f_1 factorizes through a map

$$f : D(R) \to H, \quad f(c_{ij}) = c'_{ij}$$

Obviously we have, for any $l \in \{1, \cdots, n\}$,

$$(I \otimes f)\rho(m_l) = \sum_v m_v \otimes f(c_{vl}) = \sum_v m_v \otimes c'_{vl} = \rho'(m_l)$$

so $\rho' = (I \otimes f)\rho$.

Conversely, $(I \otimes f)\rho = \rho'$ necessarily implies $f(c_{ij}) = c'_{ij}$, proving the uniqueness of f. This completes the proof of the theorem.

Remark 23. In the graded algebra $T(\mathcal{M}^n(k))$ the obstruction elements o^{ij}_{kl} are of degree one, i.e. are elements of the comatrix coalgebra $\mathcal{M}^n(k)$. This will lead us in the next section to the study of some special functions defined only for a coalgebra, which will also play an important role in solving the Long equation.

Examples 15. 1. Let $a, b, c \in k$ and $R \in \mathcal{M}_4(k)$ given by equation (5.10). R is then a solution of the quantum Yang-Baxter equation and the Long equation.

We will describe the bialgebra $D(R)$, which is obtained considering R as a solution for the Long equation. If $(b, c) = (0, 0)$ then $R = 0$ and $D(R) = T(\mathcal{M}^4(k))$. Suppose now that $(b, c) \neq (0, 0)$, and write

$$R(m_u \otimes m_v) = \sum_{i,j=1}^{2} x_{uv}^{ij} m_i \otimes m_j$$

the only x_{uv}^{ij} that are different from zero are

$$x_{11}^{11} = ab, \quad x_{12}^{11} = ac, \quad x_{12}^{12} = ab, \quad x_{21}^{11} = b, \qquad (7.16)$$
$$x_{21}^{21} = ab, \quad x_{12}^{11} = c, \quad x_{22}^{12} = b, \quad x_{22}^{21} = ac, \quad x_{22}^{22} = ab$$

The sixteen relations $o_{kl}^{ij} = 0$ are the following ones, written in the lexicografical in (i, j, k, l).

$$abc_1^1 + bc_1^2 = abc_1^1, \quad abc_2^1 + bc_2^2 = bc_1^1 + abc_2^1$$

$$acc_1^1 + cc_1^2 = acc_1^1, \quad acc_2^1 + cc_2^2 = cc_1^1 + acc_2^1$$

$$0 = 0, \quad 0 = 0, \quad abc_1^1 + bc_1^2 = abc_1^1, \quad abc_2^1 + bc_2^2 = bc_1^1 + abc_2^1$$

$$abc_1^2 = abc_1^2, \quad abc_2^2 = abc_2^2, \quad acc_1^2 = acc_1^2, \quad acc_2^2 = cc_1^2 + acc_2^2$$

$$0 = 0, \quad 0 = 0, \quad abc_1^2 = abc_1^2, \quad abc_2^2 = abc_2^2 + bc_1^2$$

The only nontrivial relations are

$$bc_1^2 = 0, \quad bc_2^2 = bc_1^1, \quad cc_1^2 = 0, \quad cc_2^2 = cc_1^1$$

As, $(b, c) \neq (0, 0)$, these reduce to

$$c_1^2 = 0, \qquad c_2^2 = c_1^1$$

Now, if we denote $c_1^1 = x$, $c_2^1 = y$ we obtain that $D(R)$ can be described as follows:

– As an algebra $D(R) = k < x, y >$, the free algebra generated by x and y.
– The comultiplication Δ and the counit ε are given by

$$\Delta(x) = x \otimes x, \quad \Delta(y) = x \otimes y + y \otimes x, \quad \varepsilon(x) = 1, \quad \varepsilon(y) = 0.$$

We observe that the bialgebra $D(R)$ does not depend on the parameters a, b, c.

2. Take $q \in k$, and consider the solution $R_q \in \mathcal{M}_4(k)$ of the Hopf equation constructed in Proposition 135. R_q is also a solution of the Long equation, as R_q has the form $f \otimes g$ with $fg = gf$. In Proposition 135 we have described the bialgebra $B(R_q)$, using the FRT Theorem for the Hopf equation. The description of $D(R_q)$, using the FRT Theorem for the Long equation is much simpler, and does not depend on the parameter q.

– As an algebra $D(R_q) = k < x, y >$, the free algebra generated by x and y.
– x and y are grouplike elements. This describes the coalgebra structure.

This construction follows from a computation similar to the one in the previous example. The only surviving obstruction relations are

$$c_1^2 = 0, \qquad c_2^1 = q(c_1^1 - c_2^2)$$

We obtain the above construction after we put $c_{11} = x$ and $c_{22} = y$.

7.3 Long coalgebras

In this Section we will define *Long maps* on coalgebras: if C is a coalgebra and I a coideal of C, then a Long map is a k-linear map $\sigma : C \otimes C/I \to k$ satisfying (7.17). This condition ensures that, for any right C-comodule (M, ρ), the natural map $R_{(\sigma, M, \rho)}$ is a solution of the Long equation. Conversely, over a finite dimensional vector space M, any solution of the Long equation arises in this way. Hence Long maps on a coalgebra may be viewed as a Long equation analog of coquasitriangular bialgebras.

The image of $c \in C$ in the quotient coalgebra will be denoted by $\bar{c}$. If (M, ρ) is a right C-comodule, then then $(M, \bar{\rho})$ is a right C/I-comodule via $\bar{\rho}(m) = m_{[0]} \otimes \overline{m_{[1]}}$, for all $m \in M$.

Definition 17. *Let C be a coalgebra and I be a coideal of C. A k-linear map $\sigma : C \otimes C/I \to k$ is called a Long map if*

$$\sigma(c_{(1)} \otimes \bar{d}) \, \overline{c_{(2)}} = \sigma(c_{(2)} \otimes \bar{d}) \, \overline{c_{(1)}} \tag{7.17}$$

for all $c, d \in C$. If $I = 0$, then σ is called proper Long map and (C, σ) is called a Long coalgebra.

Examples 16. 1. If C is cocommutavive then any k-linear map $\sigma : C \otimes C/I \to k$ is a Long map. In particular, any cocommutative coalgebra is a Long coalgebra for any $\sigma : C \otimes C \to k$.
2. Let C be a coalgebra, I a coideal, and $f \in \mathrm{Hom}_k(C/I, k)$. Then

$$\sigma_f : C \otimes C/I \to k, \quad \sigma_f(c \otimes \bar{d}) = \varepsilon(c) f(\bar{d})$$

for all $c, d \in C$, is a Long map. In particular, any coalgebra has a trivial structure of Long coalgebra via $\sigma : C \otimes C \to k$, $\sigma(c \otimes d) = \varepsilon(c)\varepsilon(d)$.
3. Let $C = \mathcal{M}^n(k)$ be the comatrix coalgebra of order n. For any $a \in k$, the map

$$\sigma : C \otimes C \to k, \quad \sigma(c_j^i \otimes c_q^p) = \delta_j^i a$$

is a proper Long map.
Indeed, for $c = c_j^i$, $d = c_q^p$, we have

$$\sigma(c_{(1)} \otimes d)c_{(2)} = \sigma(c_t^i \otimes c_q^p)c_j^t = a c_j^i,$$

and

$$\sigma(c_{(2)} \otimes d)c_{(1)} = \sigma(c_j^t \otimes c_q^p)c_t^i = a c_j^i,$$

Proposition 143. *Let $\sigma : C \otimes C/I \to k$ a Long map, and consider a right C-comodule (M, ρ). Then the map*

$$R_{(\sigma, M, \rho)} : M \otimes M \to M \otimes M, \quad R_{(\sigma, M, \rho)}(m \otimes n) = \sigma(m_{[1]} \otimes \overline{n_{[1]}}) m_{[0]} \otimes n_{[0]}$$

is a solution of the Long equation.

Proof. A straightforward computation. We write $R = R_{(\sigma,M,\rho)}$. For all l, m, $n \in M$, we have

$$R^{12}R^{23}(l \otimes m \otimes n) = R^{12}\left(\sigma(m_{[1]} \otimes \overline{n_{[1]}})l \otimes m_{[0]} \otimes n_{[0]}\right)$$

$$= \sigma(m_{[2]} \otimes \overline{n_{[1]}})\sigma(l_{[1]} \otimes \overline{m_{[1]}})l_{[0]} \otimes m_{[0]} \otimes n_{[0]}$$

$$(7.17) \quad = \sigma(l_{[1]} \otimes \overline{m_{[2]}})\sigma(m_{[1]} \otimes \overline{n_{[1]}})l_{[0]} \otimes m_{[0]} \otimes n_{[0]}$$

$$= R^{23}\left(\sigma(l_{[1]} \otimes \overline{m_{[1]}})l_{[0]} \otimes m_{[0]} \otimes n\right)$$

$$= R^{23}R^{12}(l \otimes m \otimes n)$$

so R is a solution of the Long equation.

Theorem 75. *Let M be an n-dimensional vector space and $R \in \mathrm{End}(M \otimes M)$ a solution of the Long equation.*

1. *There exists a coideal $I(R)$ of the comatrix coalgebra $\mathcal{M}^n(k)$, a coaction ρ of $\mathcal{M}^n(k)$ on M, and a unique Long map*

$$\sigma : \mathcal{M}^n(k) \otimes \mathcal{M}^n(k)/I(R) \to k$$

   such that $R = R_{(\sigma,M,\rho)}$. Furthermore, if R is bijective, σ is convolution invertible in $\mathrm{Hom}_k(\mathcal{M}^n(k) \otimes \mathcal{M}^n(k)/I(R), k)$.
2. *If R is commutative or $R\tau = \tau R$, then there exists a Long coalgebra $(L(R), \tilde{\sigma})$ and a coaction $\tilde{\rho}$ of $L(R)$ on M such that $R = R_{(\tilde{\sigma},M,\tilde{\rho})}$.*

Proof. 1. As before, let $\{m_1, \cdots, m_n\}$, and write R in matrix form

$$R(m_u \otimes m_v) = x_{uv}^{ij} m_i \otimes m_j \tag{7.18}$$

The comatrix coalgebra $\mathcal{M}^n(k)$ coacts on M:

$$\rho(m_l) = m_v \otimes c_l^v$$

Let $I(R)$ be the k-subspace of $\mathcal{M}^n(k)$ generated by the o_{kl}^{ij}. We know from (7.14) that $I(R)$ is a coideal of $\mathcal{M}^n(k)$.

We will first prove that σ is unique. Let $\sigma : \mathcal{M}^n(k) \otimes \mathcal{M}^n(k)/I(R) \to k$ be a Long map such that $R = R_\sigma$. Then

$$R_\sigma(m_v \otimes m_u) = \sigma\left((m_v)_{[1]} \otimes \overline{(m_u)_{[1]}}\right)(m_v)_{[0]} \otimes (m_u)_{[0]}$$

$$= \sigma(c_v^i \otimes \overline{c_u^j})m_i \otimes m_j$$

and it follows using (7.18) that

$$\sigma(c_v^i \otimes \overline{c_u^j}) = x_{uv}^{ij} \tag{7.19}$$

and σ is completely determined.

We will now prove the existence of σ. First define

$$\sigma_0 : \ \mathcal{M}^n(k) \otimes \mathcal{M}^n(k) \to k, \qquad \sigma_0(c_v^i \otimes c_u^j) = x_{uv}^{ij}$$

We have to show that σ_0 factorizes through a map

$$\sigma : \ \mathcal{M}^n(k) \otimes \mathcal{M}^n(k)/I(R) \to k$$

To this end, it suffices to show that $\sigma_0(\mathcal{M}^n(k) \otimes I(R)) = 0$. This can be seen as follows

$$\sigma_0(c_q^p \otimes o_{kl}^{ij}) = x_{vk}^{ij}\sigma_0(c_q^p \otimes c_l^v) - x_{lk}^{\alpha j}\sigma_0(c_q^p \otimes c_\alpha^i)$$
$$= x_{vk}^{ij}x_{ql}^{pv} - x_{lk}^{\alpha j}x_{q\alpha}^{pi} = 0$$

We still have to show that σ is a Long map. For $c = c_j^i$ and $d = c_q^p$, we find

$$\sigma(c_{(1)} \otimes \overline{d})\,\overline{c_{(2)}} = \sigma(c_v^i \otimes \overline{c_q^p})\,\overline{c_j^v} = x_{vq}^{ip}\overline{c_j^v}$$

and

$$\sigma(c_{(2)} \otimes \overline{d})\,\overline{c_{(1)}} = \sigma(c_j^\alpha \otimes \overline{c_q^p})\,\overline{c_\alpha^i} = x_{jq}^{\alpha p}\overline{c_\alpha^i}$$

so

$$\sigma(c_{(1)} \otimes \overline{d})\,\overline{c_{(2)}} - \sigma(c_{(2)} \otimes \overline{d})\,\overline{c_{(1)}} = \overline{o_{qj}^{ip}} = 0$$

and it follows that σ is a Long-map.

Suppose now that R is bijective and let $S = R^{-1}$. Let (y_{uv}^{ij}) be a family of scalars of k such that

$$S(m_u \otimes m_v) = y_{uv}^{ij}m_i \otimes m_j,$$

S is the inverse of R, so

$$x_{\alpha\beta}^{pi}y_{qj}^{\alpha\beta} = \delta_q^p\delta_j^i = y_{\alpha\beta}^{pi}x_{qj}^{\alpha\beta}$$

We define

$$\sigma_0' : \ \mathcal{M}^n(k) \otimes \mathcal{M}^n(k) \to k, \quad \sigma_0'(c_v^i \otimes c_u^j) = y_{vu}^{ij}$$

First we prove that σ_0' is a convolution inverse of σ_0. We easily compute that

$$\sigma_0\Big((c_q^p)_{(1)} \otimes (c_j^i)_{(1)}\Big)\sigma_0'\Big((c_q^p)_{(2)} \otimes (c_j^i)_{(2)}\Big)$$
$$= \sigma_0(c_\alpha^p \otimes c_\beta^i)\sigma_0'(c_q^\alpha \otimes c_j^\beta) = x_{\alpha\beta}^{pi}y_{qj}^{\alpha\beta} = \delta_q^p\delta_j^i = \varepsilon(c_j^i)\varepsilon(c_q^p)$$

and

$$\sigma_0'\Big((c_q^p)_{(1)} \otimes (c_j^i)_{(1)}\Big)\sigma_0\Big((c_q^p)_{(2)} \otimes (c_j^i)_{(2)}\Big)$$
$$= \sigma_0'(c_\alpha^p \otimes c_\beta^i)\sigma_0(c_q^\alpha \otimes c_j^\beta) = y_{\alpha\beta}^{pi}x_{qj}^{\alpha\beta} = \delta_q^p\delta_j^i = \varepsilon(c_j^i)\varepsilon(c_q^p)$$

We next show that σ_0' factorizes through a map $\sigma' : \ \mathcal{M}^n(k)\otimes\mathcal{M}^n(k)/I(R) \to k$. We have

$$\sigma'_0(c_q^p \otimes o_{kl}^{ij}) = x_{vk}^{ij} y_{ql}^{pv} - x_{lk}^{\alpha j} y_{q\alpha}^{pi}$$

$S = R^{-1}$ and R is a solution of the Long equation, so $R^{23}S^{12} = S^{12}R^{23}$. Using (7.2) we obtain that $\sigma'_0(c_q^p \otimes o_{kl}^{ij}) = 0$, hence σ'_0 factorizes through a map σ'. σ' is a convolution inverse of σ.

2. Suppose first that $R\tau = \tau R$. Then

$$x_{uv}^{ji} = x_{vu}^{ij} \tag{7.20}$$

Let

$$L(R) = \mathcal{M}^n(k)/I(R)$$

The rest of the proof is similar to the proof of part 1), we only have to prove that the map $\sigma : \mathcal{M}^n(k) \otimes \mathcal{M}^n(k)/I(R) \to k$ factorizes through a map $\tilde{\sigma} : L(R) \otimes L(R) \to k$. This can be seen as follows

$$\sigma(o_{kl}^{ij} \otimes \overline{c_q^p}) = x_{vk}^{ij}\sigma(c_l^v \otimes \overline{c_q^p}) - x_{lk}^{\alpha j}\sigma(c_\alpha^i \otimes \overline{c_q^p})$$
$$= x_{vk}^{ij} x_{lq}^{vp} - x_{lk}^{\alpha j} x_{\alpha q}^{pi}$$
$$(7.20) \qquad = x_{vk}^{ij} x_{ql}^{pv} - x_{lk}^{\alpha j} x_{q\alpha}^{pi} = 0$$

If R is commutative, then we can prove that $R^{12}R^{13} = R^{13}R^{12}$ if and only if

$$x_{vk}^{ij} x_{lq}^{vp} = x_{lk}^{\alpha j} x_{\alpha q}^{ip}$$

Thus,

$$\sigma(o_{kl}^{ij} \otimes \overline{c_q^p}) = x_{vk}^{ij} x_{lq}^{vp} - x_{lk}^{\alpha j} x_{\alpha q}^{ip} = 0$$

and again σ factorizes through a map $\tilde{\sigma} : L(R) \otimes L(R) \to k$.

As an immediate consequence, we obtain

Corollary 39. Let M be a finite dimensional vector space and $R \in \text{End}(M \otimes M)$. The following statements are equivalent:

1. R is a solution of the system

$$\begin{cases} R^{12}R^{13} = R^{13}R^{12} \\ R^{12}R^{23} = R^{23}R^{12} \end{cases} \tag{7.21}$$

   in $\text{End}(M \otimes M \otimes M)$;
2. there exists a Long coalgebra $(L(R), \sigma)$ and a structure of right $L(R)$-comodule (M, ρ) such that $R = R_{(\sigma, M, \rho)}$.

Remark 24. If R is a commutative solution of the Long equation, then R satisfies the integrability condition

$$[R^{12}, R^{13} + R^{23}] = 0$$

which appears in the study of the *Knizhnik-Zamolodchikov equation* (see [108], [166]).

Examples 17. 1. Let $C = \mathcal{M}^4(k)$ and I the two dimensional k-subspace of C with k-basis $\{c_{21}, c_{22} - c_{11}\}$. Then I is a coideal of C. Take scalars $a, b, c \in k$ and let (x_{uv}^{ij}) be given by the formulas (7.16). Then

$$\sigma : \mathcal{M}^4(k) \otimes \mathcal{M}^4(k)/I \to k, \quad \sigma(c_u^i \otimes \overline{c_v^j}) = x_{uv}^{ij}$$

is a Long map.

2. Let $C = \mathcal{M}^4(k)$, $q \in k$ and I the two-dimensional coideal of C with k-basis $\{c_{21}, c_{12} + qc_{22} - qc_{11}\}$. Let

$$x_{12}^{11} = -q, \quad x_{12}^{12} = 1, \quad x_{22}^{11} = -q^2, \quad x_{22}^{12} = q$$

and $(x_{uv}^{ij}) = 0$ in all other situations. Then

$$\sigma : \mathcal{M}^4(k) \otimes \mathcal{M}^4(k)/I \to k, \quad \sigma(c_u^i \otimes \overline{c_v^j}) = x_{uv}^{ij}$$

is a Long map.

3. Let $C = \mathcal{M}^4(k)$ and I the two-dimensional coideal of C with basis $\{c_{12}, c_{21}\}$. Let $a, b \in k$ and

$$\sigma : \mathcal{M}^4(k) \otimes \mathcal{M}^4(k)/I \to k$$

such that

$$\sigma(c_1^1 \otimes \overline{c_1^1}) = a, \quad \sigma(c_1^1 \otimes \overline{c_1^2}) = b$$

and all others are zero. Then σ is a Long map.

4. Let n be a positive integer and $\phi : \{1, \cdots, n\} \to \{1, \cdots, n\}$ a function with $\phi^2 = \phi$. It is easy to see that $R = (x_{ij}^{kl})$ given by

$$x_{uv}^{ji} = \delta_{uv} \delta_{\phi(i)v} \delta_{\phi(j)v}$$

for all $i, j, u, v = 1, \cdots, n$ is a commutative solution of the Long equation. Hence we can construct the Long coalgebra $L(R)$. By a long but trivial computation we can show that the coalgebra $L(R)$ is a quotient of the comatrix coalgebra $\mathcal{M}^n(k)$ through the coideal I generated by elements of the form

$$\begin{cases} c_{\phi(j)l}, & \text{for all } l \neq \phi(j) \\ \sum_{\alpha \in \phi^{-1}(\phi(j))} c_{i\alpha}, & \text{for all } \phi(i) \neq \phi(j) \\ \sum_{\alpha \in \phi^{-1}(\phi(i))} c_{i\alpha} - c_{\phi(i)\phi(i)}, & \text{for all } i = 1, \cdots, n \end{cases} \tag{7.22}$$

We will now construct a specific example. Let $n = 4$ and ϕ given by

$$\phi(1) = 1, \quad \phi(2) = \phi(3) = \phi(4) = 2.$$

Then I is the coideal generated by (cf.(7.22)): c_{1l}, c_{2j}, c_{i1}, $c_{32} + c_{33} + c_{34} - c_{22}$, $c_{42} + c_{43} + c_{44} - c_{22}$ for all $l \neq 1$, $j \neq 2$, $i \neq 1$. Now, if we denote $c_{11} = x_1$, $c_{22} = x_2$, $c_{32} = x_3$, $c_{33} = x_4$, $c_{42} = x_5$, $c_{44} = x_6$ we obtain the description of the corresponding Long coalgebra $L(R)$: $L(R)$ is the six dimensional vector space with $\{x_1, \cdots, x_6\}$ as a basis. The comultiplication Δ and the couinty ε are given by

$$\Delta(x_1) = x_1 \otimes x_1, \quad \Delta(x_2) = x_2 \otimes x_2,$$

$$\Delta(x_3) = x_3 \otimes x_2 + x_4 \otimes x_3 + (x_2 - x_3 - x_4) \otimes x_5$$

$$\Delta(x_4) = x_4 \otimes x_4 + (x_2 - x_3 - x_4) \otimes (x_2 - x_5 - x_6),$$

$$\Delta(x_5) = x_5 \otimes x_2 + (x_2 - x_5 - x_6) \otimes x_3 + x_6 \otimes x_5$$

$$\Delta(x_6) = (x_2 - x_5 - x_6) \otimes (x_2 - x_3 - x_4) + x_6 \otimes x_6$$

$$\varepsilon(x_1) = \varepsilon(x_2) = \varepsilon(x_4) = \varepsilon(x_6) = 1, \quad \varepsilon(x_3) = \varepsilon(x_5) = 0.$$

8 The Frobenius-Separability equation

We introduce and study the Frobenius-separability equation (or FS-equation) $R^{12}R^{23} = R^{23}R^{13} = R^{13}R^{12}$; we will see that it implies the braid equation, in the sense that all solutions of the FS-equation are solutions of the braid equation. The FS-equation can be used to determine the structure of separable algebras and Frobenius algebras. Given a solution R of the FS-equation satisfying a certain normalizing condition, we construct a Frobenius or a separable algebra $\mathcal{A}(R)$ that can be described using generators and relations. Furthermore, any finite dimensional Frobenius or separable algebra is isomorphic to such an $\mathcal{A}(R)$.

It is remarkable that the same equation can be used to describe two different kinds of algebras, namely separable algebras and Frobenius algebras. We had a similar phenomenon in Chapter 3, where we gave a categorical explanation for the relation between separability properties and Frobenius type properties. Here also we will see that the difference lies in the normalizing properties.

8.1 Frobenius algebras and separable algebras

Let A be a k-algebra. An element $e = \sum e^1 \otimes e^2 = e^1 \otimes e^2 \in A \otimes A$ will be called A-central if for any $a \in A$ we have

$$a \cdot e = e \cdot a \tag{8.1}$$

where $A \otimes A$ is viewed as an A-bimodule in the usual way

$$a_1 \cdot (b \otimes c) \cdot a_2 = a_1 b \otimes c a_2$$

for all $a_1, a_2, b, c \in A$. Of course, there exists a bijection between the set of all A-central elements and the set of all A-bimodule maps $\Delta : A \to A \otimes A$. Recall that A is called a *separable algebra* if there exists a separability idempotent, that is an A-central element $e = e^1 \otimes e^2 \in A \otimes A$ satisfying the normalizing separability condition

$$e^1 e^2 = 1. \tag{8.2}$$

A is separable if and only if the multiplication map $m : A \otimes A \to A$ splits in the category $_A\mathcal{M}_A$ of A-bimodules.

Remark 25. We have proved in Proposition 12 that any separable k-algebra is finite dimensional over k, and in Proposition 13 we have show that any separable algebra is semisimple.

Also recall that a k-coalgebra C is called *coseparable* if there exists a coseparability idempotent, that is a k-linear map $\sigma : C \otimes C \to k$ such that

$$\sigma(c \otimes d_{(1)})d_{(2)} = \sigma(c_{(2)} \otimes d)c_{(1)} \quad \text{and} \quad \sigma(c_{(1)} \otimes c_{(2)}) = \varepsilon(c)$$

for all c, $d \in C$.

For a k-algebra A, the dual $A^* = \operatorname{Hom}(A, k)$ is an (A, A)-bimodule with A-actions given by the formulas

$$(a \cdot f \cdot b)(x) = f(bxa)$$

for all a, b, $x \in A$ and $f \in A^*$. We recall that a k-algebra A is called a *Frobenius algebra* if A is finite dimensional over k and there exists an isomorphism of left A-modules $A \cong A^*$. Furthermore, A is called a *symmetric algebra* if $A \cong A^*$ as A-bimodules. Of course, a symmetric algebra is always a Frobenius algebra.

Remark 26. Using Wedderburn's Theorem, Eilenberg and Nakayama proved that any semisimple k-algebra is symmetric (cf. e.g. [113, p. 443]). In particular, any separable k-algebra is Frobenius. There exist necessary and sufficient conditions for a Frobenius algebra to be separable; for a detailed discussion, we refer to [100]. We will see below that a Frobenius algebra is separable if its characteristic element ω_A is invertible.

Proposition 144. *Let A be a k-algebra, $\Delta : A \to A \otimes A$ an A-bimodule map and $e = \Delta(1_A)$.*

1. *In $A \otimes A \otimes A$, we have the equality*

$$e^{12}e^{23} = e^{23}e^{13} = e^{13}e^{12} \tag{8.3}$$

2. *Δ is coassociative;*
3. *If (A, Δ, ε) is a coalgebra structure on A, then A is finite dimensional over k.*

Proof. 1. From the fact that Δ is an A-bimodule map, it follows immediately that e A-central. Write $E = E^1 \otimes E^2 = e$. Then

$$\begin{aligned}
e^{12}e^{23} &= (e^1 \otimes e^2 \otimes 1)(1 \otimes E^1 \otimes E^2) \\
&= e^1 \otimes \underline{e^2 E^1} \otimes E^2 \\
&= E^1 e^1 \otimes e^2 \otimes E^2 = e^{13}e^{12}
\end{aligned}$$

and

$$\begin{aligned}
e^{12}e^{23} &= e^1 \otimes \underline{e^2 E^1 \otimes E^2} \\
&= e^1 \otimes E^1 \otimes E^2 e^2 = e^{23}e^{13}
\end{aligned}$$

2. follows from 1. and the formulas

$$(\Delta \otimes I)\Delta(a) = e^{12}e^{23} \cdot (1_A \otimes 1_A \otimes a)$$

$$(I \otimes \Delta)\Delta(a) = e^{23}e^{13} \cdot (1_A \otimes 1_A \otimes a)$$

3. Let $a \in A$. Applying $\varepsilon \otimes I_A$ and $I_A \otimes \varepsilon$ to (8.1), we obtain, using the fact that ε is a counit map,

$$a = \varepsilon(ae^1)e^2 = e^1\varepsilon(e^2a)$$

and it follows that $\{e^1, \varepsilon(e^2\bullet)\}$ (or $\{e^2, \varepsilon(\bullet e^1)\}$) are dual bases of A as a vector space over k.

The next Corollary is a special case of the item 3) of Theorem 27: the equivalence 1) $\Leftrightarrow$ 3) has been proved by Abrams (see [4, Theorem 2.1]).

Corollary 40. *For a k-algebra A, the following statements are equivalent:*

1. *A is a Frobenius algebra;*
2. *there exist $e = e^1 \otimes e^2 \in A \otimes A$ and $\varepsilon \in A^*$ such that e is A-central and the normalizing Frobenius condition*

$$\varepsilon(e^1)e^2 = e^1\varepsilon(e^2) = 1_A \tag{8.4}$$

   is satisfied. (ε, e) is called a Frobenius pair.
3. *there exist a coalgebra structure $(A, \Delta_A, \varepsilon_A)$ on A such that the comultiplication $\Delta_A : A \to A \otimes A$ is an A-bimodule map.*

Proof. 1. $\Leftrightarrow$ 2. is a special case of Theorem 27.
2. $\Leftrightarrow$ 3.: observe that an A-bimodule map $\Delta : A \to A \otimes A$ is completely determined by $e = \Delta(1_A)$, and that there is a bijective corespondence between the set of all A-central elements and the set of all A-bimodule maps $\Delta : A \to A \otimes A$, and use the second statement in Proposition 144. It is easy to see that the counit property is satisfied if and only if (8.4) holds.

Let A be a Frobenius algebra over a field k. The element $w_A = (m_A \circ \Delta)(1_A) \in A$ is called the *characteristic element* of A and it generalizes the classical Euler class $e(X)$ of a connected orientated compact manifold X ([5]). If the characteristic element w_A is invertible in A then A is a separable k-algebra. Indeed, let $\Delta(1_A) = e^1 \otimes e^2$. Then

$$ae^1 \otimes e^2 = e^1 \otimes e^2a$$

for all $a \in A$. Hence, $w_A \in Z(A)$, the center of A. It follows that its inverse w_A^{-1} is also an element in $Z(A)$. Now,

$$R = w_A^{-1}(e^1 \otimes e^2)$$

is a separability idempotent, so A is a separable k-algebra.
Our next result is the coalgebra version of Corollary 40.

Theorem 76. *For a coalgebra C over a field k, the following statements are equivalent:*

1. *C is a co-Frobenius coalgebra, i.e. the forgetful functor $F : {}^C\mathcal{M} \to {}_k\mathcal{M}$ is Frobenius;*
2. *C is finite dimensional and there exists an associative and unitary algebra structure $(C, m_C, 1_C)$ on C such that the multiplication $m_C : C \otimes C \to C$ is left and right C-colinear.*

8.2 The Frobenius-separability equation

Proposition 144 leads us to the following definition.

Definition 18. *Let A be a k-algebra and $R = R^1 \otimes R^2 \in A \otimes A$.*

1. *R is called a solution of the FS-equation (or Frobenius-separability equation) if*
$$R^{12}R^{23} = R^{23}R^{13} = R^{13}R^{12} \tag{8.5}$$
   in $A \otimes A \otimes A$.
2. *R is called a solution of the S-equation (or separability equation) if R is a solution of the FS-equation and the normalizing separability condition holds:*
$$R^1 R^2 = 1_A \tag{8.6}$$
3. *(R, ε) is called a solution of the F-equation (Frobenius equation) if R is a solution of the FS-equation, and $\varepsilon \in A^*$ is such that the normalizing Frobenius condition holds:*
$$\varepsilon(R^1)R^2 = R^1\varepsilon(R^2) = 1_A. \tag{8.7}$$
4. *Two solutions of the FS-equation are called equivalent if there exists an invertible element $u \in A$ such that $S = (u \otimes u)R(u^{-1} \otimes u^{-1})$.*

Remarks 23. 1. The FS-equation appeared first in [15, Lemma 3.6]. If R is a solution of the FS-equation then R is also a solution of the braid equation

$$R^{12}R^{23}R^{12} = R^{23}R^{12}R^{23} \tag{8.8}$$

2. In many applications, the algebra A is of the form $A = \text{End}(M)$, where M is a finite dimensional vector space. Then we can view R as an element of $\text{End}(M \otimes M)$, using the canonical isomorphism $\text{End}(M) \otimes \text{End}(M) \cong \text{End}(M \otimes M)$. (8.5) can then be viewed as an equation in $\text{End}(M \otimes M \otimes M)$. This is what we will do in Proposition 145 and in most of the examples in this Section.

3. We will prove now that the FS-equation is at the same time an associativity and coassociativity constraint. First, let H be a Hopf algebra and let

$$R = \beta: \; H \otimes H \to H \otimes H, \quad R(g \otimes h) = gh_{(1)} \otimes h_{(2)}$$

be the canonical map that shows that H is a Hopf-Galois extension of k. We have proved in Examples 10 that R is a solution of the Hopf equation. Moreover, the comultiplication Δ_H and the multiplication m_H of H can be recovered from R, namely

$$\Delta_H(h) = R(1_H \otimes h) \quad \text{and} \quad m_H = (\text{Id} \otimes \varepsilon)R$$

We generalize this construction as follows. Let M be a vector space, $1_M \in M$, $\varepsilon \in M^*$, and $R \in \text{End}(M \otimes M)$. We define

$$\Delta = \Delta_R: \; M \to M \otimes M, \quad \Delta(x) = R(1_M \otimes x)$$

for all $x \in M$ and

$$m = m_R: \; M \otimes M \to M, \quad m = (\text{Id} \otimes \varepsilon)R$$

Then we have

$$(\Delta \otimes \text{Id})\Delta(x) = (R^{12}R^{23})(1_M \otimes 1_M \otimes x)$$
$$(\text{Id} \otimes \Delta)\Delta(x) = (R^{23}R^{13})(1_M \otimes 1_M \otimes x)$$

for all $x \in M$ and

$$m(\text{Id} \otimes m) = (\text{Id} \otimes \varepsilon \otimes \varepsilon)R^{12}R^{23}, \quad m(m \otimes \text{Id}) = (\text{Id} \otimes \varepsilon \otimes \varepsilon)R^{13}R^{12}.$$

We conclude that, if R is a solution of the FS-equation, then Δ_R is coassociative and m_R is an associative.

The FS-equation can be rewritten in matrix form. We follow the notation introduced in Section 5.1. Fix a basis $\{m_1, \cdots, m_n\}$ of M. A linear map $R: \; M \otimes M \to M \otimes M$ can be described by its matrix $X = (x_{uv}^{ij})$. We have

$$R(m_u \otimes m_v) = x_{uv}^{ij} m_i \otimes m_j \tag{8.9}$$

and

$$R = x_{uv}^{ij} e_i^u \otimes e_j^v \tag{8.10}$$

It is then straightforward to compute

$$R^{12}R^{23}(m_u \otimes m_v \otimes m_w) = x_{uk}^{ij} x_{vw}^{kl} m_i \otimes m_j \otimes m_l \tag{8.11}$$
$$R^{23}R^{13}(m_u \otimes m_v \otimes m_w) = x_{vk}^{jl} x_{uw}^{ik} m_i \otimes m_j \otimes m_l \tag{8.12}$$
$$R^{13}R^{12}(m_u \otimes m_v \otimes m_w) = x_{kw}^{il} x_{uv}^{kj} m_i \otimes m_j \otimes m_l \tag{8.13}$$

Proposition 145. *For $R \in \text{End}(M \otimes M)$, we have, with notation as above.*

1. R is a solution of the FS-equation if and only if

$$x_{uk}^{ij}x_{vw}^{kl} = x_{vk}^{jl}x_{uw}^{ik} = x_{kw}^{il}x_{uv}^{kj} \tag{8.14}$$

for all $i, j, l, u, v, w \in \{1, \cdots, n\}$.
2. R satisfies (8.6) if and only if

$$x_{ik}^{kj} = \delta_i^j \tag{8.15}$$

for all $i, j \in \{1, \cdots, n\}$.
3. Let ε be the trace map. Then (R, ε) satisfies (8.7) if and only if

$$x_{ki}^{kj} = x_{ik}^{jk} = \delta_i^j \tag{8.16}$$

for all $i, j \in \{1, \cdots, n\}$.

Proof. 1. follows immediately from (8.11-8.13). 2. and 3. follow from (8.10), using the multiplication rule

$$e_j^i e_l^k = \delta_l^i e_j^k$$

and the formula for the trace

$$\varepsilon(e_j^i) = \delta_j^i$$

As a consequence of Proposition 145, we can show that $R \in \text{End}(M \otimes M)$ is a solution of the equation $R^{12}R^{23} = R^{13}R^{12}$ if and only if a certain multiplication on $M \otimes M$ is associative.

Corollary 41. *Let $\{m_1, \cdots, m_n\}$ be a basis of M and $R \in \text{End}(M \otimes M)$, given by (8.9). Then R is a solution of the equation $R^{12}R^{23} = R^{13}R^{12}$ if and only if the multiplication on $M \otimes M$ given by*

$$(m_k \otimes m_l) \cdot (m_r \otimes m_j) = x_{jl}^{ak} m_r \otimes m_a$$

$(k, l, r, j = 1, \cdots, n)$ is associative.
In this case M is a left $M \otimes M$-module with structure

$$(m_k \otimes m_l) \bullet m_j = \sum_a x_{jl}^{ak} m_a$$

for all $k, l, j = 1, \cdots, n$.

Proof. Write $m_{kl} = m_k \otimes m_l$. Then

$$m_{pq} \cdot (m_{kl} \cdot m_{rj}) = x_{jl}^{ak} m_{pq} m_{ra} = x_{jl}^{ak} x_{aq}^{ip} m_{ri}$$

and

$$(m_{pq} \cdot m_{kl}) \cdot m_{rj} = x_{lq}^{ap} m_{ka} m_{rj} = x_{ja}^{ik} x_{lq}^{ap} m_{ri}$$

such that

$$(m_{pq} \cdot m_{kl}) \cdot m_{rj} - m_{pq} \cdot (m_{kl} \cdot m_{rj}) = \left(x_{ja}^{ik} x_{lq}^{ap} - x_{jl}^{ak} x_{aq}^{ip} \right) m_{ri}$$

and the right hand side is zero for all indices p, q, k, l, r and j if and only if (8.14) holds, and in Proposition 145, we have seen that (8.14) is equivalent to $R^{12} R^{23} = R^{13} R^{12}$.

The last statement follows from

$$(m_{pq} \cdot m_{kl}) \bullet m_j - m_{pq} \bullet (m_{kl} \bullet m_j) = \left(x_{ja}^{ik} x_{lq}^{ap} - x_{jl}^{ak} x_{aq}^{ip} \right) m_i = 0$$

where we used (8.14) at the last step.

Examples 18. 1. The identity $I_{M \otimes M}$ and the switch map τ_M are trivial solutions of the FS-equation in $\text{End}(M \otimes M \otimes M)$.

2. Let $A = \mathcal{M}_n(k)$. Then

$$R_j = \sum_{i=1}^{n} e_i^j \otimes e_j^i$$

is a solution of the S-equation and $(R = \sum_{j=1}^{n} R_j, \text{trace})$ is a solution of the F-equation.

3. Let $R \in A \otimes A$ be a solution of FS-equation and $u \in A$ invertible. Then

$$^u R = (u \otimes u) R (u^{-1} \otimes u^{-1}) \tag{8.17}$$

is also a solution of the FS-equation. Let $\mathbf{FS}(A)$ be the set of all solutions of the FS-equation, and $U(A)$ the multiplicative group of invertible elements in A. Then (8.17) defines an action of $U(A)$ on $\mathbf{FS}(A)$.

4. If $a \in A$ is an idempotent, then $a \otimes a$ is a solution of the FS-equation.

5. Let A be a k-algebra, and $e \in A \otimes A$ an A-central element. Then for any left A-module M, the homotety $R = R_e : M \otimes M \to M \otimes M$ given by

$$R(m \otimes n) = e^1 \cdot m \otimes e^2 \cdot n \tag{8.18}$$

$(m, n \in M)$ is a solution of the FS-equation in $\text{End}(M \otimes M \otimes M)$. This is an easy consequence of (8.3). Moreover, if e is a separability idempotent (respectively (e, ε) is a Frobenius pair), then R is a solution of the S-equation (respectively a solution of the F-equation).

6. Let G be a finite group, and $A = kG$. Then $e = \sum_{g \in G} g \otimes g^{-1}$ is an A-central element and (e, p_1) is a Frobenius pair ($p_1 : kG \to k$ is the map defined by $p_1(g) = \delta_{1,g}$, for all $g \in G$). Hence, kG is a Frobenius algebra. Furthermore, if $(\text{char}(k), |G|) = 1$, then $e' = |G|^{-1} e$ is a separability idempotent.

7. Using a computer, Bogdan Ichim computed for us that $\mathbf{FS}(\mathcal{M}_2(Z_2))$ (resp. $\mathbf{FS}(\mathcal{M}_2(Z_3))$) consists of exactly 38 (resp. 187) solutions of the FS-equation. We will present only two of them. Let k be a field of characteristic 2 (resp. 3). Then

$$R = \begin{pmatrix} 1 & 1 & 1 & 0 \\ 1 & 0 & 0 & 1 \\ 1 & 0 & 0 & 1 \\ 0 & 1 & 1 & 1 \end{pmatrix}, \quad (\text{resp.} \quad R = \begin{pmatrix} 1 & 0 & 0 & 1 \\ 0 & 1 & 1 & 2 \\ 0 & 1 & 1 & 2 \\ 1 & 2 & 2 & 1 \end{pmatrix})$$

are solutions of FS-equation.

8. Let $\{m_1, m_2, \cdots, m_n\}$ be a basis of M, a^i_j scalars of k and let R be given by

$$R(m_u \otimes m_v) = \sum_j a^j_u m_v \otimes m_j$$

Thus $x^{ij}_{uv} = \delta^i_v a^j_u$, with notation as in (8.9). An immediate verification shows that R is a solution of the FS-equation. If ε is the trace map, then (R, ε) is a solution of the F-equation if and only if $a^j_i = \delta^j_i$. R is a solution of the S-equation if and only if n is invertible in k, and $na^j_i = \delta^j_i$.

If H is a finite dimensional unimodular involutory Hopf algebra, and t is a two-sided integral in H, then $R = t_{(1)} \otimes S(t_{(2)})$ is a solution of the quantum Yang-Baxter equation (cf. [114, Theorem 8.3.3]). In our next Proposition, we will show that, for an arbitrary Hopf algebra H, R is a solution of the FS-equation and the braid equation.

Proposition 146. *Let H be a Hopf algebra and $t \in H$ a left integral of H. Then $R = t_{(1)} \otimes S(t_{(2)}) \in H \otimes H$ is H-central, and therefore a solution of the FS-equation and the braid equation.*

Proof. For all $h \in H$, we have that $ht = \varepsilon(h)t$ and, subsequently,

$$h_{(1)}t \otimes h_{(2)} = t \otimes h$$

$$h_{(1)}t_{(1)} \otimes S(h_{(2)}t_{(2)}) \otimes h_{(3)} = t_{(1)} \otimes S(t_{(2)}) \otimes h$$

Multiplying the second and the third factor, we obtain

$$ht_{(1)} \otimes S(t_{(2)}) = t_{(1)} \otimes S(t_{(2)})h$$

proving that R is H-central.

Remarks 24. 1. If $\varepsilon(t) = 1$, then R is a separability idempotent and H is separable over k, and we recover Maschke's Theorem for Hopf algebras (see [1]).

2. If t is a right integral, then a similar argument shows that $S(t_{(1)}) \otimes t_{(2)}$ is an H-central element.

Let H be a Hopf algebra and $\sigma : H \otimes H \to k$ be a normalized Sweedler 2-cocycle, i.e.

$$\sigma(h \otimes 1_H) = \sigma(1_H \otimes h) = \varepsilon(h)$$

$$\sigma(k_{(1)} \otimes l_{(1)})\sigma(h \otimes k_{(2)}l_{(2)}) = \sigma(h_{(1)} \otimes k_{(1)})\sigma(h_{(2)}k_{(2)} \otimes l) \qquad (8.19)$$

for all h, k, $l \in H$. The *crossed product* algebra H_σ is equal to H as k-module and the (associative) multiplication is given by

$$g \cdot h = \sigma(g_{(1)} \otimes h_{(1)})g_{(2)}h_{(2)}$$

for all $g, h \in H_\sigma = H$.

Proposition 147. *Let H be a cocommutative Hopf algebra over a commutative ring k, $t \in H$ a right integral, and $\sigma : H \otimes H \to k$ a normalized convolution invertible Sweedler 2-cocycle. Then*

$$R_\sigma = \sigma^{-1}\Big(S(t_{(2)}) \otimes t_{(3)}\Big)S(t_{(1)}) \otimes t_{(4)} \in H_\sigma \otimes H_\sigma \qquad (8.20)$$

is H_σ-central, and a solution of the FS-equation. Consequently, if H is a separable algebra, then H_σ is also a separable algebra.

Proof. The method of proof is the same as in Proposition 146, but the situation is more complicated. For all $h \in H$, we have $th = \varepsilon(h)t$, and

$$h \otimes t = h_{(1)} \otimes th_{(2)}$$

and

$$h \otimes S(t_{(1)}) \otimes S(t_{(2)}) \otimes t_{(3)} \otimes t_{(4)}$$
$$= h_{(1)} \otimes S(h_{(2)})S(t_{(1)}) \otimes S(h_{(3)})S(t_{(2)}) \otimes t_{(3)}h_{(4)} \otimes t_{(4)}h_{(5)}$$

We now compute easily that

$$h \cdot R_\sigma = \sigma^{-1}(S(t_{(2)}) \otimes t_{(3)})h \cdot S(t_{(1)}) \otimes t_{(4)}$$
$$= \sigma^{-1}(S(h_{(3)})S(t_{(2)} \otimes t_{(3)}h_{(4)})h_{(1)} \cdot (S(h_{(2)})S(t_{(1)})) \otimes t_{(4)}h_{(5)}$$
$$= \sigma^{-1}(S(h_{(3)})S(t_{(2)} \otimes t_{(3)}h_{(4)})\sigma((h_{(1)})_{(1)}$$
$$\qquad \otimes S(h_{(2)})_{(1)}S(t_{(1)})_{(1)})(h_{(1)})_{(2)}S(h_{(2)})_{(2)}S(t_{(1)})_{(2)} \otimes t_{(4)}h_{(5)}$$
$$= \sigma^{-1}(S(h_{(3)})S(t_{(3)} \otimes t_{(4)}h_{(4)})\sigma(h_{(1)} \otimes S(h_{(2)}S(t_{(2)})))S(t_{(1)}) \otimes t_{(5)}h_{(5)}$$

On the other hand

$$R_\sigma \cdot h = \sigma^{-1}(S(t_{(2)}) \otimes t_{(3)})S(t_{(1)}) \otimes t_{(4)} \cdot h$$
$$= \sigma^{-1}(S(t_{(2)}) \otimes t_{(3)})\sigma(t_{(4)} \otimes h_{(1)})S(t_{(1)}) \otimes t_{(5)}h_{(2)}$$

In order to prove that R_σ is H_σ-central, it suffices to show that, for all $f, g \in H$:

$$\sigma^{-1}(S(h_{(3)})S(g_{(2)}) \otimes g_{(3)}h_{(4)})\sigma(h_{(1)} \otimes S(h_{(2)})S(g_{(1)}))$$
$$= \sigma^{-1}(S(g_{(1)}) \otimes g_{(2)})\sigma(g_{(3)} \otimes h) \qquad (8.21)$$

So far, we have not used the cocommutativity of H. If H is cocommutative, then we can omit the Sweedler indices, since they contain no information. Hence we can write

$$\Delta(h) = \sum h \otimes h = h \otimes h$$

The cocycle relation (8.19) can then be rewritten as

$$\sigma(h \otimes kl) = \sigma^{-1}(k \otimes l)\sigma(h \otimes k)\sigma(kh \otimes l) \tag{8.22}$$
$$\sigma(hk \otimes l) = \sigma^{-1}(h \otimes k)\sigma(k \otimes l)\sigma(h \otimes kl) \tag{8.23}$$

Using (8.22), (8.23) and the fact that σ is normalized, we compute

$$\sigma^{-1}(S(h)S(g) \otimes gh)\sigma(h \otimes S(h)S(g))$$
$$= \sigma(S(h) \otimes S(g))\sigma^{-1}(S(g) \otimes gh)\sigma^{-1}(S(h) \otimes S(g)gh)$$
$$\sigma^{-1}(S(h) \otimes S(g))\sigma(h \otimes S(h))\sigma(hS(h) \otimes g)$$
$$= \sigma(g \otimes h)\sigma^{-1}(S(g) \otimes g)\sigma^{-1}(S(g)g \otimes h)$$
$$= \sigma^{-1}(S(g) \otimes g)\sigma(g \otimes h)$$

proving (8.21). We also used that

$$\sigma(h \otimes S(h)) = \sigma(S(h) \otimes h)$$

which follows from the cocycle condition and the fact that σ is normalized. Finally, if H is separable, then we can find a right integral t such that $\varepsilon(t) = 1$, and we easily see that $m_{H_\sigma}(R_\sigma) = 1$, proving that R_σ is a solution of the S-equation.

In [84] solutions of the braid equation are constructed starting from 1-cocycles on a group G. The interesting point in this construction is that, at set theory level, any "nondegenerate symmetric" solution of the braid equation arises in this way (see [84, Theorem 2.9]). Now, taking G a finite group and $H = kG$ in Proposition 147, we find a large class of solutions to the braid equation, arising from 2-cocycles $\sigma : G \times G \to k^*$. These solutions R can be described using a family of scalars (x_{uv}^{ij}), as in Proposition 145, where the indices now run through G. Let $n = |G|$, and write $M_n(k) \cong M_G(k)$, with entries indexed by $G \times G$.

Corollary 42. *Let G be a finite group of order n, and $\sigma : G \times G \to k^*$ a normalized 2-cocycle. Then $R_\sigma = (x_{uv}^{ij})_{i,j,u,v \in G}$ given by*

$$x_{uv}^{ij} = \delta_{j,\, ui^{-1}v}\; \sigma^{-1}(iu^{-1}, ui^{-1})\sigma(iu^{-1}, u)\sigma(ui^{-1}, v) \tag{8.24}$$

(i, j, u, $v \in G$) is a solution of the FS-equation. If n is invertible in k, then $n^{-1}R$ is a solution of the S-equation.

Proof. The twisted group algebra $k_\sigma G$ is the k-module with basis G, and multiplication given by $g \cdot h = \sigma(g, h)gh$, for any $g, h \in G$. $t = \sum_{g \in G} g$ is a left integral in kG and the element R_σ defined in (8.20) takes the form

$$R_\sigma = \sum_{g \in G} \sigma^{-1}(g^{-1}, g)g^{-1} \otimes g$$

Using the multiplication rule on $k_\sigma G$, we find that the map

$$\tilde{R}_\sigma : k_\sigma G \otimes k_\sigma G \to k_\sigma G \otimes k_\sigma G, \quad \tilde{R}_\sigma(u \otimes v) = R_\sigma \cdot (u \otimes v)$$

is given by

$$\tilde{R}_\sigma(u \otimes v) = \sum_{i \in G} \sigma^{-1}(iu^{-1}, ui^{-1})\sigma(iu^{-1}, u)\sigma(ui^{-1}, v) \, i \otimes ui^{-1}v$$

If we write

$$\tilde{R}_\sigma(u \otimes v) = \sum_{i,j \in G} x_{uv}^{ij} \, i \otimes j$$

then x_{uv}^{ij} is given by (8.24).

We will now present a coalgebra version of Example 18.3. First we adapt an old definition of Larson ([116]).

Definition 19. *Let C be a k-coalgebra. A map $\sigma : C \otimes C \to k$ is called an FS-map if*

$$\sigma(c \otimes d_{(1)})d_{(2)} = \sigma(c_{(2)} \otimes d)c_{(1)}. \tag{8.25}$$

If, in addition, σ satisfies the normalizing condition

$$\sigma(c_{(1)} \otimes c_{(2)}) = \varepsilon(c) \tag{8.26}$$

then σ is called a coseparability idempotent.
If there exists an $f \in C$ such that the FS-map σ satisfies the normalizing condition

$$\sigma(f \otimes c) = \sigma(c \otimes f) = \varepsilon(c) \tag{8.27}$$

for all $c \in C$, then we call (σ, f) an F-map.

The following Corollary is a special case of Theorem 35.

Corollary 43. *Let C be a coalgebra.*

1. *The forgetful functor $F : \mathcal{M}^C \to \mathcal{M}_k$ is separable (or equivalently C is a coseparable coalgebra) if and only if there exists a coseparability idempotent σ.*
2. *Let $G = - \otimes C : \mathcal{M}_k \to \mathcal{M}^C$ be the right adjoint of F. Then (F, G) is a Frobenius pair if and only if there exists an F-map (σ, f).*

Using FS-maps, we can construct solutions of the FS-equation.

Examples 19. 1. The comatrix coalgebra $\mathcal{M}^n(k)$ is coseparable and

$$\sigma : \mathcal{M}^n(k) \otimes \mathcal{M}^n(k) \to k, \quad \sigma(c_i^j \otimes c_k^l) = \delta_k^j \delta_l^i$$

is a coseparability idempotent.

2. Let k be a field of characteristic zero, and consider the Hopf algebra $C = k[Y]$, with $\Delta(Y) = Y \otimes 1 + 1 \otimes Y$ and $\varepsilon(Y) = 0$. Then there is only one FS-map $\sigma : C \otimes C \to k$, namely the zero map. Indeed, σ is completely described by $\sigma(Y^i \otimes Y^j) = a_{ij}$. We will show that all $a_{ij} = 0$. Using the fact that $\Delta(Y^n) = \Delta(Y)^n$, we easily find that (8.25) is equivalent to

$$\sum_{j=0}^{m} \binom{m}{j} a_{n,m-j} Y^j = \sum_{i=0}^{n} \binom{n}{i} a_{n-i,m} Y^i \qquad (8.28)$$

for all positive integers n and m. Taking $n > m$, and identifying the coefficients in Y^n, we find $a_{nm} = 0$. If $m > n$, we also find $a_{nm} = 0$, now identifying coefficients in Y^m. We can now write $a_{nm} = a_n \delta_{nm}$. Take $m > n$. The right-hand side of (8.28) amounts to zero, while the left-hand side is

$$\binom{m}{m-n} a_n Y^{n-m}$$

It follows that $a_n = 0$ for all n, and $\sigma = 0$.

Proposition 148. *Let C be a coalgebra, $\sigma : C \otimes C \to k$ an FS-map and M a right C-comodule. Then the map*

$$R_\sigma : M \otimes M \to M \otimes M, \quad R_\sigma(m \otimes n) = \sigma(m_{[1]} \otimes n_{[1]}) m_{[0]} \otimes n_{[0]}$$

is a solution of the FS-equation in $\text{End}(M \otimes M \otimes M)$.

Proof. Write $R = R_\sigma$ and take $l, m, n \in M$.

$$R^{12} R^{23} (l \otimes m \otimes n) = R^{12} \Big(\sigma(m_{[1]} \otimes n_{[1]}) l \otimes m_{[0]} \otimes n_{[0]} \Big)$$
$$= \sigma(m_{[2]} \otimes n_{[1]}) \sigma(l_{[1]} \otimes m_{[1]}) l_{[0]} \otimes m_{[0]} \otimes n_{[0]} \qquad (8.29)$$

Applying (8.25), with $m = c$, $n = d$, we obtain

$$R^{12} R^{23} (l \otimes m \otimes n) = \sigma(m_{[1]} \otimes n_{[1]}) \sigma(l_{[1]} \otimes n_{[2]}) l_{[0]} \otimes m_{[0]} \otimes n_{[0]}$$
$$= R^{23} \Big(\sigma(l_{[1]} \otimes n_{[1]}) l_{[0]} \otimes m \otimes n_{[0]} \Big)$$
$$= R^{23} R^{13} (l \otimes m \otimes n)$$

Applying (8.25), with $m = d$, $l = c$, we obtain

$$R^{12}R^{23}(l \otimes m \otimes n) = \sigma(l_{[1]} \otimes n_{[1]})\sigma(l_{[2]} \otimes m_{[1]})l_{[0]} \otimes m_{[0]} \otimes n_{[0]}$$
$$= R^{13}\Big(\sigma(l_{[1]} \otimes n_{[1]})l_{[0]} \otimes m_{[0]} \otimes n\Big)$$
$$= R^{13}R^{12}(l \otimes m \otimes n)$$

proving that R is a solution of the FS-equation.

Remark 27. If C is finite dimensional and $A = C^*$ is its dual algebra, then there is a one-to-one correspondence between FS-maps $\sigma : C \otimes C \to k$ and A-central elements $e \in A \otimes A$. The correspondence is given by the formula

$$\sigma(c \otimes d) = \sum \langle c, e^1 \rangle \langle d, e^2 \rangle.$$

In this situation, the map R_e from Example 18.5 is equal to R_σ. Indeed,

$$R_\sigma(m \otimes n) = \sigma(m_{[1]} \otimes n_{[1]})m_{[0]} \otimes n_{[0]} = \langle m_{(1)}, e^1 \rangle \langle n_{(1)}, e^2 \rangle$$
$$= e^1 \cdot m \otimes e^2 \cdot n = R_e(m \otimes n)$$

We will now present two more classes of solutions of the FS-equation.

Proposition 149. *Take $a \in k$, $X = \{1, \dots, n\}$, and $\theta : X^3 \to X$ a map satisfying*

$$\theta(u, i, j) = v \iff \theta(v, j, i) = u \tag{8.30}$$
$$\theta(i, u, j) = v \iff \theta(j, v, i) = u \tag{8.31}$$
$$\theta(i, j, u) = v \iff \theta(j, i, v) = u \tag{8.32}$$
$$\theta(i, j, k) = \theta(u, v, w) \iff \theta(j, i, u) = \theta(k, w, v) \tag{8.33}$$

1. $R = (x_{ij}^{uv}) \in M_{n^2}(k)$ *given by*

$$x_{ij}^{uv} = a\delta_{\theta(i,v,j)}^u \tag{8.34}$$

   is a solution of the FS-equation.
2. *Assume that n is invertible in k, and take $a = n^{-1}$.*
   2a. R is a solution of the S-equation if and only if

$$\theta(k, k, i) = i$$

   for all $i, k \in X$.
   2b. Let ε be the trace map. (R, ε) is a solution of the F-equation if and only if

$$\theta(i, k, k) = i$$

   for all $i, k \in X$.

Proof. 1. We have to verify (8.14). Using (8.32), we compute

$$x_{uk}^{ij} x_{vw}^{kl} = a^2 \delta_{\theta(u,j,k)}^i \delta_{\theta(v,l,w)}^k$$
$$= a^2 \delta_{\theta(j,u,i)}^k \delta_{\theta(v,l,w)}^k = a^2 \delta_{\theta(v,l,w)}^{\theta(j,u,i)} \tag{8.35}$$

In a similar way, we find

$$x_{vk}^{jl} x_{uw}^{ik} = a^2 \delta_{\theta(l,v,j)}^{\theta(w,i,u)} \tag{8.36}$$

$$x_{kw}^{il} x_{uv}^{kj} = a^2 \delta_{\theta(u,j,v)}^{\theta(i,w,l)} \tag{8.37}$$

Using (8.33), we find that (8.35), (8.36), and (8.37) are equal, proving (8.14).
2a. We easily compute that

$$x_{ik}^{kj} = n^{-1} \delta_{\theta(i,j,k)}^k = n^{-1} \delta_{\theta(k,k,i)}^j$$

and it follows that R is a solution of the S-equation if and only if $\theta(k,k,i) = i$
for all i and k.
2b. We compute

$$x_{ki}^{kj} = n^{-1} \delta_{\theta(k,j,i)}^k = n^{-1} \delta_{\theta(i,k,k)}^j$$
$$x_{ik}^{jk} = n^{-1} \delta_{\theta(i,k,k)}^j$$

and it follows that (R, ε) is a solution of the F-equation if and only if
$\theta(i,k,k) = i$ for all i, k.

Examples 20. 1. Let G be a finite group. Then the map

$$\theta: \ G \times G \times G \to G, \quad \theta(i,j,k) = ij^{-1}k$$

satisfies conditions (8.30-8.33).
2. Let G be a group of order n acting on $X = \{1, 2, \cdots, n\}$, and assume that
the action of G is transitive and free, which means that for every $i, j \in X$,
there exists a unique $g \in G$ such that $g(i) = j$. Then the map $\theta: \ X \times X \times X \to X$ defined by

$$\theta(i,j,k) = g^{-1}(k)$$

where $g \in G$ is such that $g(i) = j$, satisfies conditions (8.30-8.33).

Proposition 150. *Let n be a positive integer, $\phi: \ \{1, \cdots, n\} \to \{1, \cdots, n\}$
a function with $\phi^2 = \phi$ and $\{m_1, \cdots, m_n\}$ a basis of M. Then*

$$R^\phi: \ M \otimes M \to M \otimes M, \quad R^\phi(m_i \otimes m_j) = \delta_{ij} \sum_{a,b \in \phi^{-1}(i)} m_a \otimes m_b \tag{8.38}$$

is a solution of the FS-equation.

Proof. Write $R = R^\phi$, and take $p, q, r \in \{1, \ldots, n\}$. Then

$$R^{12}R^{23}(m_p \otimes m_q \otimes m_r) = R^{12}\Big(\delta_{qr} \sum_{a,b\in\phi^{-1}(q)} m_p \otimes m_a \otimes m_b\Big)$$

$$= \delta_{qr} \sum_{a,b\in\phi^{-1}(q)} \delta_{ap} \sum_{c,d\in\phi^{-1}(p)} m_c \otimes m_d \otimes m_b$$

$$= \delta_{qr}\delta_{\phi(p)q} \sum_{b\in\phi^{-1}(q)} \sum_{c,d\in\phi^{-1}(p)} m_c \otimes m_d \otimes m_b.$$

If $\phi^{-1}(p)$ is nonempty (take $x \in \phi^{-1}(p)$), and $\phi(p) = q$, then $q = \phi(p) = \phi^2(x) = \phi(x) = p$, so we can write

$$R^{12}R^{23}(m_p \otimes m_q \otimes m_r) = \delta_{pqr\phi(p)} \sum_{a,b,c\in\phi^{-1}(p)} m_a \otimes m_b \otimes m_c.$$

In a similar way, we can compute that

$$R^{23}R^{13}(m_p \otimes m_q \otimes m_r) = R^{13}R^{12}(m_p \otimes m_q \otimes m_r)$$

$$= \delta_{pqr\phi(p)} \sum_{a,b,c\in\phi^{-1}(p)} m_a \otimes m_b \otimes m_c.$$

Now we will generalize Example 18.3 to algebras without unit. Recall that a left A-module M is called unital (or regular) if the natural map $A \otimes_A M \to M$ is an isomorphism.

Proposition 151. *Let M be a unital A-module, and $f : A \to A \otimes A$ an A-bimodule map. Then the map $R : M \otimes M \to M \otimes M$ mapping $m \otimes a \cdot n$ to $f(a)(m \otimes n)$ is a solution of the FS-equation.*

Proof. Observe first that it suffices to define R on elements of the form $m \otimes a \cdot n$, since the map $A \otimes_A M \to M$ is surjective. R is well-defined since

$$R(m \otimes a \cdot (b \cdot n)) = f(a)(m \otimes b \cdot n)$$
$$= f(a)(I_M \otimes b)(m \otimes n) = f(ab)(m \otimes n)$$

Write $f(a) = a^1 \otimes a^2$, for all $a \in A$. Then

$$f(ab) = a^1 \otimes a^2b = ab^1 \otimes b^2.$$

Now

$$R^{12}(R^{23}(m \otimes bn \otimes ap)) = R^{12}(m \otimes a^1bn \otimes a^2p)$$
$$= a^1b^1m \otimes b^2n \otimes a^2p = R^{13}(b^1m \otimes b^2n \otimes p)$$
$$= R^{13}(R^{12}(m \otimes bn \otimes ap))$$

In a similar way, we prove that $R^{12}R^{23} = R^{23}R^{13}$.

8.3 The structure of Frobenius algebras and separable algebras

The first statement of Proposition 144 can be restated as follows: for a k-algebra A, any A-central element $R \in A \otimes A$ is a solution of the FS-equation. Over a field k, we can prove the converse: any solution of the FS-equation arises in this way.

Theorem 77. *Let A be a k-algebra and $R = R^1 \otimes R^2 \in A \otimes A$ a solution of the FS-equation.*

1. *Then there exists a k-subalgebra $\mathcal{A}(R)$ of A such that $R \in \mathcal{A}(R) \otimes \mathcal{A}(R)$ and R is $\mathcal{A}(R)$-central.*
2. *If $R \sim S$ are equivalent solutions of the FS-equation, then $\mathcal{A}(R) \cong \mathcal{A}(S)$.*
3. *$(\mathcal{A}(R), R)$ satisfies the following universal property: if $(B, m_B, 1_B)$ is an algebra, and $e \in B \otimes B$ is an B-central element, then any algebra map $\alpha : B \to A$ with $(\alpha \otimes \alpha)(e) = R$ factors through an algebra map $\tilde{\alpha} : B \to \mathcal{A}(R)$.*
4. *If $R \in A \otimes A$ is a solution of the S-equation (resp. the F-equation), then $\mathcal{A}(\mathcal{R})$ is a separable (resp. Frobenius) algebra.*

Proof. 1. Let $\mathcal{A}(R) = \{a \in A \mid a \cdot R = R \cdot a\}$. Obviously $\mathcal{A}(R)$ is a k-subalgebra of A and $1_A \in \mathcal{A}(R)$. We also claim that $R \in \mathcal{A}(R) \otimes \mathcal{A}(R)$. To this end, we observe first that $\mathcal{A}(R) = \mathrm{Ker}(\varphi)$, with $\varphi : A \to A \otimes A^{\mathrm{op}}$ defined by the formula

$$\varphi(a) = (a \otimes 1_A - 1_A \otimes a)R.$$

A is flat as a k-algebra, so

$$\mathcal{A}(R) \otimes A = \mathrm{Ker}(\varphi \otimes \mathrm{Id}_A).$$

Now,

$$(\varphi \otimes I_A)(R) = r^1 R^1 \otimes R^2 \otimes r^2 - R^1 \otimes R^2 r^1 \otimes r^2$$
$$= R^{13} R^{12} - R^{12} R^{23} = 0$$

and it follows that $R \in \mathcal{A}(R) \otimes A$. In a similar way, using that $R^{12} R^{23} = R^{23} R^{13}$, we get that $R \in A \otimes \mathcal{A}(R)$, and we find that $R \in \mathcal{A}(R) \otimes \mathcal{A}(R)$. Indeed, k is a field, so

$$\mathcal{A}(R) \otimes \mathcal{A}(R) = A \otimes \mathcal{A}(R) \cap \mathcal{A}(R) \otimes A$$

Hence, R is an $\mathcal{A}(R)$-central element of $\mathcal{A}(R) \otimes \mathcal{A}(R)$.

2. Let $u \in U(A)$ such that $S = (u \otimes u)R(u^{-1} \otimes u^{-1})$. Then

$$f_u : \mathcal{A}(R) \to \mathcal{A}(S), \quad f_u(a) = uau^{-1}$$

for all $a \in \mathcal{A}(R)$ is a well-defined isomorphism of k-algebras.

3. Let $b \in B$. If we apply $\alpha \otimes \alpha$ to the equality $(b \otimes 1_B)e = e(1_B \otimes b)$ we find

that the image of α is contained in $\mathcal{A}(R)$, and the universal property follows.
4. The first statement follows from the definition of separable algebras and the second one follows from 3) of Corollary 40.

Remark 28. The converse of 2. does not hold: let A be a separable k-algebra and $R \in A \otimes A$ a separability idempotent. Then R and $S = 0 \otimes 0$ are solutions of the FS-equation, $\mathcal{A}(R) = \mathcal{A}(S) = A$ and, of course, R and S are not equivalent.

If A is finite dimensional, then we can describe the algebra $\mathcal{A}(R)$ using generators and relations. Let $\{m_1, m_2, \ldots, m_n\}$ be a basis of a finite dimensional vector space M. We have seen that an endomorphism $R \in \text{End}(M \otimes M)$ can be written in matrix form (see (8.9) and (8.10)). Suppose that R is a solution of FS-equation. Identifying $\text{End}(M)$ and $\mathcal{M}_n(k)$, we will write $\mathcal{A}(n, R)$ for the subalgebra of $\mathcal{M}_n(k)$ corresponding to $\mathcal{A}(R)$. An easy computation shows that

$$\mathcal{A}(n, R) = \{(a_j^i) \in \mathcal{M}_n(k) \mid x_{\alpha v}^{ij} a_u^\alpha = x_{uv}^{i\alpha} a_\alpha^j, \text{ for all } i, j, u, v = 1, \cdots, n\}$$
(8.39)

where R is a matrix satisfying (8.14).
We can now prove the main result of this Chapter.

Theorem 78. *For an n-dimensional k-algebra A, the following statements are equivalent:*

1. *A is a Frobenius (resp. separable) algebra.*
2. *There exists an algebra isomorphism*

$$A \cong \mathcal{A}(n, R),$$

where $R = (x_{uv}^{ij}) \in \mathcal{M}_n(k) \otimes \mathcal{M}_n(k) \cong \text{End}(A) \otimes \text{End}(A)$ is a solution of the Frobenius (resp. separability) equation.

Proof. 1. $\Rightarrow$ 2. Both Frobenius and separable algebras are characterized by the existence of an A-central element with some normalizing properties. Let $e = e^1 \otimes e^2 \in A \otimes A$ be such an A-central element. Then the map

$$R = R_e : A \otimes A \to A \otimes A, \qquad R(a \otimes b) = e^1 a \otimes e^2 b$$

$(a, b \in A)$, is a solution to the FS-equation. Here we view $R_e \in \text{End}_k(A \otimes A) \cong \text{End}_k(A) \otimes \text{End}_k(A)$ (A is finite dimensional over k). Consequently, we can construct the algebra $\mathcal{A}(R) \subseteq \text{End}_k(A)$. We will prove that A and $\mathcal{A}(R)$ are isomorphic when A is a Frobenius algebra, or a separable algebra.
First we consider the injection $i : A \to \text{End}_k(A)$, with $i(a)(b) = ab$, for $a, b \in A$. Then image of i is included in $\mathcal{A}(R)$. Indeed,

$$\mathcal{A}(R) = \{f \in \text{End}_k(A) \mid (f \otimes I_A) \circ R = R \circ (I_A \otimes f)\}$$

Using the fact that e is an A-central element, it follows easily that $(i(a) \otimes I_A) \circ R = R \circ (I_A \otimes i(a))$, for all $a \in A$, proving that $\mathrm{Im}(i) \subseteq \mathcal{A}(R)$.
If $f \in \mathcal{A}(R)$, then $(f \otimes I_A) \circ R = R \circ (I_A \otimes f)$, and, evaluating this equality at $1_A \otimes a$, we find

$$f(e^1) \otimes e^2 a = e^1 \otimes e^2 f(a) \qquad (8.40)$$

Now assume that A is a Frobenius algebra. Then there exists $\varepsilon : A \to k$ such that $\varepsilon(e^1)e^2 = e^1\varepsilon(e^2) = 1_A$. Applying $\varepsilon \otimes I_A$ to (8.40) we obtain that

$$f(a) = (\varepsilon(f(e^1))e^2)a$$

for all $a \in A$. Thus, $f = i(\varepsilon(f(e^1))e^2)$. This proves that $\mathrm{Im}(i) = \mathcal{A}(R)$, and the corestriction of i to $\mathcal{A}(R)$ is an isomorphism of algebras.
If A is separable, then $e^1 e^2 = 1_A$. Applying m_A to (8.40) we find

$$f(a) = (f(e^1)e^2)a$$

for all $a \in A$. Consequently $f = i(f(e^1)e^2)$, proving again that A and $\mathcal{A}(R)$ are isomorphic.
$2. \Rightarrow 1$. This is the last statement of Theorem 77.

Remark 29. Over a field k that is algebraically closed or of characteristic zero, the structure of finite dimensional separable k-algebras is given by the classical Wedderburn-Artin Theorem: a finite dimensional algebra A is separable if and only if is semisimple, if and only if it is a direct product of matrix algebras.

As we have seen in Lemma 27, the dual A^* of a separable finite dimensional algebra is a coseparable finite dimensional coalgebra. Thus we obtain a structure Theorem for coseparable coalgebras, by using duality arguments.
More precisely, let C be a coseparable k-coalgebra which of dimension n over k. Then there exists an FS-map $\sigma : C \otimes C \to k$ satisfying the normalizing condition (8.26). Let $A = C^*$ and $A \otimes A \cong C^* \otimes C^*$. Then σ is an A-central element of A, satisfying the normalizing condition (8.6). It follows from Theorem 78 and (8.39) that $A \cong \mathcal{A}(n, \sigma)$, and therefore $C \cong \mathcal{A}(n, \sigma)^*$. Now $\mathcal{A}(n, \sigma)^*$ can be described as a quotient of the comatrix coalgebra: $\mathcal{A}(n, \sigma)^* = \mathcal{M}^n(k)/I$, where I is the coideal of $\mathcal{M}^n(k)$ that annihilates $\mathcal{A}(n, \sigma)$; I is generated by

$$\{o_{uv}^{ij} = c_{u\alpha}x_{\alpha v}^{ij} - x_{uv}^{i\alpha}c_{\alpha j} \mid i, j, k, l = 1 \ldots, m\}$$

This coalgebra will be denoted by

$$\mathcal{C}(n, R) = \mathcal{M}^n(k)/I$$

We have obtain the following

Theorem 79. *For an n-dimensional C, the following statements are equivalent.*

1. C is a co-Frobenius (resp. coseparable) coalgebra.
2. There exists a coalgebra isomorphism

$$A \cong C(n, R),$$

where $R = (x_{uv}^{ij}) \in M_n(k) \otimes M_n(k) \cong \mathrm{End}_k(A) \otimes \mathrm{End}_k(A)$ is a solution of the Frobenius (resp. separability) equation.

Examples 21. 1. Let M be a n-dimensional vector space and $R = I_{M \otimes M}$. Then

$$\mathcal{A}(R) = \{f \in \mathrm{End}(M) \mid f \otimes I_M = I_M \otimes f\} = k$$

2. Now let $R = \tau_M$ be the switch map. For all $f \in \mathrm{End}(M)$ and $m, n \in M$, we have

$$((f \otimes I_M) \circ \tau)(m \otimes n) = f(n) \otimes m = (\tau \circ (I_M \otimes f))(m \otimes n)$$

and, consequently, $\mathcal{A}(\tau) \cong M_n(k)$.

3. Let M be a finite dimensional vector space over a field k, and $f \in \mathrm{End}(M)$ an idempotent. Then

$$\mathcal{A}(f \otimes f) = \{g \in \mathrm{End}(M) \mid f \circ g = g \circ f = \alpha f \text{ for some } \alpha \in k\}$$

Indeed, $g \in \mathcal{A}(f \otimes f)$ if and only if $g \circ f \otimes f = f \otimes f \circ g$. Multiplying the two factors, we find that $g \circ f = f \circ g$. Now $g \circ f \otimes f = f \otimes g \circ f$ implies that $g \circ f = \alpha f$ for some $\alpha \in k$. The converse is obvious.

In particular, assume that M has dimension 2 and let $\{m_1, m_2\}$ be a basis of M. Let f be the idempotent endomorphism with matrix

$$\begin{pmatrix} 1 - rq & q \\ r(1 - rq) & rq \end{pmatrix}$$

Assume first that $rq \neq 1$ and $r \neq 0$, and take $g \in \mathrm{End}(M)$ with matrix

$$\begin{pmatrix} a & b \\ c & d \end{pmatrix}$$

$g \in \mathcal{A}(f \otimes f)$ if and only if

$$\alpha = a + br \quad ; \quad c + dr = r(a + br) \quad ; \quad br(1 - rq) = qc$$

The two last equations can be easily solved for b and c in terms of a and d, and we see that $\mathcal{A}(f \otimes f)$ has dimension two. We know from the proof of Theorem 77 that $f \in \mathcal{A}(f \otimes f)$. Another solution of the above system is I_M, and we find that $\mathcal{A}(f \otimes f)$ is the two-dimensional subalgebra of $\mathrm{End}(M)$ with basis $\{f, I_M\}$. Put $f' = I_M - f$. Then $\{f, f'\}$ is also a basis for $\mathcal{A}(f \otimes f)$, and $\mathcal{A}(f \otimes f) = k \times k$. $C(f \otimes f)$ is the grouplike coalgebra of dimension two. We find the same result if $rq = 1$ or $q = 0$.

4. Let $R \in \mathcal{M}_{n^2}(k)$ given by equation (8.34) as in Proposition 149. Then the algebra $\mathcal{A}(n, R)$ is given by

$$\mathcal{A}(n, R) = \{(a_j^i) \in \mathcal{M}_n(k) \mid a_u^{\theta(j,v,i)} = a_{\theta(u,i,v)}^j, \quad (\forall)\ i, j, u, v \in X\} \quad (8.41)$$

Proposition 149 tells us when this algebra is separable or Frobenius over k. Assume now that G is a finite group with $|G| = n$ invertible in k and θ is given as in Example 20.2. Then the above algebra $\mathcal{A}(n, R)$ equals

$$\mathcal{A} = \{(a_j^i) \in \mathcal{M}_n(k) \mid a_{g(i)}^i = a_{g(j)}^j, \quad (\forall)\ i, j \in X \text{ and } g \in G\}$$

Indeed, if $(a_j^i) \in \mathcal{A}$, then $(a_j^i) \in \mathcal{A}(n, R)$, since $g(\theta(j, v, i)) = u$ implies that $g(j) = \theta(u, i, v)$. Conversely, for $(a_j^i) \in \mathcal{A}(n, R)$, we choose $i, j \in X$ and $g \in G$. Let $u = g(i)$ and $v = g(j)$. Then $\theta(j, v, u) = i$, hence

$$a_{g(i)}^i = a_u^{\theta(j,v,u)} = a_{\theta(u,u,v)}^j = a_v^j = a_{g(j)}^j,$$

showing that $\mathcal{A} = \mathcal{A}(n, R)$. From the fact that G acts transitively, it follows that a matrix in $\mathcal{A}(n, R)$ is completely determined by its top row. For every $g \in G$, we define $A_g \in \mathcal{A}(n, R)$ by $(A_g)_i^1 = \delta_{g(1),i}$. Then $\mathcal{A}(n, R) = \{A_g \mid g \in G\}$, and we have an algebra isomorphism

$$f:\ \mathcal{A}(n, R) \to kG, \quad f(A_g) = g$$

For example, take the cyclic group of order n, $G = C_n$.

$$\mathcal{A}(n, R) = \left\{ \begin{pmatrix} x_1 & x_2 & \cdots & x_n \\ x_n & x_1 & \cdots & x_{n-1} \\ x_{n-1} & x_n & \cdots & x_{n-2} \\ \cdot & \cdot & \cdots & \cdot \\ x_2 & x_3 & \cdots & x_1 \end{pmatrix} \mid x_1, \cdots, x_n \in k \right\} \cong kC_n$$

5. Let G be a finite group of order n and $\sigma:\ G \times G \to k^*$ a 2-cocycle. Let R_σ be the solution of the FS-equation given by (8.24). We then obtain directly from (8.39) that $\mathcal{A}(n, R_\sigma)$ consists of all $G \times G$-matrices $(a_j^i)_{i,j \in G}$ satisfying the relations

$$a_u^{jv^{-1}i}\, \sigma^{-1}(ui^{-1}vj^{-1}, jv^{-1})\sigma(vj^{-1}, jv^{-1}i)\sigma(jv^{-1}, v) =$$

$$a_{ui^{-1}v}^j\, \sigma^{-1}(iu^{-1}, ui^{-1})\sigma(iu^{-1}, u)\sigma(ui^{-1}, v)$$

for all $i, j, u, v \in G$. This algebra is separable if n is invertible in k.

We will now present some new classes of examples, starting from the solution R^ϕ of the FS-equation discussed in Proposition 150. In this case, we find easily that

$$x_{uv}^{ij} = \delta_{uv}\delta_{\phi(i)u}\delta_{\phi(j)v}$$

and, according to (8.39), $\mathcal{A}(R^\phi)$ consists of matrices $\left(a_j^i\right)$ satisfying

$$\sum_{\alpha=1}^{n} a_u^\alpha x_{\alpha v}^{ij} = \sum_{\alpha=1}^{n} x_{uv}^{i\alpha} a_\alpha^j$$

or

$$a_u^v \delta_{\phi(i)v}\delta_{\phi(j)v} = \sum_{\alpha\in\phi^{-1}(v)} \delta_{uv}\delta_{\phi(i)u}a_\alpha^j \qquad (8.42)$$

for all $i, j, v, u = 1, \ldots, n$. The left hand side of (8.42) is nonzero if and only if $\phi(i) = \phi(j) = v$. If $\phi(i) = \phi(j) = v = u$, then (8.42) amounts to

$$a_{\phi(i)}^{\phi(i)} = \sum_{\{\alpha|\phi(\alpha)=\phi(i)\}} a_\alpha^i.$$

If $\phi(i) = \phi(j) = v \neq u$, then (8.42) takes the form

$$a_u^{\phi(i)} = 0.$$

Now assume that the left hand side of (8.42) is zero. If $\phi(i) \neq \phi(j)$, then the right hand side of (8.42) is also zero, except when $u = v = \phi(i)$. Then (8.42) yields

$$\sum_{\{\alpha|\phi(\alpha)=\phi(i)\}} a_\alpha^j = 0.$$

If $\phi(i) = \phi(j) \neq u$, then (8.42) reduces to $0 = 0$. We summarize our results as follows.

Proposition 152. *Consider an idempotent map $\phi : \{1, \ldots, n\} \to \{1, \ldots, n\}$, and the corresponding solution R^ϕ of the FS-equation. Then $\mathcal{A}(R^\phi)$ is the subalgebra of $\mathcal{M}_n(k)$ consisting of all matrices (a_j^i) satisfying*

$$a_{\phi(i)}^{\phi(i)} = \sum_{\{\alpha|\phi(\alpha)=\phi(i)\}} a_\alpha^i \qquad (i = 1, \ldots, n) \qquad (8.43)$$

$$a_j^{\phi(i)} = 0 \qquad (\phi(i) \neq j) \qquad (8.44)$$

$$0 = \sum_{\{\alpha|\phi(\alpha)=\phi(i)\}} a_\alpha^j \qquad (\phi(i) \neq \phi(j)) \qquad (8.45)$$

$\mathcal{A}(R^\phi)$ is a separable k-algebra if and only if $(R^\phi, \varepsilon = \mathrm{trace})$ is a solution of the F-equation if and only if ϕ is the identity map. In this case, $\mathcal{A}(R^\phi)$ is the direct sum of n copies of k.

Proof. The first part was done above. $\mathcal{A}(R)$ is separable if and only if (8.15) holds. This comes down to

$$\delta_{ju} = \sum_{v=1}^{n} \delta_{uv}\delta_{\phi(u)u}\delta_{\phi(j)v} = \delta_{\phi(u)u}\delta_{\phi(j)u}$$

and this implies that $\phi(u) = u$ for all u. (8.43,8.44,8.45) reduce to $a^i_j = 0$ for $i \neq j$, and $\mathcal{A}(R^I)$ consists of all diagonal matrices.

$(R^\phi, \varepsilon = \text{trace})$ is a solution of the F-equation if and only if (8.16) holds, and a similar computation shows that this also implies that ϕ is the identity.

Examples 22. 1. Take $n = 4$, and ϕ given by

$$\phi(1) = \phi(2) = 2, \ \ \phi(3) = \phi(4) = 4$$

(8.43,8.44,8.45) take the following form

$$a^2_2 = a^1_1 + a^1_2 \qquad a^4_4 = a^3_3 + a^3_4$$
$$a^2_1 = a^2_2 = a^2_4 = 0 \ \ a^4_1 = a^4_2 = a^4_3 = 0$$
$$a^3_1 = -a^3_2 \qquad a^1_3 = -a^1_4$$

and

$$\mathcal{A}(4, R^\phi) = \left\{ \begin{pmatrix} x & y-x & u & -u \\ 0 & y & 0 & 0 \\ v & -v & z & t-z \\ 0 & 0 & 0 & t \end{pmatrix} \ \Big|\ x, y, z, t, u, v \in k \right\}$$

The dual coalgebra can also be described easily. Write $x_i = c^i_i$ $(i = 1, \ldots, 4)$, $x_5 = c^1_3$ and $x_6 = c^3_1$. Then $\mathcal{C}(R^\phi)$ is the six dimensional coalgebra with basis $\{x_1, \ldots, x_6\}$ and

$$\Delta(x_1) = x_1 \otimes x_1 + x_5 \otimes x_6, \quad \Delta(x_2) = x_2 \otimes x_2,$$
$$\Delta(x_3) = x_3 \otimes x_3 + x_6 \otimes x_5, \quad \Delta(x_4) = x_4 \otimes x_4,$$
$$\Delta(x_5) = x_1 \otimes x_5 + x_5 \otimes x_3, \quad \Delta(x_6) = x_6 \otimes x_1 + x_3 \otimes x_6,$$
$$\varepsilon(x_i) = 1 \quad (i = 1, 2, 3, 4), \quad \varepsilon(x_i) = 0 \quad (i = 5, 6).$$

2. Again, take $n = 4$, but let ϕ be given by the formula

$$\phi(1) = 1, \ \phi(2) = \phi(3) = \phi(4) = 2$$

Then (8.43,8.44,8.45) reduce to

$$a^2_2 = a^3_2 + a^3_3 + a^3_4 = a^4_2 + a^4_3 + a^4_4$$

$$a^1_2 = a^1_3 = a^1_4 = a^2_1 = a^2_3 = a^2_4 = a^3_1 = a^4_1 = 0,$$

hence

$$\mathcal{A}(4,R^\phi) = \left\{ \begin{pmatrix} x & 0 & 0 & 0 \\ 0 & y & 0 & 0 \\ 0 & u & z & y-z-u \\ 0 & v & y-t-v & t \end{pmatrix} \;\middle|\; x,y,z,t,u,v \in k \right\}$$

Putting $c_i^i = x_i$ $(i = 1,\ldots,4)$, $x_5 = c_2^3$ and $x_6 = c_2^4$, we find that $\mathcal{C}(R^\phi)$ is the six dimensional coalgebra with structure maps

$$\begin{aligned}
\Delta(x_1) &= x_1 \otimes x_1, \quad \Delta(x_2) = x_2 \otimes x_2, \\
\Delta(x_3) &= x_3 \otimes x_3 + (x_2 - x_3 - x_5) \otimes (x_2 - x_4 - x_6), \\
\Delta(x_4) &= x_4 \otimes x_4 + (x_2 - x_4 - x_6) \otimes (x_2 - x_3 - x_5), \\
\Delta(x_5) &= x_5 \otimes x_2 + x_3 \otimes x_5 + (x_2 - x_3 - x_5) \otimes x_6, \\
\Delta(x_6) &= x_6 \otimes x_2 + x_4 \otimes x_6 + (x_2 - x_4 - x_6) \otimes x_5, \\
\varepsilon(x_i) &= 1 \quad (i = 1,2,3,4), \quad \varepsilon(x_i) = 0 \quad (i = 5,6).
\end{aligned}$$

8.4 The category of FS-objects

We have seen in Corollary 41 that the equation $R^{12}R^{23} = R^{13}R^{12}$ is equivalent to the fact that a certain multiplication on $M \otimes M$ is associative. We will now prove that the other equation, namely $R^{12}R^{23} = R^{23}R^{13}$, is equivalent to the fact that a certain comultiplication is coassociative.

Proposition 153. *Let $(A, m_A, 1_A)$ be an algebra, $R = R^1 \otimes R^2 \in A \otimes A$ and*

$$\delta : A \to A \otimes A, \quad \delta(a) = R^1 \otimes R^2 a$$

for all $a \in A$. The following statements are equivalent:

1. *(A, δ) is a coassociative coalgebra (not necessarily with counit);*
2. *$R^{12}R^{23} = R^{23}R^{13}$ in $A \otimes A \otimes A$.*

In this case any left A-module $(M, \cdot)$ has a structure of left comodule over the coalgebra (A, δ) via

$$\rho : M \to A \otimes M, \quad \rho(m) = R^1 \otimes R^2 \cdot m$$

for all $m \in M$.

Proof. The equivalence of 1. and 2. follows from the formulas

$$\begin{aligned}
(\delta \otimes I)\delta(a) &= R^{12}R^{23} \cdot (1_A \otimes 1_A \otimes a) \\
(I \otimes \delta)\delta(a) &= R^{23}R^{13} \cdot (1_A \otimes 1_A \otimes a)
\end{aligned}$$

for all $a \in A$. The final statement follows from

$$(\delta \otimes I)\rho(m) = R^{12}R^{23} \cdot (1_A \otimes 1_A \otimes m)$$
$$(I \otimes \rho)\rho(m) = R^{23}R^{13} \cdot (1_A \otimes 1_A \otimes m)$$

for all $m \in M$.

Suppose now that $(A, m_A, 1_A)$ is an algebra over k and let $R \in A \otimes A$ be a A-central element. Then $R^{12}R^{23} = R^{23}R^{13}$ in $A \otimes A \otimes A$. It follows that $(A, \Delta_R = \delta^{\mathrm{cop}})$ is also a coassociative coalgebra, where $\Delta_R(a) = \delta^{\mathrm{cop}}(a) = R^2 a \otimes R^1$, for all $a \in A$. We remark that Δ_R is not an algebra map, i.e. (A, m_A, Δ_R) is not a bialgebra. Any left A-module $(M, \cdot)$ has a structure of right comodule over the coalgebra (A, Δ_R) via

$$\rho_R : \ M \to M \otimes A, \quad \rho_R(m) = R^2 \cdot m \otimes R^1$$

for all $m \in M$. Moreover, for any $a \in A$ and $m \in M$ we have that

$$\rho_R(a \cdot m) = a_{(1)} \cdot m \otimes a_{(2)} = m_{[0]} \otimes am_{[1]}.$$

Indeed, from the definition of the coaction on M and the comultiplication on A, we have immediately

$$\rho_R(a \cdot m) = R^2 a \cdot m \otimes R^1 = a_{(1)} \cdot m \otimes a_{(2)}$$

On the other hand

$$m_{[0]} \otimes am_{[1]} = R^2 \cdot m \otimes aR^1 = R^2 a \cdot m \otimes R^1$$

where in the last equality we used that R is a A-central element. These considerations lead us to the following

Definition 20. *Let (A, m_A, Δ_A) be at once an algebra and a coalgebra (but not necessarily a bialgebra). An FS-object over A is a k-module M that is at once a left A-module and a right A-comodule such that*

$$\rho(a \cdot m) = a_{(1)} \cdot m \otimes a_{(2)} = m_{[0]} \otimes am_{[1]} \tag{8.46}$$

for all $a \in A$ and $m \in M$.

The category of FS-objects and A-linear A-colinear maps will be denoted by $_A\mathcal{FS}^A$. This category measures how far A is from a bialgebra. If A has not a unit (resp. a counit), then the objects in $_A\mathcal{FS}^A$ will be assumed to be unital (resp. counital).

Proposition 154. *If $(A, m_A, 1_A, \Delta_A, \varepsilon_A)$ is a bialgebra with unit and counit, then the forgetful functor*

$$F : \ _A\mathcal{FS}^A \to \ _k\mathcal{M}$$

is an isomorphism of categories.

Proof. Define $G : {}_k\mathcal{M} \to {}_A\mathcal{FS}^A$ as follows: $G(M) = M$ as a k-module, with trivial A-action and A-coaction:

$$\rho(m) = m \otimes 1_A \quad \text{and} \quad a \cdot m = \varepsilon_A(a)m$$

for all $a \in A$ and $m \in M$. It is clear that $G(M) \in {}_A\mathcal{FS}^A$.
Now, assume that M is an FS-object over A. Applying $I \otimes \varepsilon_A$ to (8.46), we find that

$$a \cdot m = m_{[0]}\varepsilon_A(a)\varepsilon_A(m_{[1]}) = \varepsilon_A(a)m$$

Taking $a = 1_A$ in (8.46), we find

$$\rho(m) = 1_A \cdot m \otimes 1_A = m \otimes 1_A.$$

Hence, G is an inverse for the forgetful functor $F : {}_A\mathcal{FS}^A \to {}_k\mathcal{M}$.

Definition 21. *A triple* (A, m_A, Δ_A) *is called a weak Frobenius algebra (WF-algebra, for short) if* (A, m_A) *is an algebra (not necessarily with unit),* (A, Δ_A) *is a coalgebra (not necessarily with counit) and* $(A, m_A, \Delta_A) \in {}_A\mathcal{FS}^A$*, that is*

$$\Delta(ab) = a_{(1)}b \otimes a_{(2)} = b_{(1)} \otimes ab_{(2)}. \tag{8.47}$$

for all $a, b \in A$.

Remarks 25. 1. Assume that A is an WF-algebra with unit, and write $\Delta(1_A) = e^2 \otimes e^1$. From (8.47), it follows that

$$\Delta(a) = e^2 a \otimes e^1 = e^2 \otimes ae^1 \tag{8.48}$$

and this implies that $\Delta^{\mathrm{cop}}(1_A) = e^1 \otimes e^2$ is an A-central element. Conversely, if A is an algebra with unit, and $e = e^1 \otimes e^2$ is an A-central element, then $e^{12}e^{23} = e^{23}e^{13}$ (see (8.3)), and it is easy to prove that this last statement is equivalent to the fact that $\Delta : A \to A \otimes A$ given by $\Delta(a) = e^2 a \otimes e^1$ is coassociative. Thus A is a WF-algebra. We have proved that WF-algebras with unit correspond to algebras with unit together with an A-central element.
2. From (8.47), it follows immediately that $f := \Delta^{\mathrm{cop}} : A \to A \otimes A$ is an A-bimodule map. Conversely, if f is an A-bimodule map, then it is easy to prove that $\Delta = \tau \circ f$ defines a coassociative comultiplication on A, making A into a WF-algebra. Now, using Corollary 40, we obtain that a finitely generated projective and unitary k-algebra $(A, m_A, 1_A)$ is Frobenius if and only if A is an unitary and counitary WF-algebra. Thus, we can view WF-algebras as a generalization of Frobenius algebras.

Proposition 155. *Let* $(A, m_A, 1_A, \Delta_A)$ *be a WF-algebra with unit. Then the forgetful functor*

$$F : {}_A\mathcal{FS}^A \to {}_A\mathcal{M}$$

is an isomorphism of categories.

Proof. We define a functor $G : {}_A\mathcal{M} \to {}_A\mathcal{FS}^A$ as follows: $G(M) = M$ as an A-module, with right A-coaction given by the formula

$$\rho(m) = e^2 \cdot m \otimes e^1$$

where $\Delta(1_A) = e^2 \otimes e^1$. ρ is a coaction because $e^{12}e^{23} = e^{23}e^{13}$, and, using (8.48), we see that

$$\rho(a \cdot m) = e^2 a \cdot m \otimes e^1 = a_{(1)} \cdot m \otimes a_{(2)}$$
$$= e^2 \cdot m \otimes ae^1 = m_{[0]} \otimes am_{[1]}$$

as needed. Now, G and F are each others inverses.

Now, we will give the coalgebra version of Proposition 155. Consider a WF-algebra $(A, m_A, \Delta_A, \varepsilon_A)$, with a counit ε_A, and consider $\sigma = \varepsilon \circ m_A \circ \tau : A \otimes A \to k$, that is,

$$\sigma(c \otimes d) = \varepsilon(dc)$$

for all $c, d \in A$. Now

$$\Delta(cd) = c_{(1)}d \otimes c_{(2)} = d_{(1)} \otimes cd_{(2)}$$

so

$$\sigma(d \otimes c_{(1)})c_{(2)} = (\varepsilon \otimes I_C)(\Delta(cd)) = (I_C \otimes \varepsilon)(\Delta(cd)) = \sigma(d_{(2)} \otimes c)d_{(1)}$$

and σ is an FS-map. Conversely, let (A, Δ_A) be a coalgebra with counit, and assume that $\sigma : A \otimes A \to k$ is an FS-map. A straightforward computation shows that the formula

$$c \cdot d = \sigma(d_{(2)} \otimes c)d_{(1)}$$

defines an associative multiplication on A and that $(A, \cdot, \Delta_A)$ is a WF-algebra. Thus, WF-algebras with counit correspond to coalgebras with counit together with an FS-map.

Proposition 156. *Let $(A, m_A, \Delta_A, \varepsilon_A)$ be a WF-algebra with counit. Then the forgetful functor*

$$F : {}_A\mathcal{FS}^A \to \mathcal{M}^A$$

is an isomorphism of categories.

Proof. The inverse of F is the functor $G : \mathcal{M}^A \to {}_A\mathcal{FS}^A$ defined as follows: $G(M) = M$ as a A-comodule, with A-action given by

$$a \cdot m = \sigma(m_{[1]} \otimes a)m_{[0]}$$

for all $a \in A$, $m \in M$. Further details are left to the reader.

As an immediate consequence of Proposition 155 and Proposition 156 we obtain the following generalization of Abrams' result [4, Theorem 3.3]

Corollary 44. *Let* $(A, m_A, 1_A, \Delta_A, \varepsilon_A)$ *be a* WF-*algebra with unit and counit. Then we have an equivalence of categories*

$$_A\mathcal{FS}^A \cong {}_A\mathcal{M} \cong \mathcal{M}^A$$

Let us finally show that Proposition 155 also holds over WF-algebras that are unital as modules over themselves.

Proposition 157. *Let* A *be a* WF-*alegebra that is unital as a module over itself. We have an equivalence between the categories* $_A\mathcal{M}$ *and* $_A\mathcal{FS}^A$.

Proof. For a unital A-module M, we define $F(M)$ as the A-module M with A-coaction given by

$$\rho(a \cdot m) = a_{(1)} \cdot m \otimes a_{(2)}$$

It is clear that ρ defines an A-coaction. One equality in (8.46) is obvious, and the other one follows from (8.47): for all $a, b \in A$ and $m \in M$, we have that

$$(b \cdot m)_{[0]} \otimes a(b \cdot m)_{[1]} = b_{(1)} \cdot m \otimes ab_{(2)}$$
$$= (ab)_{(1)} \cdot m \otimes (ab)_{(2)} = \rho(a \cdot (b \cdot m))$$

It follows that $F(M)$ is an FS-object, and F defines the desired category equivalence.

References

1. Abe, E. (1977): Hopf Algebras. Cambridge University Press, Cambridge
2. Anderson, F.W., Fuller, K.R. (1974): Rings and categories of modules. Springer-Verlag, Berlin
3. Abrams, L. (1996): Two-dimensional topological quantum field theories and Frobenius algebras. J. Knot Theory Ramifications, **5**, 569–587
4. Abrams, L. (1999): Modules, comodules and cotensor products over Frobenius algebras. J. Algebra, **219**, 201–213
5. Abrams, L. (2000): The quantum Euler class and the quantum cohomology of the Grassmanians. Israel J. Math., **117**, 335–352
6. Andruskiewitsch, N., Grana, M. (2002): From racks to pointed Hopf algebras. Preprint
7. Andruskiewitsch, N., Schneider, H.-J. (1998): Lifting of quantum linear space and pointed Hopf algebras of order p^3. J. Algebra, **209**, 658–691
8. Atiyah, M. (1997): An introduction to topological quantum field theories. Turkish J. Math, **21**, 1–7
9. Baaj, S., Blanchard, E., Skandalis, G. (1999): Unitaries multiplicatifs en dimension finie et leurs sous-objets. Ann. Inst. Fourier (Grenoble), **49**, 1305–1344
10. Baaj, S., Skandalis, G. (1993): Unitaries multiplicatifs et dualite pour les produits croises de C^*-algebres. Ann. Sci. Ecole Norm. Sup., **26**, 425–488
11. Bass, H. (1968): Algebraic K-Theory. Benjamin, New York
12. Beattie, M., Dăscălescu, S., Grünenfelder, L. (1999): On the number of types of finite dimensional Hopf algebras. Inv. Math., **136**, 1–7
13. Beattie, M., Dăscălescu, S., Grünenfelder, L. (2000): Constructing pointed Hopf algebras by Ore extensions. J. Algebra, **225**, 743–770
14. Beattie, M., Dăscălescu, S., Raianu, Ş., Van Oystaeyen, F. (1998): The categories of Yetter-Drinfeld modules, Doi-Hopf modules and two-sided two-cosided Hopf modules. Appl. Categorical Structures, **6**, 223–237
15. Beidar, K.I., Fong, Y., Stolin, A. (1997): On Frobenius algebras and the Yang-Baxter equation. Trans. Amer. Math. Soc., **349**, 3823–3836
16. Bichon, J., Street, R. (2001): Militaru's D-equation in monoidal category. Appl. Categorical Structures, to appear
17. Bespalov, Y., Drabant, B. (1998): Cross product bialgebras III. Preprint
18. Bespalov, Y., Drabant, B. (1999): Cross product bialgebras I. J. Algebra, **219**, 466–506
19. Bespalov, Y., Drabant, B. (2001): Cross product bialgebras II. J. Algebra, **240**, 445–504
20. Böhm, G. (2000): Doi-Hopf modules over weak Hopf algebras. Comm. Algebra, **28**, 4687–4698

21. Borceux, F. (1994): Handbook of categorical algebra 1. Cambridge University Press, Cambridge
22. Borceux, F. (1994): Handbook of categorical algebra 2. Cambridge University Press, Cambridge
23. Brzeziński, T. (1999): On modules associated to coalgebra-Galois extensions. J. Algebra, **215**, 290–317
24. Brzeziński, T. (2000): Frobenius properties and Maschke-type theorems for entwined modules. Proc. Amer. Math. Soc., **128**, 2261–2270
25. Brzeziński, T. (2000): Coalgebra-Galois extensions from the extension point of view. In: Caenepeel, S., Van Oystaeyen. F. (eds) Hopf algebras and quantum groups. Marcel Dekker, New York
26. Brzeziński, T. (2001): The structure of corings. Induction functors, Maschke-type theorem, and Frobenius and Galois properties. Algebras and Representation Theory, to appear
27. Brzeziński, T. (2001): Towers of corings. Preprint
28. Brzeziński, T. (2001) The structure of corings with a group-like element. Preprint
29. Brzeziński, T., Caenepeel, S., Militaru, G. (2001): Doi-Koppinen modules for quantum groupoids. J. Pure Appl. Algebra, to appear
30. Brzeziński, T., Caenepeel, S., Militaru, G., Zhu, S. (2001): Frobenius and Maschke type Theorems for Doi-Hopf modules and entwined modules revisited: a unified approach. In: Granja, A., Hermida Alonso, J., Verschoren, A. (eds) Ring theory and Algebraic Geometry. Marcel Dekker, New York
31. Brzeziński, T., Hajac, P.M. (1999): Coalgebra extensions and algebra coextensions of Galois type. Comm. Algebra, **27**, 1347–1367
32. Brzeziński, T., Majid, S. (1998): Coalgebra bundles. Comm. Math. Phys., **191**, 467–492
33. Brzeziński, T., Majid, S. (2000): Quantum geometry of algebra factorizations and coalgebra bundles. Comm. Math. Phys., **213**, 491–521
34. Brzeziński, T., Militaru, G. (2000): Bialgebroids, x_R-bialgebras and duality. J. Algebra, to appear
35. Caenepeel, S. (1998): Brauer groups, Hopf algebras and Galois theory. Kluwer Academic Publishers, Dordrecht
36. Caenepeel, S., Dăscălescu, S. (1999): On pointed Hopf algebras of dimension 2^n. Bull. London Math. Soc., **31**, 17–24
37. Caenepeel, S., Dăscălescu, S., Militaru, G., Panaite, F. (1997): Coalgebra deformation of bialgebras by Harrison cocycles, copairings of Hopf algebras and double crosscoproducts. Bull. Belg. Math. Soc., **4**, 647–671
38. Caenepeel, S., Dăscălescu, S., Raianu, Ş. (1996): Cosemisimple Hopf algebras coacting on coalgebras. Comm. Algebra, **24**, 1649–1677
39. Caenepeel, S., De Groot, E. (2000): Modules over weak entwining structures. Contemp. Math. **267**, 31–54
40. Caenepeel, S., De Groot, E., Militaru, G. (2001): Frobenius functors of the second kind. Comm. Algebra, to appear
41. Caenepeel, S., Ion, B., Militaru, G., Zhu, S. (1999): Separable functors for the category of Doi-Hopf modules, Applications. Adv. Math., **145**, 239–290
42. Caenepeel, S., Ion, B., Militaru, G. (2000): The structure of Frobenius algebras and separable algebras. K-Theory, **19**, 365–402

43. Caenepeel, S., Ion, B., Militaru, G., Zhu, S. (2000): Separable functors for the category of Doi-Hopf modules II. In: Caenepeel, S., Van Oystaeyen, F. (eds) Hopf algebras and quantum groups. Marcel Dekker, New York

44. Caenepeel, S., Ion, B., Militaru, G., Zhu, S. (2000): Smash biproducts of algebras and coalgebras. Algebras and Representation Theory, **3**, 19–42

45. Caenepeel, S., Jiao, Z. (2000): Pairings and the twisted product. Arab J. Math. Sci., **6**, 17-36

46. Caenepeel, S., Kadison, L. (2001): Are biseparable extensions Frobenius?. K-Theory, **24**, 361–383

47. Caenepeel, S., Militaru, G. (2001): Maschke functors, semisimple functors and separable functors of the second kind. Applications. Preprint

48. Caenepeel, S., Militaru, G., Zhu, S. (1997): A Maschke type theorem for Doi-Hopf modules. J. Algebra, **187**, 388–412

49. Caenepeel, S., Militaru, G., Zhu, S. (1997): Doi-Hopf modules, Yetter-Drinfeld modules and Frobenius type properties. Trans. Amer. Math. Soc., **349**, 4311–4342

50. Caenepeel, S., Militaru, G., Zhu, S. (1997): Crossed modules and Doi-Hopf modules. Israel J. Math., **100**, 221–247

51. Caenepeel, S., Raianu, Ş. (1995): Induction functors for the Doi-Koppinen unified Hopf modules. In: Facchini, A., Menini, C. (eds) Abelian groups and Modules. Kluwer Academic Publishers, Dordrecht

52. Caenepeel, S., Van Oystaeyen, F., Zhang, Y. (1994): Quantum Yang-Baxter module algebras. K-Theory, **8**, 231–255

53. Caenepeel, S., Van Oystaeyen, F., Zhang, Y. (1997): The Brauer group of Yetter-Drinfeld module algebras. Trans. Amer. Math. Soc., **349**, 3737–3771

54. Caenepeel, S., Van Oystaeyen, F., Zhou, B. (1998): Making the category of Doi-Hopf modules into a braided monoidal category. Algebras and Representation Theory, **1**, 75–96

55. Cartan, H., Eilenberg, S. (1956): Homological Algebra. Princeton Univ. Press, Princeton

56. Castaño Iglesias, F., Gómez Torrecillas, J., Năstăsescu, C. (1997): Separable functors in coalgebras. Tsukuba J. Math., **12**, 329–344

57. Castaño Iglesias, F., Gómez Torrecillas, J., Năstăsescu, C. (1998): Separable functors in graded rings. J. Pure Appl. Algebra, **127**, 219–230

58. Castaño Iglesias, F., Gómez Torrecillas, J., Năstăsescu, C. (1999): Frobenius functors; Applications. Comm. Algebra, **27**, 4879–4900

59. Cipolla, M. (1976): Discesa fedelmente piatta dei moduli. Rendiconti del Circolo Matematico di Palermo, Serie II, **25**

60. Cohen, M., Fischman, D. (1986): Hopf Algebra Actions. J. Algebra, **100**, 363–379

61. Cohen, M., Fischman, D. (1992): Semisimple Extensions and Elements of Trace 1. J. Algebra, **149**, 419–437

62. Chase, S., Sweedler, M.E. (1969): Hopf algebras and Galois theory. Springer Verlag, Berlin

63. Dăscălescu, S., Năstăsescu, C., Raianu, Ş. (2001): Hopf algebras: an Introduction. Marcel Dekker, New York

64. Dăscălescu, S., Nichita, F. (1999): Yang-Baxter operators arising from (co)algebra structures. Comm. Algebra, **27**, 5833–5845

65. Davydov, A.A. (2001): Pentagon equation and matrix bialgebras. Comm Algebra, **29**, 2627–2650

66. DeMeyer, F., Ingraham, E. (1971): Separable algebras over commutative rings. Springer Verlag, Berlin

67. Doi, Y. (1983): On the structure of relative Hopf modules. Comm. Algebra, **11**, 243–255

68. Doi, Y. (1984): Cleft comodule algebras and Hopf modules. Comm. Algebra, **12**, 1155–1169

69. Doi, Y. (1985): Algebras with total integral. Comm. Algebra, **13**, 2137–2159

70. Doi, Y. (1990): Hopf extensions of algebras and Maschke type theorems. Israel J. Math., **72**, 99–108

71. Doi, Y. (1992): Unifying Hopf modules. J. Algebra, **153**, 373–385

72. Doi, Y. (1993): Braided bialgebras and quadratic bialgebras. Comm. Algebra, **21**, 1731–1749

73. Doi, Y., Takeuchi, M. (1989): Hopf-Galois extensions of algebras, the Miyashita-Ulbrich action and Azumaya algebras. J. Algebra, **121**, 488–516

74. Drinfeld, V.G. (1997): Quantum groups. In: Proc. ICM at Berkeley. Amer. Math. Soc., Providence

75. Drinfeld, V.G. (1990): On almost cocommutative Hopf algebras. Leningrad Math. J., **1**, 321–342

76. Drinfeld, V.G. (1990): Quasi Hopf algebras. Leningrad Math. J., **1**, 1419–1457

77. Drinfeld, V.G. (1991): On quasitriangular quasi-Hopf algebras and on a group that is closely connected with $\mathrm{Gal}(\overline{Q}/Q)$. Leningrad Math. J., **2**, 829–860

78. Drinfeld, V.G. (1992): On some unsolved problems in quantum group theory. Lecture Notes in Mathematics, **1510**, 1–8

79. Dubrovin, B., (1996): Geometry of 2D topological field theories. In: Integrable systems and quantum groups (Montecatini Terme), Lecture Notes in Math., Springer, Berlin

80. El Kaoutit, L., Gómez Torrecillas, J., Lobillo, F.J. (2002) Semisimple corings. preprint

81. Etingof, P., Gelaki, S. (1998): A method of construction of finite dimensional triangular semisimple Hopf algebras. Math. Res. Lett., **5**, 551–561

82. Etingof, P., Gelaki, S. (2000): The classification of triangular semisimple and cosemisimple Hopf algebras over an algebraically closed field. Internat. Math. Res. Notices, **5**, 223–234

83. Etingof, P., Kazhdan, D. (1998): Quantization of Lie bialgebras II. Sel. Math.(New Ser.), **2**, 213–231

84. Etingof, P., Schedler, T., Soloviev, A. (1999): Set-theoretical solutions to the quantum Yang-Baxter equation. Duke Math. J., **100**, 169–209

85. Faddeev, F.D., Reshetikhin, N.Y., Takhtajan, L.A. (1989): Quantization of Lie groups and Lie algebras. Algebraic Analysis, **1**, 129–139

86. Faith, C. (1973): Algebra: Rings, Modules and Categories I. Springer-Verlag, Berlin

87. Farnsteiner, R. (1994): On Frobenius extensions defined by Hopf algebras. J. Algebra, **166**, 130–141

88. Fischman, D., Montgomery, S., Schneider, H.-J. (1997): Frobenius extensions of subalgebras of Hopf algebras. Trans. Amer. Math. Soc. **349**, 4857–4895

89. Gamst, J., Hoechstman, K. (1969): Quaternions généralisés. C.R. Acad. Sci. Paris, **269**, 560–562

90. Gelaki, S. (1998): On pointed Hopf algebras and Kaplansky's 10th conjecture. J. Algebra, **209**, 635–657

91. Gomez Torrecillas, J. (2001): Separable functors in corings. Inter. J. Math. and Math. Sci., to appear.

92. Graña, M. (2000): On Nichols algebras of low dimension. Contemp. Math., **267**, 111–134

93. Grothendieck, A. (1959): Technique de Descente I. Sém. Bourbaki, **190**

94. Hirata, K., Sugano, K. (1966): On semi-simple and Separable Extensions over noncommutative Rings. J. Math. Soc. Japan, **18**, 360–373

95. Hobst, D., Pareigis, B. (2001): Double quantum groups. J. Algebra, **242**, 460–494

96. Ion, B., Stănciulescu, M. (1999): Several examples of noncommutative and noncocommutative bialgebras arising from the Hopf equation. Rev. Roum. Math. Pures Appl., **44**, 385–404

97. Janelidze, G., Tholen, W., (2002): Facets of descent III. in preparation

98. Kadison, L. (1996): The Jones polynomial and certain separable Frobenius extensions. J. Algebra, **186**, 461–475

99. Kadison, L. (1999): Separability and the twisted Frobenius bimodules. Algebras and Representation Theory, **2**, 397–414

100. Kadison, L. (1999): New examples of Frobenius extensions. Amer. Math. Soc., Providence

101. Kan, D.M. (1958): Adjoint functors. Trans. Amer. Math. Soc., **87**, 294–329

102. Kan, H.B. (1999): Injective envelopes, cogenerators and direct products in the Doi-Hopf modules. Chinese Sci. Bull., **44**, 1350–1356

103. Kaplansky, I. (1975): Bialgebras. University of Chicago Press, Chicago

104. Kasch, F. (1954): Grundlagen einer Theorie der Frobenius-Erweiterungen. Math. Ann., **127**, 453-474

105. Kasch, F. (1961): Projektive Frobenius-Erweiterungen; Sitzungsber. Heidelberger Akad. Wiss.(Math.-Naturw. Kl.), **61**, 89–109

106. Kashaev, R.M. (1997): Heisenberg Double and the Pentagon relation. Algebra i Analiz, **8**, 63–74; translation in St. Petersburg Math. J., **8**, 585–592

107. Kashaev, R.M., Sergeev, S.M. (1998): On pentagon, ten-term and tetraedron relations. Comm. Math. Phys., **195**, 309–319

108. Kassel, C. (1995): Quantum Groups. Springer Verlag, Berlin

109. Knus, M.A., Ojanguren, M. (1974): Théorie de la descente et algèbres d'Azumaya. Springer Verlag, Berlin

110. Koppinen, M. (1992): A duality theorem for crossed products of Hopf algebras. J. Algebra, **146**, 153–174

111. Koppinen, M. (1995): Variations on the smash product with applications to group-graded rings. J. Pure Appl. Algebra, **104**, 61–80

112. Kulish, P.P., Mudrov, A.I. (1998): On twisting solutions to the Yang-Baxter equation. Preprint math. QA/9811044

113. Lam, T.Y. (1998): Lectures on modules and rings. Springer Verlag, Berlin

114. Lambe. L.A., Radford, D. (1992): Algebraic aspects of the quantum Yang-Baxter equation. J. Algebra, **54**, 228–288

115. Lambe, L.A., Radford, D. (1997): Introduction to the quantum Yang-Baxter equation and quantum groups: an algebraic approach. Kluwer Academic Publishers, Dordrecht

116. Larson, R.G. (1973): Coseparable coalgebras, J. Pure and Appl. Algebra, **3**, 261–267

117. Larson, R.G., Sweedler, M. (1969): An associative orthogonal bilinear form for Hopf algebras. Amer. J. Math., **91**, 75–93

350   References

118. Long, F. (1974): The Brauer group of dimodule algebras, J. Algebra, **30**, 559–601
119. Lu, D.M. (2001): Tannaka duality and the FRT-Construction. Comm. Algebra, **29**, 5717–5731
120. Lu, J.H. (1994): On the Drinfeld double and the Heisenberg double of a Hopf algebra. Duke Math. J., **74**, 763–776
121. Lu, J.H., Yan, M., Zhu, Y.C. (2000): On the set-theoretical Yang-Baxter equation. Duke J. Math., **104**, 1–18
122. Mac Lane, S. (1963): Natural associativity and commutativity. Rice Univ. Studies, **49**, 4–28
123. Mac Lane, S. (1997): Categories for the working mathematician (second edition). Springer Verlag, Berlin
124. Madar, A., Marcus, A. (2001): Frobenius functors and transfer. Publ. Math. Debrecen, to appear
125. Majid, S. (1990): Physics for algebraists: non-commutative and non-cocommutative Hopf algebras by a bicrossproduct construction. J. Algebra, **130**, 17–64
126. Majid, S. (1991): Doubles of quasitriangular Hopf algebras. Comm. Algebra, **19**, 3061–3073
127. Majid, S. (1993): Quantum random walks and time reversal. Int. J. Mod. Phys., **8**, 4521–4545
128. Majid, S. (1995): Foundations of quantum group theory. Cambridge Univ. Press, Cambridge
129. Manin, Y.I. (1999): Frobenius manifolds, quantum cohomology, and moduli spaces. American Mathematical Society, Providence
130. Menini, C., Militaru, G. (2002): Integrals, quantum Galois extensions and the affineness criterion for quantum Yetter-Drinfeld modules. J. Algebra, **247**, 467–508
131. Menini, C., Năstăsescu, C. (1994): When are induction and coinduction functors isomorphic? Bull. Belgian Math. Soc., **1**, 521–558
132. Mesablishvili, B. (2000): Pure morphisms of commutative rings are effective descent morphisms for modules - a new proof. Theory Appl. Categories, **7**, 38–42
133. Militaru, G. (1998): The Hopf modules category and the Hopf equation. Comm. Algebra, **26**, 3071–3097
134. Militaru, G. (1998): New types of bialgebras arising from the Hopf equation. Comm. Algebra, **26**, 3099–3117
135. Militaru, G. (1999): The Long dimodules category and nonlinear equations. Algebra Representation Theory, **2**, 177–200
136. Militaru, G. (1999): A class of solutions for the integrability condition of the Knizhnik-Zamolodchikov equation: a Hopf algebraic approach. Comm. Algebra, **27**, 2393–2407
137. Militaru, G. (2000): Heisenberg double, pentagon equation, structure and classification of finite dimensional Hopf algebras. Preprint math. QA/0009141
138. Militaru, G., Ştefan, D. (1994): Extending modules for Hopf Galois extensions. Comm. Algebra, **14**, 5657–5678
139. Molnar, R.K. (1977): Semi-direct products of Hopf algebras. J. Algebra, **47**, 29–51
140. Montgomery, S. (1993): Hopf algebras and their actions on rings. American Mathematical Society, Providence

141. Morita, K. (1965): Adjoint pairs of functors and Frobenius extensions. Sci. Rep. Tokyo Kyoiku Daigaku (Sect. A), **9**, 40–71
142. Nakayama, T., Tsuzuku, T. (1960): On Frobenius extensions I. Nagoya Math. J., **17**, 89–110
143. Năstăsescu, C., Raianu, Ş., Van Oystaeyen, F. (1990): Modules graded by G-sets. Math. Z., **203**, 605–627
144. Năstăsescu, C., Torrecillas, B. (1993): Graded coalgebras. Tsukuba J. Math., **17**, 461–479
145. Năstăsescu, C., Van den Bergh, M., Van Oystaeyen, F. (1989): Separable functors applied to graded rings. J. Algebra, **123**, 397–413
146. Năstăsescu, C., Van Oystaeyen, F. (1982): Graded ring theory. North Holland, Amsterdam
147. Năstăsescu, C., Van Oystaeyen, F., Shaoxue, L. (1991): Graded modules over G-sets II. Math. Z., **207**, 341–358
148. Nichita, F. (1999): Self-Inverse Yang-Baxter operators from (co)algebra structures. J. Algebra, **218**, 738–759
149. Nuss, F. (1997): Noncommutative descent and nonabelian cohomology. K-Theory, **12**, 23–74
150. Panaite, F., Van Oystaeyen, F. (1999): Quasitriangular structures for some pointed Hopf algebras of dimension 2^n. Comm. Algebra, **27**, 4929–4942
151. Pareigis, B. (1971): When Hopf algebras are Frobenius Algebras. J. Algebra, **18**, 588–596
152. Parshall, B., Wang, J. (1991): Quantum Linear Groups. Memoirs Amer. Math. Soc., **89**
153. Pierce, R. (1982): Associative algebras. Springer Verlag, Berlin
154. Radford, D. (1993): Minimal quasi-triangular Hopf algebras. J. Algebra, **157**, 285–315
155. Radford, D. (1933): Solutions to the quantum Yang-Baxter equation and the Drinfeld double. J. Algebra, **161**, 20–32
156. Radford, D. (1994): Solutions to the quantum Yang-Baxter equation arising from pointed bialgebras, Trans. Amer. Math. Soc., **343**, 455–477
157. Radford, D., Towber, J. (1993): Yetter-Drinfeld categories associated to an arbitrary bialgebra. J. Pure and Appl. Algebra, **87**, 259–279
158. Rafael, D.M. (1990): Separable functors revisited. Comm. in Algebra, **18**, 1445–1459
159. del Río, A. (1990): Categorical methods in graded ring theory. Publicacions Math., **72**, 489–531
160. Soloviev, A. (2000): Non-unitary set-theoretical solutions of the quantum Yang-Baxter equation. Math. Res. Lett., **7**, 577–596
161. Schauenburg, P. (1994): Hopf module and Yetter-Drinfeld modules. J. Algebra, **169**, 874–890
162. Schauenburg, P. (1998): Examples of equivalences of Doi-Koppinen Hopf module categories, including Yetter-Drinfeld modules. Bull. Belgian Math. Soc.-Simon Stevin, **5**, 91–98
163. Schauenburg, P. (2000): Doi-Koppinen modules versus entwined modules. New York J. Math., **6**, 325–329
164. Schneider, H.-J. (1990): Representation theory for Hopf Galois extensions. Israel J. Math., **70**, 196–231
165. Schneider, H.-J. (1990): Principal homogeneous spaces for arbitrary Hopf algebras. Israel J. Math., **70**, 167–195

166. Shnider, S., Sternberg, S. (1993): Quantum groups-from coalgebras to Drinfeld algebras: a guided tour. International Press Company, Boston
167. Ştefan, D. (1996): Cohomology of Hopf algebras and the Clifford's extension problem. J. Algebra, **182**, 165–182
168. Ştefan, D. (1997): The set of types of n-dimensional semisimple and cosemisimple Hopf algebras is finite. J. Algebra, **193**, 571–580
169. Ştefan, D., Van Oystaeyen, F. (1999): The Wedderburn-Malcev theorem for comodule algebras. Comm. Algebra, **27**, 3569–3581
170. Street, R. (1998): Fusion operators and cocycloids in monoidal categories. Appl. Categorical Structures, **6**, 177–191
171. Sugano, K. (1971): Note on separability of endomorphism rings. J. Fac. Sci. Hokkaido Univ., **21**, 196–208
172. Sweedler, M.E. (1969): Hopf algebras. Benjamin, New York
173. Sweedler, M.E. (1975): The predual Theorem to the Jacobson-Bourbaki theorem. Trans. Amer. Math. Soc., **213**, 391–406
174. Takesaki, M. (1972): Duality and von Neumann algebras. In: Lectures Operator algebras, Lecture Notes in Math., **247**, 665–779, Springer Verlag, Berlin
175. Takeuchi, M. (1972): A correspondence between Hopf ideals and sub-Hopf algebras. Manuscripta Math., **7**, 251–270
176. Takeuchi, M. (1980): $\mathrm{Ext}_{\mathrm{ad}}(SpR, -\mu^A) \cong \widehat{\mathrm{Br}}(A/k)$. J. Algebra, **67**, 436–475
177. Takeuchi, M. (1992): Finite dimensional representations of the quantum Lorentz group. Commun. Math. Phys., **144**, 557–580
178. Tambara, D. (1990): The bialgebra of coendomorphisms of an algebra. J. Fac. Sci. Univ. Tokyo, Sect. IA, Math., **37**, 425–456
179. Taylor, J.L. (1982): A bigger Brauer group. Pacific J. Math., **103**, 163–203
180. Ulbrich, K.H. (1990): Smash products and comodules of linear maps. Tsukuba J. Math., **2**, 371–378
181. Van Daele, A., Van Keer, S. (1994): The Yang-Baxter and pentagon equation. Compositio Math., **91**, 201–221
182. Van Oystaeyen, F., Zhang, Y. (1998): The Brauer group of a braided category, J. Algebra, **202**, 96–128
183. Wang, D., Wang, Y. (2001): Twistings, crossed products and Hopf Galois coextensions. Science in China (Series A), to appear
184. Wang, S.H. (2001): Doi-Koppined Hopf bimodules are modules. Comm. Algebra, **29**, 4671–4682
185. Wang, S.H., Li, J.Q. (1998): On twisted smash product for bimodule algebras and the Drinfeld double. Comm. Algebra, **26**, 2435–2444
186. Weinstein, A., Xu, P. (1992): Classical solutions of the quantum Yang-Baxter equation, Comm. Math. Phys., **148**, 309–343
187. Wisbauer, R. (1997): Semiperfect coalgebras over rings. In: Algebras and combinatorics. Papers from the international congress ICAC. Hong Kong
188. Wisbauer, R. (2001): On the category of comodules over corings. Preprint
189. Yetter, D.N. (1990): Quantum groups and representations of monoidal categories. Math. Proc. Cambridge Philos. Soc., **108**, 261–290
190. Zhu, Y. (1994): Hopf algebras in prime dimension. International Math. Research Notices, **1**, 53–59

Index

Druck: Strauss Offsetdruck, Mörlenbach
Verarbeitung: Schäffer, Grünstadt